Undergraduate Texts in Mathematics

Springer Books on Elementary Mathematics by Serge Lang

MATH! Encounters with High School Students
1985, ISBN 96129-1

The Beauty of Doing Mathematics
1985, ISBN 96149-6

Geometry. A High School Course (with G. Murrow), **Second Edition**
1989, ISBN 96654-4

Basic Mathematics
1988, ISBN 96787-7

A First Course in Calculus
1986, ISBN 96201-8

Calculus of Several Variables
1987, ISBN 96405-3

Introduction to Linear Algebra
1986, ISBN 96205-0

Linear Algebra
1987, ISBN 96412-6

Undergraduate Algebra, Third Edition
2005, ISBN 22025-9

Undergraduate Analysis
1983, ISBN 90800-5

Complex Analysis
1985, ISBN 96085-6

Math Talks for Undergraduates
1999, ISBN 98749-5

Serge Lang

Undergraduate Algebra

Third Edition

Serge Lang
Department of Mathematics
Yale University
New Haven, CT 06520
USA

Mathematics Subject Classification (2000): 13-01, 15-01

Library of Congress Cataloging-in-Publication Data
Lang, Serge, 1927–
Undergraduate algebra / Serge Lang. — 3rd ed.
p. cm. — (Undergraduate texts in mathematics)
Includes bibliographical references and index.

1. Algebra. I. Title. II. Series.
QA152.3.L36 2004
512—dc22 2004049194

ISBN 978-1-4419-1959-5 Printed on acid-free paper.
e-ISBN 978-0-387-27475-1

Printed in the United States of America. (EB/ASCO)

9 8 7 6 5 4 3 2 1

springeronline.com

Foreword

This book, together with *Linear Algebra*, constitutes a curriculum for an algebra program addressed to undergraduates.

The separation of the linear algebra from the other basic algebraic structures fits all existing tendencies affecting undergraduate teaching, and I agree with these tendencies. I have made the present book self contained logically, but it is probably better if students take the linear algebra course *before* being introduced to the more abstract notions of groups, rings, and fields, and the systematic development of their basic abstract properties. There is of course a little overlap with the book *Linear Algebra*, since I wanted to make the present book self contained. I define vector spaces, matrices, and linear maps and prove their basic properties.

The present book could be used for a one-term course, or a year's course, possibly combining it with *Linear Algebra*. I think it is important to do the field theory and the Galois theory, more important, say, than to do much more group theory than we have done here. There is a chapter on finite fields, which exhibit both features from general field theory, and special features due to characteristic p. Such fields have become important in coding theory.

There is also a chapter on some of the group-theoretic features of matrix groups. Courses in linear algebra usually concentrate on the structure theorems, quadratic forms, Jordan form, etc. and do not have the time to mention, let alone emphasize, the group-theoretic aspects of matrix groups. I find that the basic algebra course is a good place to introduce students to such examples, which mix abstract group theory with matrix theory. The groups of matrices provide concrete examples for the more abstract properties of groups listed in Chapter II.

The construction of the real numbers by Cauchy sequences and null sequences has no standard place in the curriculum, depending as it does on mixed ideas from algebra and analysis. Again, I think it belongs in a basic algebra text. It illustrates considerations having to do with rings, and also with ordering and absolute values. The notion of completion is partly algebraic and partly analytic. Cauchy sequences occur in mathematics courses on analysis (integration theory for instance), and also number theory as in the theory of p-adic numbers or Galois groups.

For a year's course, I would also regard it as appropriate to introduce students to the general language currently in use in mathematics concerning sets and mappings, up to and including Zorn's lemma. In this spirit, I have included a chapter on sets and cardinal numbers which is much more extensive than is the custom. One reason is that the statements proved here are not easy to find in the literature, disjoint from highly technical books on set theory. Thus Chapter X will provide attractive extra material if time is available. This part of the book, together with the Appendix, and the construction of the real and complex numbers, also can be viewed as a short course on the naive foundations of the basic mathematical objects.

If all these topics are covered, then there is enough material for a year's course. Different instructors will choose different combinations according to their tastes. For a one-term course, I would find it appropriate to cover the book up to the chapter on field theory, or the matrix groups. Finite fields can be treated as optional.

Elementary introductory texts in mathematics, like the present one, should be simple and always provide concrete examples together with the development of the abstractions (which explains using the real and complex numbers as examples before they are treated logically in the text). The desire to avoid encyclopedic proportions, and specialized emphasis, and to keep the book short explains the omission of some theorems which some teachers will miss and may want to include in the course. Exceptionally talented students can always take more advanced classes, and for them one can use the more comprehensive advanced texts which are easily available.

New Haven, Connecticut, 1987 S. LANG

Acknowledgments

I thank Ron Infante and James Propp for assistance with the proofreading, suggestions, and corrections.

S.L.

Foreword to the Third Edition

In this new edition I have added new material in Chapters IV and VI, first on polynomials, and second on linear algebra in combination with group theory. The additions to Chapter VI describe various product structures for SL_n (Iwasawa and other decompositions). These also have to do with the conjugation action and the decomposition of the Lie algebra under this action. The algebra involved comes from deeper theories, but the parts I have extracted on SL_n belong to an elementary level. Students are then put into contact with some algebra used as a backdrop for analysis on groups, starting with SL_n.

A new section in Chapter IV gives a complete account of the Mason-Stothers theorem about polynomials, with Noah Snyder's beautifully simple proof. It is worth emphasizing that the derivative for polynomials is a purely algebraic operation, for which limits are not required. A Springer pamphlet has been published to present a self-contained treatment of polynomials (from scratch) culminating with this topic. Here it takes its place as a section in the general chapter on polynomials. It occurs as a natural twin for the section on the *abc* conjecture.

I have tried on several occasions to put students in contact with genuine research mathematics, by selecting instances of conjectures which can be formulated in language at the level of this course. I have stated more than half a dozen such conjectures, of which the *abc* conjecture provides one spectacular example. Usually students have to wait years before they realize that mathematics is a live activity, sustained by its open problems. I have found it very effective to break down this obstacle whenever possible.

Acknowledgment

I thank Keith Conrad for his suggestions and help with the proofreading in previous editions. I also thank Allen Altman for numerous additional corrections.

New Haven 2004 SERGE LANG

Contents

CHAPTER I

The Integers

I, §1. TERMINOLOGY OF SETS

A collection of objects is called a **set**. A member of this collection is also called an **element** of the set. It is useful in practice to use short symbols to denote certain sets. For instance, we denote by $\mathbf{Z}$ the set of all integers, i.e. all numbers of the type $0, \pm 1, \pm 2, \ldots$. Instead of saying that x is an element of a set S, we shall also frequently say that x **lies in** S, and write $x \in S$. For instance, we have $1 \in \mathbf{Z}$, and also $-4 \in \mathbf{Z}$.

If S and S' are sets, and if every element of S' is an element of S, then we say that S' is a **subset** of S. Thus the set of positive integers $\{1, 2, 3, \ldots\}$ is a subset of the set of all integers. To say that S' is a subset of S is to say that S' is part of S. Observe that our definition of a subset does not exclude the possibility that $S' = S$. If S' is a subset of S, but $S' \neq S$, then we shall say that S' is a **proper** subset of S. Thus $\mathbf{Z}$ is a subset of $\mathbf{Z}$, and the set of positive integers is a proper subset of $\mathbf{Z}$. To denote the fact that S' is a subset of S, we write $S' \subset S$, and also say that S' is **contained** in S.

If S_1, S_2 are sets, then the **intersection** of S_1 and S_2, denoted by $S_1 \cap S_2$, is the set of elements which lie in both S_1 and S_2. For instance, if S_1 is the set of integers $\geqq 1$ and S_2 is the set of integers $\leqq 1$, then

$$S_1 \cap S_2 = \{1\}$$

(the set consisting of the number 1).

The **union** of S_1 and S_2, denoted by $S_1 \cup S_2$, is the set of elements which lie in S_1 or in S_2. For instance, if S_1 is the set of integers $\leqq 0$

and S_2 is the set of integers $\geqq 0$, then $S_1 \cup S_2 = \mathbf{Z}$ is the set of all integers.

We see that certain sets consist of elements described by certain properties. If a set has no elements, it is called the **empty** set. For instance, the set of all integers x such that $x > 0$ and $x < 0$ is empty, because there is no such integer x.

If S, S' are sets, we denote by $S \times S'$ the set of all pairs (x, x') with $x \in S$ and $x' \in S'$.

We let $\# S$ denote the number of elements of a set S. If S is finite, we also call $\# S$ the **order** of S.

I, §2. BASIC PROPERTIES

The integers are so well known that it would be slightly tedious to axiomatize them immediately. Hence we shall assume that the reader is acquainted with the elementary properties of arithmetic, involving addition, multiplication, and inequalities, which are taught in all elementary schools. In the appendix and in Chapter III, the reader will see how one can axiomatize such rules concerning addition and multiplication. For the rules concerning inequalities and ordering, see Chapter IX.

We mention explicitly one property of the integers which we take as an axiom concerning them, and which is called **well-ordering**.

Every non-empty set of integers $\geqq 0$ has a least element.

(This means: If S is a non-empty set of integers $\geqq 0$, then there exists an integer $n \in S$ such that $n \leqq x$ for all $x \in S$.)

Using this well-ordering, we shall prove another property of the integers, called induction. It occurs in several forms.

Induction: First Form. *Suppose that for each integer $n \geqq 1$ we are given an assertion $A(n)$, and that we can prove the following two properties:*

(1) *The assertion $A(1)$ is true.*
(2) *For each integer $n \geqq 1$, if $A(n)$ is true, then $A(n + 1)$ is true.*

Then for all integers $n \geqq 1$, the assertion $A(n)$ is true.

Proof. Let S be the set of all positive integers n for which the assertion $A(n)$ is false. We wish to prove that S is empty, i.e. that there is no element in S. Suppose there is some element in S. By well-ordering, there exists a least element n_0 in S. By assumption, $n_0 \neq 1$, and hence $n_0 > 1$. Since n_0 is least, it follows that $n_0 - 1$ is not in S, in other words the assertion $A(n_0 - 1)$ is true. But then by property (2), we conclude that $A(n_0)$ is also true because

$$n_0 = (n_0 - 1) + 1.$$

This is a contradiction, which proves what we wanted.

Example. We wish to prove that for each integer $n \geqq 1$,

$$A(n):\quad 1 + 2 + \cdots + n = \frac{n(n+1)}{2}.$$

This is certainly true when $n = 1$, because

$$1 = \frac{1(1+1)}{2}.$$

Assume that our equation is true for an integer $n \geqq 1$. Then

$$\begin{aligned} 1 + \cdots + n + (n+1) &= \frac{n(n+1)}{2} + (n+1) \\ &= \frac{n(n+1) + 2(n+1)}{2} \\ &= \frac{n^2 + n + 2n + 2}{2} \\ &= \frac{(n+1)(n+2)}{2}. \end{aligned}$$

Thus we have proved the two properties (1) and (2) for the statement denoted by $A(n+1)$, and we conclude by induction that $A(n)$ is true for all integers $n \geqq 1$.

Remark. In the statement of induction, we could replace 1 by 0 everywhere, and the proof would go through just as well.

The second form is a variation on the first.

Induction: Second Form. *Suppose that for each integer $n \geqq 0$ we are given an assertion $A(n)$, and that we can prove the following two properties:*

(1′) *The assertion $A(0)$ is true.*
(2′) *For each integer $n > 0$, if $A(k)$ is true for every integer k with $0 \leqq k < n$, then $A(n)$ is true.*

Then the assertion $A(n)$ is true for all integers $n \geqq 0$.

Proof. Again let S be the set of integers $\geqq 0$ for which the assertion is false. Suppose that S is not empty, and let n_0 be the least element of S. Then $n_0 \neq 0$ by assumption (1′), and since n_0 is least, for every integer k with $0 \leqq k < n_0$, the assertion $A(k)$ is true. By (2′) we conclude that $A(n_0)$ is true, a contradiction which proves our second form of induction.

As an example of well ordering, we shall prove the statement known as the **Euclidean algorithm**.

Theorem 2.1. *Let m, n be integers and $m > 0$. Then there exist integers q, r with $0 \leqq r < m$ such that*

$$n = qm + r.$$

The integers q, r are uniquely determined by these conditions.

Proof. The set of integers q such that $qm \leqq n$ is bounded from above proof?), and therefore by well ordering has a largest element satisfying

$$qm \leqq n < (q + 1)m = qm + m.$$

Hence

$$0 \leqq n - qm < m.$$

Let $r = n - qm$. Then $0 \leqq r < m$. This proves the existence of the integers q and r as desired.

As for uniqueness, suppose that

$$n = q_1 m + r_1, \qquad 0 \leqq r_1 < m,$$

$$n = q_2 m + r_2, \qquad 0 \leqq r_2 < m.$$

If $r_1 \neq r_2$, say $r_2 > r_1$. Subtracting, we obtain

$$(q_1 - q_2)m = r_2 - r_1.$$

But $r_2 - r_1 < m$, and $r_2 - r_1 > 0$. This is impossible because $q_1 - q_2$ is an integer, and so if $(q_1 - q_2)m > 0$ then $(q_1 - q_2)m \geqq m$. Hence we conclude that $r_1 = r_2$. But then $q_1 m = q_2 m$, and thus $q_1 = q_2$. This proves the uniqueness, and concludes the proof of our theorem.

Remark. The result of Theorem 2.1 is nothing but an expression of the result of long division. We call r the **remainder** of the division of n by m.

I, §2. EXERCISES

1. If n, m are integers $\geqq 1$ and $n \geqq m$, define the **binomial coefficients**

$$\binom{n}{m} = \frac{n!}{m!\,(n-m)!}.$$

As usual, $n! = n \cdot (n-1) \cdots 1$ is the product of the first n integers. We define $0! = 1$ and $\binom{n}{0} = 1$. Prove that

$$\binom{n}{m-1} + \binom{n}{m} = \binom{n+1}{m}.$$

2. Prove by induction that for any integers x, y we have

$$(x+y)^n = \sum_{i=0}^{n} \binom{n}{i} x^i y^{n-i} = y^n + \binom{n}{1} x y^{n-1} + \binom{n}{2} x^2 y^{n-2} + \cdots + x^n.$$

3. Prove the following statements for all positive integers:
   (a) $1 + 3 + 5 + \cdots + (2n-1) = n^2$
   (b) $1^2 + 2^2 + \cdots + n^2 = n(n+1)(2n+1)/6$
   (c) $1^3 + 2^3 + 3^3 + \cdots + n^3 = [n(n+1)/2]^2$

4. Prove that

$$\left(1 + \frac{1}{1}\right)^1 \left(1 + \frac{1}{2}\right)^2 \cdots \left(1 + \frac{1}{n-1}\right)^{n-1} = \frac{n^{n-1}}{(n-1)!}$$

5. Let x be a real number. Prove that there exists an integer q and a real number s with $0 \leqq s < 1$ such that $x = q + s$, and that q, s are uniquely determined. Can you deduce the euclidean algorithm from this result without using induction?

I, §3. GREATEST COMMON DIVISOR

Let n be a non-zero integer, and d a non-zero integer. We shall say that d **divides** n if there exists an integer q such that $n = dq$. We then write $d \mid n$. If m, n are non-zero integers, by a **common divisor** of m and n we mean an integer $d \neq 0$ such that $d \mid n$ and $d \mid m$. By a **greatest common divisor** or **g.c.d.** of m and n, we mean an integer $d > 0$ which is a common divisor, and such that, if e is a divisor of m and n then e divides d. We shall see in a moment that a greatest common divisor always exists. It is immediately verified that a greatest common divisor is uniquely determined. We define the g.c.d. of several integers in a similar way.

Let J be a subset of the integers. We shall say that J is an **ideal** if it has the following properties:

The integer 0 is in J. If m, n are in J, then $m + n$ is in J. If m is in J, and n is an arbitrary integer, then nm is in J.

Example. Let $m_1,\ldots,m_r$ be integers. Let J be the set of all integers which can be written in the form

$$x_1 m_1 + \cdots + x_r m_r$$

with integers $x_1,\ldots,x_r$. Then it is immediately verified that J is an ideal. Indeed, if $y_1,\ldots,y_r$ are integers, then

$$(x_1 m_1 + \cdots + x_r m_r) + (y_1 m_1 + \cdots + y_r m_r) = (x_1 + y_1)m_1 + \cdots + (x_r + y_r)m_r$$

lies in J. If n is an integer, then

$$n(x_1 m_1 + \cdots + x_r m_r) = n x_1 m_1 + \cdots + n x_r m_r$$

lies in J. Finally, $0 = 0m_1 + \cdots + 0m_r$ lies in J, so J is an ideal. We say that J is **generated** by $m_1,\ldots,m_r$ and that $m_1,\ldots,m_r$ are **generators**.

We note that $\{0\}$ itself is an ideal, called the **zero ideal**. Also, $\mathbf{Z}$ is an ideal, called the **unit ideal.**

Theorem 3.1. *Let J be an ideal of $\mathbf{Z}$. Then there exists an integer d which is a generator of J. If $J \neq \{0\}$ then one can take d to be the smallest positive integer in J.*

Proof. If J is the zero ideal, then 0 is a generator. Suppose $J \neq \{0\}$. If $n \in J$ then $-n = (-1)n$ is also in J, so J contains some positive integer. Let d be the smallest positive integer in J. We contend that d is a generator of J. To see this, let $n \in J$, and write $n = dq + r$ with $0 \leqq r < d$. Then $r = n - dq$ is in J, and since $r < d$, it follows that $r = 0$. This proves that $n = dq$, and hence that d is a generator, as was to be shown.

Theorem 3.2. *Let m_1, m_2 be positive integers. Let d be a positive generator for the ideal generated by m_1, m_2. Then d is a greatest common divisor of m_1 and m_2.*

Proof. Since m_1 lies in the ideal generated by m_1, m_2 (because $m_1 = 1m_1 + 0m_2$), there exists an integer q_1 such that

$$m_1 = q_1 d,$$

whence d divides m_1. Similarly, d divides m_2. Let e be a non-zero integer dividing both m_1 and m_2, say

$$m_1 = h_1 e \quad \text{and} \quad m_2 = h_2 e$$

with integers h_1, h_2. Since d is in the ideal generated by m_1 and m_2, there are integers s_1, s_2 such that $d = s_1 m_1 + s_2 m_2$, whence

$$d = s_1 h_1 e + s_2 h_2 e = (s_1 h_1 + s_2 h_2)e.$$

Consequently, e divides d, and our theorem is proved.

Remark. Exactly the same proof applies when we have more than two integers. For instance, if $m_1, \ldots, m_r$ are non-zero integers, and d is a positive generator for the ideal generated by $m_1, \ldots, m_r$, then d is a greatest common divisor of $m_1, \ldots, m_r$.

Integers $m_1, \ldots, m_r$ whose greatest common divisor is 1 are said to be **relatively prime.** If that is the case, then there exist integers $x_1, \ldots, x_r$ such that

$$x_1 m_1 + \cdots + x_r m_r = 1,$$

because 1 lies in the ideal generated by $m_1, \ldots, m_r$.

I, §4. UNIQUE FACTORIZATION

We define a **prime number** p to be an integer $\geqq 2$ such that, given a factorization $p = mn$ with positive integers m, n, then $m = 1$ or $n = 1$. The first few primes are 2, 3, 5, 7, 11,....

Theorem 4.1. *Every positive integer $n \geqq 2$ can be expressed as a product of prime numbers (not necessarily distinct),*

$$n = p_1 \cdots p_r,$$

uniquely determined up to the order of the factors.

Proof. Suppose that there is at least one integer $\geqq 2$ which cannot be expressed as a product of prime numbers. Let m be the smallest such integer. Then in particular m is not prime, and we can write $m = de$ with integers $d, e > 1$. But then d and e are smaller than m, and since m was chosen smallest, we can write

$$d = p_1 \cdots p_r \quad \text{and} \quad e = p'_1 \cdots p'_s$$

with prime numbers $p_1, \ldots, p_r, p'_1, \ldots, p'_s$. Thus

$$m = de = p_1 \cdots p_r p'_1 \cdots p'_s$$

is expressed as a product of prime numbers, a contradiction, which proves that every positive integer $\geqq 2$ can be expressed as a product of prime numbers.

We must now prove the uniqueness, and for this we need a lemma.

Lemma 4.2. *Let p be a prime number, and m, n non-zero integers such that p divides mn. Then $p|m$ or $p|n$.*

Proof. Assume that p does not divide m. Then the greatest common divisor of p and m is 1, and there exist integers a, b such that

$$1 = ap + bm.$$

(We use Theorem 3.2.) Multiplying by n yields

$$n = nap + bmn.$$

But $mn = pc$ for some integer c, whence

$$n = (na + bc)p,$$

and p divides n, as was to be shown.

This lemma will be applied when p divides a product of prime numbers $q_1 \cdots q_s$. In that case, p divides q_1 or p divides $q_2 \cdots q_s$. If p divides q_1, then $p = q_1$. Otherwise, we can proceed inductively, and we conclude that in any case, there exists some i such that $p = q_i$.

Suppose now that we have two products of primes

$$p_1 \cdots p_r = q_1 \cdots q_s.$$

By what we have just seen, we may renumber the primes $q_1, \ldots, q_s$ and then we may assume that $p_1 = q_1$. Cancelling q_1, we obtain

$$p_2 \cdots p_r = q_2 \cdots q_s.$$

We may then proceed by induction to conclude that after a renumbering of the primes $q_1,\dots,q_s$ we have $r = s$, and $p_i = q_i$ for all i. This proves the desired uniqueness.

In expressing an integer as a product of prime numbers, it is convenient to bunch together all equal factors. Thus let n be an integer > 1, and let $p_1,\dots,p_r$ be the *distinct* prime numbers dividing n. Then there exist unique integers $m_1,\dots,m_r > 0$ such that $n = p_1^{m_1}\cdots p_r^{m_r}$. We agree to the usual convention that for any non-zero integer x, $x^0 = 1$. Then given any positive integer n, we can write n as a product of prime powers with distinct primes $p_1,\dots,p_r$:

$$n = p_1^{m_1}\cdots p_r^{m_r},$$

where the exponents $m_1,\dots,m_r$ are integers $\geqq 0$, and uniquely determined.

The set of quotients of integers m/n with $n \neq 0$ is called the **rational numbers**, and denoted by $\mathbf{Q}$. We assume for the moment that the reader is familiar with $\mathbf{Q}$. We show later how to construct $\mathbf{Q}$ from $\mathbf{Z}$ and how to prove its properties.

Let $a = m/n$ be a rational number, $n \neq 0$ and assume $a \neq 0$, so $m \neq 0$. Let d be the greatest common divisor of m and n. Then we can write $m = dm'$ and $n = dn'$, and m', n' are relatively prime. Thus

$$a = \frac{m'}{n'}.$$

If we now express $m' = p_1^{i_1}\cdots p_r^{i_r}$ and $n' = q_1^{j_1}\cdots q_s^{j_s}$ as products of prime powers, we obtain a factorization of a itself, and we note that no p_k is equal to any q_l.

If a rational number is expressed in the form m/n where m, n are integers, $n \neq 0$, and m, n are relatively prime, then we call n the **denominator** of the rational number, and m its **numerator**. Occasionally, by abuse of language, when one writes a quotient m/n where m, n are not necessarily relatively prime, one calls n a denominator for the fraction.

I, §4. EXERCISES

1. Prove that there are infinitely many prime numbers. [*Hint from Euclid*: Let 2, 3, 5,...,P be the set of primes up to P. Show that there is another prime as follows. Let

$$N = 2\cdot 3\cdot 5\cdot 7\cdots P + 1,$$

the product being taken over all primes $\leqq P$. Show that any prime dividing N is not among the primes up to P.]

2. Define a **twin prime** to be a prime p such that $p+2$ is also prime. For instance (3, 5), (5, 7), (11, 13) are twin primes.
   (a) Write down all the twin primes less than 100.
   (b) Are there infinitely many twin primes? Use a computer to compute more twin primes and see if there is any regularity in their occurrence.
3. Observe that $5 = 2^2 + 1$, $17 = 4^2 + 1$, $37 = 6^2 + 1$ are primes. Are there infinitely many primes of the form $n^2 + 1$ where n is a positive integer? Compute all the primes less than 100 which are of the form $n^2 + 1$. Use a computer to compute further primes and see if there is any regularity of occurrence for these primes.
4. Start with a positive odd integer n. Then $3n + 1$ is even. Divide by the largest power of 2 which is a factor of $3n + 1$. You obtain an odd integer n_1. Iterate these operations. In other words, form $3n_1 + 1$, divide by the maximal power of 2 which is a factor of $3n_1 + 1$, and iterate again. What do you think happens? Try it out, starting with $n = 1$, $n = 3$, $n = 5$, and go up to $n = 41$. You will find that at some point, for each of these values of n, the iteration process comes back to 1. There is a conjecture which states that the above iteration procedure will always yield 1, no matter what odd integer n you started with. For an expository article on this problem, see J. C. Lagarias, "The $3x + 1$ problem and its generalizations", *American Mathematical Monthly*, Vol. 92, No. 1, 1985. The problem is traditionally credited to Lothar Collatz, dating back to the 1930's. The problem has a reputation for getting people to think unsuccessfully about it, to the point where someone once made the joke that "this problem was part of a conspiracy to slow down mathematical research in the U.S.". Lagarias gives an extensive bibliography of papers dealing with the problem and some of its offshoots.

Prime numbers constitute one of the oldest and deepest areas of research in mathematics. Fortunately, it is possible to state the greatest problem of mathematics in simple terms, and we shall now do so. The problem is part of the more general framework to describe how the primes are distributed among the integers. There are many refinements to this question. We start by asking approximately how many primes are there $\leqq x$ when x becomes arbitrarily large? We want first an asymptotic formula. We recall briefly a couple of definitions from the basic terminology of functions. Let f, g be two functions of a real variable and assume g positive. We say that $f(x) = O(g(x))$ for $x \to \infty$ if there is a constant $C > 0$ such that $|f(x)| \leqq Cg(x)$ for all x sufficiently large. We say that $f(x)$ is **asymptotic** to $g(x)$ and we write $f \sim g$ if

$$\lim_{x \to \infty} \frac{f(x)}{g(x)} = 1.$$

Let $\pi(x)$ denote the number of primes $p \leqq x$. At the end of the 19th century Hadamard and de la Vallée-Poussin proved the **prime number theorem**, that

$$\pi(x) \sim \frac{x}{\log x}.$$

Thus $x/\log x$ gives a first-order approximation to count the primes. But although the formula $x/\log x$ has the attractiveness of being a closed formula, and of being very simple, it does not give a very good approximation, and conjecturally, there is a much better one, as follows.

Roughly speaking, the idea is that the probability for a positive integer n to be prime is $1/\log n$. What does this mean? It means that $\pi(x)$ should be given with very good approximation by the sum

$$L(x) = \frac{1}{\log 2} + \frac{1}{\log 3} + \cdots + \frac{1}{\log n} = \sum_{k=2}^{n} \frac{1}{\log k},$$

where n is the largest integer $\leqq x$. If x is taken to be an integer, then we take $n = x$. For those who have had calculus, you will see immediately that the above sum is a Riemann sum for the integral usually denoted by $Li(x)$, namely

$$Li(x) = \int_2^x \frac{1}{\log t}\, dt,$$

and that the sum differs from the integral by a small error, bounded independently of x; in other words, $L(x) = Li(x) + O(1)$.

The question is: How good is the approximation of $\pi(x)$ by the sum $L(x)$, or for that matter by the integral $Li(x)$? That's where the big problem comes from. The following conjecture was made by Riemann around 1850.

Riemann Hypothesis. *We have*

$$\pi(x) = L(x) + O(x^{1/2} \log x).$$

This means that the sum $L(x)$ gives an approximation to $\pi(x)$ with an error term which has the order of magnitude $x^{1/2} \log x$ so roughly the square root of x, which is very small compared to x when x is large. You can verify this relationship experimentally by making up tables for $\pi(x)$ and for $L(x)$. You will find that the difference is quite small.

Even knowing the Riemann hypothesis, lots of other questions would still arise. For instance, consider twin primes

$$(3, 5), (5, 7), (11, 13), (17, 19), \ldots.$$

These are primes p such that $p + 2$ is also prime. Let $\pi_t(x)$ denote the number of twin primes $\leqq x$. It is not known today whether there are infinitely many twin primes, but conjecturally it is possible to give an asymptotic estimate for their number. Hardy and Littlewood conjectured that there is a constant $C_t > 0$ such that

$$\pi_t(x) \sim C_t \frac{x}{(\log x)^2}.$$

and they determined this constant explicitly.

Finally, let $\pi_s(x)$ denote the number of primes $\leqq x$ which are of the form $n^2 + 1$. It is not known whether there are infinitely many such primes, but Hardy–Littlewood have conjectured that there is a constant $C_s > 0$ such that

$$\pi_s(x) \sim C_s \frac{x^{1/2}}{\log x}.$$

and they have determined this constant explicitly. The determination of such constants as C_t or C_s is not so easy and depends on subtle relations of dependence between primes. For an informal discussion of these problems with a general audience, and some references to original papers, cf. my books: *The Beauty of Doing Mathematics*, and *Math talks for undergraduates* (the talk on prime numbers), Springer-Verlag.

I, §5. EQUIVALENCE RELATIONS AND CONGRUENCES

Let S be a set. By an **equivalence relation** in S we mean a relation, written $x \sim y$, between certain pairs of elements of S, satisfying the following conditions:

ER 1. *We have $x \sim x$ for all $x \in S$.*

ER 2. *If $x \sim y$ and $y \sim z$ then $x \sim z$.*

ER 3. *If $x \sim y$ then $y \sim x$.*

Suppose we have such an equivalence relation in S. Given an element x of S, let C_x consist of all elements of S which are equivalent to x. Then all elements of C_x are equivalent to one another, as follows at once from our three properties. (Verify this in detail.) Furthermore, you will also verify at once that if x, y are elements of S, then either $C_x = C_y$, or C_x, C_y have no element in common. Each C_x is called an **equivalence class**. We see that our equivalence relation determines a decomposition of S into disjoint equivalence classes. Each element of a class is called a **representative** of the class.

Our first example of the notion of equivalence relation will be the notion of congruence. Let n be a positive integer. Let x, y be integers. We shall say that x is **congruent to** y **modulo** n if there exists an integer m such that $x - y = mn$. This means that $x - y$ lies in the ideal generated by n. If $n \neq 0$, this also means that $x - y$ is divisible by n. We write the relation of congruence in the form.

$$x \equiv y \pmod{n}.$$

It is then immediately verified that this is an equivalence relation, namely that the following properties are verified:

(a) We have $x \equiv x \pmod{n}$.
(b) If $x \equiv y$ and $y \equiv z \pmod{n}$, then $x \equiv z \pmod{n}$.
(c) If $x \equiv y \pmod{n}$ then $y \equiv x \pmod{n}$.

Congruences also satisfy further properties:

(d) If $x \equiv y \pmod{n}$ and z is an integer, then $xz \equiv yz \pmod{n}$.
(e) If $x \equiv y$ and $x' \equiv y' \pmod{n}$, then $xx' \equiv yy' \pmod{n}$. Furthermore $x + x' \equiv y + y' \pmod{n}$.

We give the proof of the first part of (e) as an example. We can write

$$x = y + mn \qquad \text{and} \qquad x' = y' + m'n$$

with some integers m, m'. Then

$$xx' = (y + mn)(y' + m'n) = yy' + mny' + ym'n + mm'nn,$$

and the expression on the right is immediately seen to be equal to

$$yy' + wn$$

for some integer w, so that $xx' = yy' \pmod{n}$, as desired.

We define the **even** integers to be those which are congruent to 0 mod 2. Thus n is even if and only if there exists an integer m such that $n = 2m$. We define the **odd** integers to be all the integers which are not even. It is trivially shown that an odd integer n can be written in the form $2m + 1$ for some integer m.

I, §5. EXERCISES

1. Let n, d be positive integers and assume $1 < d < n$. Show that n can be written in the form

$$n = c_0 + c_1 d + \cdots + c_k d^k$$

with integers c_i such that $0 \leqq c_i < d$, and that these integers c_i are uniquely determined. [*Hint*: For the existence, write $n = qd + c_0$ by the Euclidean algorithm, and then use induction. For the uniqueness, use induction, assuming $c_0, \ldots, c_r$ are uniquely determined; show that c_{r+1} is then uniquely determined.]

2. Let m, n be non-zero integers written in the form

$$m = p_1^{i_1} \cdots p_r^{i_r} \quad \text{and} \quad n = p_1^{j_1} \cdots p_r^{j_r},$$

where i_ν, j_ν are integers $\geqq 0$ and $p_1, \ldots, p_r$ are distinct prime numbers.
   (a) Show that the g.c.d. of m, n can be expressed as a product $p_1^{k_1} \cdots p_r^{k_r}$ where $k_1, \ldots, k_r$ are integers $\geqq 0$. Express k_ν in terms of i_ν and j_ν.
   (b) Define the notion of **least common multiple**, and express the least common multiple of m, n as a product $p_1^{k_1} \cdots p_r^{k_r}$ with integers $k_\nu \geqq 0$. Express k_ν in terms of i_ν and j_ν.
3. Give the g.c.d. and l.c.m. of the following pairs of positive integers:
   (a) $5^3 2^6 3$ and 225
   (b) 248 and 28.

4. Let n be an integer $\geqq 2$.
   (a) Show that any integer x is congruent mod n to a unique integer m such that $0 \leqq m < n$.
   (b) Show that any integer $x \neq 0$, relatively prime to n, is congruent to a unique integer m relatively prime to n, such that $0 < m < n$.
   (c) Let $\varphi(n)$ be the number of integers m relatively prime to n, such that $0 < m < n$. We call φ the **Euler phi function**. We also define $\varphi(1) = 1$. If $n = p$ is a prime number, what is $\varphi(p)$?
   (d) Determine $\varphi(n)$ for each integer n with $1 \leqq n \leqq 10$.

5. **Chinese Remainder Theorem.** Let n, n' be relatively prime positive integers. Let a, b be integers. Show that the congruences

$$x \equiv a \pmod{n},$$
$$x \equiv b \pmod{n'}$$

can be solved simultaneously with some $x \in \mathbf{Z}$. Generalize to several congruences $x \equiv a_i \bmod n_i$, where $n_1, \ldots, n_r$ are pairwise relatively prime positive integers.

6. Let a, b be non-zero relatively prime integers. Show that $1/ab$ can be written in the form

$$\frac{1}{ab} = \frac{x}{a} + \frac{y}{b}$$

with some integers x, y.

7. Show that any rational number $a \neq 0$ can be written in the form

$$a = \frac{x_1}{p_1^{r_1}} + \cdots + \frac{x_n}{p_n^{r_n}},$$

where $x_1, \ldots, x_n$ are integers, $p_1, \ldots, p_n$ are distinct prime numbers, and $r_1, \ldots, r_n$ are integers $\geqq 0$.

8. Let p be a prime number and n an integer, $1 \leqq n \leqq p - 1$. Show that the binomial coefficient $\binom{p}{n}$ is divisible by p.

9. For all integers x, y and all primes p show that $(x + y)^p \equiv x^p + y^p \pmod{p}$.

10. Let n be an integer $\geqq 2$. Show by examples that the bionomial coefficient $\binom{p^n}{k}$ need not be divisible by p^n for $1 \leqq k \leqq p^n - 1$.

11. (a) Prove that a positive integer is divisible by 3 if and only if the sum of its digits is divisible by 3.
    (b) Prove that it is divisible by 9 if and only if the sum of its digits is divisible by 9.
    (c) Prove that it is divisible by 11 if and only if the alternating sum of its digits is divisible by 11. In other words, let the integer be

$$n = a_k a_{k-1} \cdots a_0 = a_0 + a_1 10 + a_2 10^2 + \cdots + a_k 10^k, \qquad 0 \leqq a_i \leqq 9.$$

    Then n is divisible by 11 if and only if $a_0 - a_1 + a_2 - a_3 + \cdots + (-1)^k a_k$ is divisible by 11.

12. A positive integer is called **palyndromic** if its digits from left to right are the same as the digits from right to left. For instance, 242 and 15851 are palyndromic. The integers 11, 101, 373, 10301 are palyndromic primes. Observe that except for 11, the others have an odd number of digits.
    (a) Is there a palyndromic prime with four digits? With an even number of digits (except for 11)?
    (b) Are there infinitely many palyndromic primes? (This is an unsolved problem in mathematics.)

CHAPTER II

Groups

II, §1. GROUPS AND EXAMPLES

A **group** G is a set, together with a rule (called a **law of composition**) which to each pair of elements x, y in G associates an element denoted by xy in G, having the following properties.

GR 1. *For all x, y, z in G we have associativity, namely*

$$(xy)z = x(yz).$$

GR 2. *There exists an element e of G such that $ex = xe = x$ for all x in G.*

GR 3. *If x is an element of G, then there exists an element y of G such that $xy = yx = e$.*

Strictly speaking, we call G a **multiplicative** group. If we denote the element of G associated with the pair (x, y) by $x + y$, then we write **GR 1** in the form

$$(x + y) + z = x + (y + z),$$

GR 2 in the form that there exists an element 0 such that

$$0 + x = x + 0 = x$$

for all x in G, and **GR 3** in the form that given $x \in G$, there exists an element y of G such that

$$x + y = y + x = 0.$$

With this notation, we call G an **additive** group and $x + y$ the **sum**. We shall use the + notation only when the group satisfies the additional rule

$$x + y = y + x$$

for all x, y in G. With the multiplicative notation, this is written $xy = yx$ for all x, y in G, and if G has this property, we call G a **commutative**, or **abelian** group.

We shall now prove various simple statements which hold for all groups.

Let G be a group. The element e of G whose existence is asserted by **GR 2** *is uniquely determined.*

Proof. If e, e' both satisfy this condition, then

$$e' = ee' = e.$$

We call this element the **unit element** of G. We call it the **zero** element in the additive case.

Let $x \in G$. The element y such that $yx = xy = e$ is uniquely determined.

Proof. If z satisfies $zx = xz = e$, then

$$z = ez = (yx)z = y(xz) = ye = y.$$

We call y the **inverse** of x, and denote it by x^{-1}. In the additive notation, we write $y = -x$.

We shall now give examples of groups. Many of these involve notions which the reader will no doubt have encountered already in other courses.

Example 1. Let **Q** denote the rational numbers, i.e. the set of all fractions m/n where m, n are integers, and $n \neq 0$. Then **Q** is a group under addition. Furthermore, the non-zero elements of **Q** form a group under multiplication, denoted by **Q***.

Example 2. The real numbers and complex numbers are groups under addition. The non-zero real numbers and non-zero complex numbers are groups under multiplication. We shall always denote the real and complex numbers by **R** and **C** respectively, and the group of non-zero elements by **R*** and **C*** respectively.

Example 3. The complex numbers of absolute value 1 form a group under multiplication.

Example 4. The set consisting of the numbers 1, -1 is a group under multiplication, and this group has 2 elements.

Example 5. The set consiting of the numbers 1, -1, i, $-i$ is a group under multiplication. This group has 4 elements.

Example 6 (The direct product). Let G, G' be groups. Let $G \times G'$ be the set consisting of all pairs (x, x') with $x \in G$ and $x' \in G'$. If (x, x') and (y, y') are such pairs, define their product to be $(xy, x'y')$. Then $G \times G'$ is a group.

It is a simple matter to verify that all the conditions **GR 1, 2, 3** are satisfied, and we leave this to the reader. We call $G \times G'$ the **direct product** of G and G'.

One may also take a direct product of a finite number of groups. Thus if $G_1, \ldots, G_n$ are groups, we let

$$\prod_{i=1}^{n} G_i = G_1 \times \cdots \times G_n$$

be the set of all n-tuples $(x_1, \ldots, x_n)$ with $x_i \in G_i$. We define multiplication componentwise, and see at once that $G_1 \times \cdots \times G_n$ is a group. If e_i is the unit element of G_i, then $(e_1, \ldots, e_n)$ is the unit element of the product.

Example 7. The Euclidean space $\mathbf{R}^n$ is nothing but the product

$$\mathbf{R}^n = \mathbf{R} \times \cdots \times \mathbf{R}$$

taken n times. In this case, we view **R** as an additive group.

A group consisting of one element is said to be **trivial**. A group in general may have infinitely many elements, or only a finite number. If G has only a finite number of elements, then G is called a **finite group**, and the number of elements of G is called its **order**. The group of Example 4 has order 2, and that of Example 5 has order 4.

In Examples 1 through 5, the groups happen to be commutative. We shall find non-commutative examples later, when we study groups of permutations. See also groups of matrices in Chapter VI.

Let G be a group. Let $x_1,\ldots,x_n$ be elements of G. We can then form their product, which we define by induction to be

$$x_1 \cdots x_n = (x_1 \cdots x_{n-1})x_n.$$

Using the associative law of **GR 1**, one can show that one gets the same value for this product no matter how parentheses are inserted around its elements. For instance for $n = 4$,

$$(x_1x_2)(x_3x_4) = x_1(x_2(x_3x_4))$$

and also

$$(x_1x_2)(x_3x_4) = ((x_1x_2)x_3)x_4.$$

We omit the proof in the general case (done by induction), because it involves slight notational complications which we don't want to go into. The above product will also be written

$$\prod_{i=1}^{n} x_i.$$

If the group is written additively, then we write the sum sign instead of the product sign, so that a sum of n terms looks like

$$\sum_{i=1}^{n} x_i = (x_1 + \cdots + x_{n-1}) + x_n = x_1 + \cdots + x_n.$$

The group G being commutative, and written additively, it can be shown by induction that the above sum is independent of the order in which $x_1,\ldots,x_n$ are taken. We shall again omit the proof. For example, if $n = 4$,

$$\begin{aligned}(x_1 + x_2) + (x_3 + x_4) &= x_1 + (x_2 + x_3 + x_4)\\ &= x_1 + (x_3 + x_2 + x_4)\\ &= x_3 + (x_1 + x_2 + x_4).\end{aligned}$$

Let G be a group, and H a subset of G. We shall say that H is a **subgroup** if it contains the unit element, and if, whenever $x, y \in H$, then xy and x^{-1} are also elements of H. (Additively, we write $x + y \in H$ and $-x \in H$.) Then H is itself a group in its own right, the law of composition in H being the same as that in G. The unit element of G constitutes a subgroup, which is called the **trivial subgroup**. Every group G is a subgroup of itself.

Example 8. The additive group of rational numbers is a subgroup of the additive group of real numbers. The group of complex numbers of absolute value 1 is a subgroup of the multiplicative group of non-zero complex numbers. The group $\{1, -1\}$ is a subgroup of $\{1, -1, i, -i\}$.

There is a general way of obtaining subgroups from a group. Let S be a subset of a group G, having at least one element. Let H be the set of elements of G consisting of all products $x_1 \cdots x_n$ such that x_i or x_i^{-1} is an element of S for each i, and also containing the unit element. Then H is obviously a subgroup of G, called the subgroup **generated** by S. We also say that S is a set of **generators** of H. If S is a set of generators for H, we shall use the notation

$$H = \langle S \rangle.$$

Thus if elements $\{x_1, \dots, x_r\}$ form a set of generators for G, we write

$$G = \langle x_1, \dots, x_r \rangle.$$

Example 9. The number 1 is a generator for the additive group of integers. Indeed, every integer can be written in the form

$$1 + 1 + \cdots + 1$$

or

$$-1 - 1 - \cdots - 1,$$

or it is the 0 integer.

Observe that in additive notation, the condition that S be a set of generators for the group is that every element of the group not 0 can be written

$$x_1 + \cdots + x_n,$$

where $x_i \in S$ or $-x_i \in S$.

Example 10. Let G be a group. Let x be an element of G. If n is a positive integer, we define x^n to be

$$xx \cdots x,$$

the product being taken n times. If $n = 0$, we define $x^0 = e$. If $n = -m$ where m is an integer > 0, we define

$$x^{-m} = (x^{-1})^m.$$

It is then routinely verified that the rule

$$\boxed{x^{m+n} = x^m x^n}$$

holds for all integers m, n. The verification is tedious but straightforward. For instance, suppose m, n are positive integers. Then

$$x^m x^n = \underbrace{x \cdots x}_{m \text{ times}} \underbrace{x \cdots x}_{n \text{ times}} = \underbrace{x \cdots x}_{m+n \text{ times}}.$$

Suppose again that m, n are positive integers, and $m < n$. Then (see Exercise 3)

$$x^{-m} x^n = \underbrace{x^{-1} \cdots x^{-1}}_{m \text{ times}} \underbrace{x \cdots x}_{n \text{ times}} = x^{n-m}$$

because the product of x^{-1} taken m times will cancel the product of x taken m times, and will leave x^{n-m} on the right-hand side. The other cases are proved similarly. One could also formalize this by induction, but we now omit these tedious steps.

Similarly, we also have the other rule of exponents, namely

$$\boxed{(x^m)^n = x^{mn}.}$$

It is also tedious to prove, but it applies to multiplication in groups just as it applies to numbers in elementary school, because one uses only the law of composition, multiplication, and its associativity, and multiplicative inverses for the proof. For instance, if m, n are positive integers, then

$$(x^m)^n = \underbrace{x^m \cdots x^m}_{n \text{ times}} = \underbrace{x \cdots x}_{mn \text{ times}} = x^{mn}.$$

If m or n is negative, then one has to go through the definitions to see that the rule applies also, and we omit these arguments.

If the group is written additively, then we write nx instead of x^n, and the rules read:

$$\boxed{(m+n)x = mx + nx \quad \text{and} \quad (mn)x = m(nx).}$$

Observe also that we have the rule

$$\boxed{(x^n)^{-1} = (x^{-1})^n.}$$

To verify this, suppose that n is a positive integer. Then (see Exercise 3)

$$\underbrace{x \cdots x}_{n \text{ times}} \underbrace{x^{-1} \cdots x^{-1}}_{n \text{ times}} = e$$

because we can use the definition $xx^{-1} = e$ repeatedly. If n is negative, one uses the definition $x^{-m} = (x^{-1})^m$ with m positive to give the proof.

Let G be a group and let $a \in G$. *Let H be the subset of elements of G consisting of all powers a^n with $n \in \mathbf{Z}$. Then H is the subgroup generated by a.* Indeed, H contains the unit element $e = a^0$. Let a^n, $a^m \in H$. Then

$$a^m a^n = a^{m+n} \in H.$$

Finally, $(a^n)^{-1} = a^{-n} \in H$. So H satisfies the conditions of a subgroup, and H is generated by a.

Let G be a group. We shall say that G is **cyclic** if there exists an element a of G such that every element x of G can be written in the form a^n for some integer n. The subgroup H above is the cyclic subgroup generated by a.

Example 11. Consider the additive group of integers $\mathbf{Z}$. Then $\mathbf{Z}$ is cyclic, generated by 1. A subgroup of $\mathbf{Z}$ is merely what we called an ideal in Chapter I. We can now interpret Theorem 3.1 of Chapter I as stating:

Let H be a subgroup of $\mathbf{Z}$. If H is not trivial, let d be the smallest positive integer in H. Then H consists of all elements nd, with $n \in \mathbf{Z}$, and so H is cyclic.

We now look more closely at cyclic groups. Let G be a cyclic group and let a be a generator. Two cases can occur.

Case 1. There is no positive integer m such that $a^m = e$. Then for every integer $n \neq 0$ it follows that $a^n \neq e$. In this case, we say that G is **infinite cyclic**, or that a has **infinite order**. In fact the elements

$$a^n \quad \text{with} \quad n \in \mathbf{Z}$$

are all distinct. To see this, suppose that $a^r = a^s$ with some integers r, $s \in \mathbf{Z}$. Then $a^{s-r} = e$ so $s - r = 0$ and $s = r$. For example, the number 2 generates an infinite cyclic subgroup of the multiplicative group of complex numbers. Its elements are

$$\ldots, 2^{-5}, 2^{-4}, \tfrac{1}{8}, \tfrac{1}{4}, \tfrac{1}{2}, 1, 2, 4, 8, 2^4, 2^5, \ldots.$$

Case 2. There exists a positive integer m such that $a^m = e$. Then we say that a has **finite order**, and we call m an **exponent** for a. Let J be the set of integers $n \in \mathbf{Z}$ such that $a^n = e$. Then J is a subgroup of $\mathbf{Z}$. This assertion is routinely verified: we have $0 \in J$ because $a^0 = e$ by definition. If $m, n \in J$ then

$$a^{m+n} = a^m a^n = ee = e,$$

so $m + n \in J$. Also $a^{-m} = (a^m)^{-1} = e$, so $-m \in J$. Thus J is a subgroup of $\mathbf{Z}$. By Theorem 3.1 of Chapter I, the smallest positive integer d in J is a generator of J. By definition, this positive integer d is the smallest positive integer such that $a^d = e$, and d is called the **period** of a. If $a^n = e$ then $n = ds$ for some integer s.

Suppose a is an element of period d. Let n be an integer. By the Euclidean algorithm, we can write

$$n = qd + r \qquad \text{with } q, r \in \mathbf{Z} \text{ and } 0 \leqq r < d.$$

Then

$$a^n = a^r.$$

Theorem 1.1. *Let G be a group and $a \in G$. Suppose that a has finite order. Let d be the period of a. Then a generates a cyclic subgroup of order d, whose elements are $e, a, \ldots, a^{d-1}$.*

Proof. The remark just before the theorem shows that this cyclic subgroup consists of the powers $e, a, \ldots, a^{d-1}$. We must now show that the elements

$$e, a, \ldots, a^{d-1}$$

are distinct. Indeed, suppose $a^r = a^s$ with $0 \leqq r \leqq d - 1$ and

$$0 \leqq s \leqq d - 1,$$

say $r \leqq s$. Then $a^{s-r} = e$. Since

$$0 \leqq s - r < d,$$

we must have $s - r = 0$, whence $r = s$. We conclude that the cyclic group generated by a in this case has order d.

Example 12. The multiplicative group $\{1, -1\}$ is cyclic of order 2.

Example 13. The complex numbers $\{1, i, -1, -i\}$ form a cyclic group of order 4. The number i is a generator.

II, §1. EXERCISES

1. Let G be a group and a, b, c be elements of G. If $ab = ac$, show that $b = c$.
2. Let G, G' be finite groups, of orders m, n respectively. What is the order of $G \times G'$?

3. Let $x_1, \ldots, x_n$ be elements of a group G. Show (by induction) that

$$(x_1 \cdots x_n)^{-1} = x_n^{-1} \cdots x_1^{-1}.$$

What does this look like in additive notation? For two elements $x, y \in G$, we have $(xy)^{-1} = y^{-1}x^{-1}$. Write this also in additive notation.

4. (a) Let G be a group and $x \in G$. Suppose that there is an integer $n \geqq 1$ such that $x^n = e$. Show that there is an integer $m \geqq 1$ such that $x^{-1} = x^m$.
(b) Let G be a finite group. Show that given $x \in G$, there exists an integer $n \geqq 1$ such that $x^n = e$.

5. Let G be a finite group and S a set of generators. Show that every element of G can be written in the form

$$x_1 \cdots x_n,$$

where $x_i \in S$.

6. Let G be a group such that $x^2 = 1$ for all $x \in G$. Prove that G is abelian.

7. There exists a group G of order 4 having two generators x, y such that $x^2 = y^2 = e$ and $xy = yx$. Determine all subgroups of G. Show that

$$G = \{e, x, y, xy\}.$$

8. There exists a group G of order 8 having two generators x, y such that $x^4 = y^2 = e$ and $xy = yx^3$. Show that every element of G can be written in the form $x^i y^j$ with integers i, j such that $i = 0, 1, 2, 3$ and $j = 0, 1$. Conclude that these elements are distinct. Make up a **multiplication table** by writing the product of two elements in the blank spaces, and expressing them in the form $x^i y^j$ with $i = 0, 1, 2, 3$ and $j = 0, 1$.

e	x	x^2	x^3	y	yx	yx^2	yx^3
x							
x^2							
x^3					yx^2		
y							
yx							
yx^2							
yx^3							

We filled one entry, namely $x^3yx = yx^2$.

9. There exists a group G of order 8 having generators denoted by i, j, k such that

$$ij = k, \quad jk = i, \quad ki = j,$$
$$i^2 = j^2 = k^2.$$

Denote i^2 by m.

(a) Show that every element of G can be written in the form

$$e, i, j, k, m, mi, mj, mk,$$

and hence that these are precisely the distinct elements of G.

(b) Make up a multiplication table as in Exercise 8.

The group G in Exercise 9 is called the **quaternion group**. One frequently writes $-1, -i, -j, -k$ instead of m, mi, mj, mk. Cf. Chapter VI, §4 for the space of quaternions.

10. There exists a group G of order 12 having generators x, y such that

$$x^6 = y^2 = e \text{ and } xy = yx^5.$$

Show that the elements $x^i y^j$ with $0 \leqq i \leqq 5$ and $0 \leqq j \leqq 1$ are the distinct elements of G. Make up a multiplication table as in the previous exercises.

11. The groups of Exercises 8 and 10 have representations as groups of symmetries. For instance, in Exercise 8, let σ be the rotation which maps each corner of the square

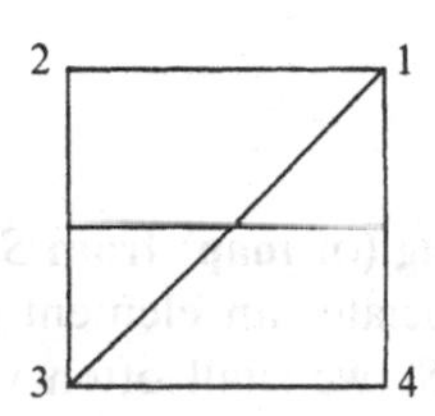

on the next corner (taking, say counterclockwise rotation), and let τ be the reflection across the indicated diagonal. Show geometrically that σ and τ satisfy the relations of Exercise 8. Express in terms of powers of σ and τ the reflection across the horizontal line as indicated on the square. Using the notation of §6, we can write $\sigma = [1234]$ and $\tau = [24]$.

12. In the case of Exercise 10, do the analogous geometric interpretation, taking a hexagon instead of a square.

(*Note*: The groups of Exercises 11 and 12 can essentially be understood as groups of permutations of the vertices. Cf. Exercises 8 and 9 of §6.)

13. Let G be a group and H a subgroup. Let $x \in G$. Let xHx^{-1} be the subset of G consisting of all elements xyx^{-1} with $y \in H$. Show that xHx^{-1} is a subgroup of G.

14. Let G be a group and let S be a set of generators of G. Assume that $xy = yx$ for all $x, y \in S$. Prove that G is abelian. Thus to test whether a group is abelian or not, it suffices to verify the commutative rule on a set of generators.

Exercises on cyclic groups

15. A **root of unity** in the complex numbers is a number ζ such that $\zeta^n = 1$ for some positive integer n. We then say that ζ is an n-th root of unity. Describe the set of n-th roots of unity in $\mathbf{C}$. Show that this set is a cyclic group of order n.

16. Let G be a finite cyclic group of order n. Show that for each positive integer d dividing n, there exists a subgroup of order d.

17. Let G be a finite cyclic group of order n. Let a be a generator. Let r be an integer $\neq 0$, and relatively prime to n.
    (a) Show that a^r is also a generator of G.
    (b) Show that every generator of G can be written in this form.
    (c) Let p be a prime number, and G a cyclic group of order p. How many generators does G have?

18. Let m, n be relatively prime positive integers. Let G, G' be cyclic groups of orders m, n respectively. Show that $G \times G'$ is cyclic, of order mn.

19. (a) Let G be a multiplicative finite abelian group. Let a be the product of all the elements of the group. Prove that $a^2 = e$.
    (b) Suppose in addition that G is cyclic. If G has odd order, show that $a = e$. If G has even order, show that $a \neq e$.

II, §2. MAPPINGS

Let S, S' be sets. A **mapping** (or **map**) **from** S **to** S' is an association which to every element of S associates an element of S'. Instead of saying that f is a mapping of S into S', we shall often write the symbols $f: S \to S'$.

If $f: S \to S'$ is a mapping, and x is an element of S, then we denote by $f(x)$ the element of S' associated to x by f. We call $f(x)$ the **value** of f at x, or also the **image** of x under f. The set of all elements $f(x)$, for all $x \in S$, is called the **image** of f. If T is a subset of S, then the set of elements $f(x)$ for all $x \in T$ is called the **image** of T, and denoted by $f(T)$.

If f is as above, we often write $x \mapsto f(x)$ to denote the image of x under f. Note that we distinguish two types of arrows, namely

$$\to \quad \text{and} \quad \mapsto.$$

Example 1. Let S and S' be both equal to $\mathbf{R}$. Let $f: \mathbf{R} \to \mathbf{R}$ be the mapping $f(x) = x^2$, i.e. the mapping whose value at x is x^2. We can also express this by saying that f is the mapping such that $x \mapsto x^2$. The image of f is the set of real numbers $\geqq 0$.

Let $f: S \to S'$ be a mapping, and T a subset of S. Then we can define a map $T \to S'$ by the same rule $x \mapsto f(x)$ for $x \in T$. In other words, we can view f as defined only on T. This map is called the **restriction** of f to T and is denoted by $f \mid T: T \to S'$.

Let S, S' be sets. A map $f: S \to S'$ is said to be **injective** if whenever $x, y \in S$ and $x \neq y$ then $f(x) \neq f(y)$. We could also write this condition in the form: if $f(x) = f(y)$ then $x = y$.

Example 2. The mapping f of Example 1 is not injective. Indeed, we have $f(1) = f(-1)$. Let $g: \mathbf{R} \to \mathbf{R}$ be the mapping $x \mapsto x + 1$. Then g is injective, because if $x \neq y$ then $x + 1 \neq y + 1$, i.e. $g(x) \neq g(y)$.

Let S, S' be sets. A map $f: S \to S'$ is said to be **surjective** if the image $f(S)$ of S is equal to all of S'. This means that given any element $x' \in S'$, there exists an element $x \in S$ such that $f(x) = x'$. One says that f is **onto** S'.

Example 3. Let $f: \mathbf{R} \to \mathbf{R}$ be the mapping $f(x) = x^2$. Then f is not surjective, because no negative number is in the image of f.

Let $g: \mathbf{R} \to \mathbf{R}$ be the mapping $g(x) = x + 1$. Then g is surjective, because given a number y, we have $y = g(y - 1)$.

Remark. Let $\mathbf{R}'$ denote the set of real numbers $\geqq 0$. One can view the association $x \mapsto x^2$ as a map of $\mathbf{R}$ into $\mathbf{R}'$. When so viewed, the map is then surjective. Thus it is a reasonable convention *not* to identify this map with the map $f: \mathbf{R} \to \mathbf{R}$ defined by the same formula. To be completely accurate, we should therefore incorporate the set of arrival and the set of departure of the map into our notation, and for instance write

$$f^S_{S'}: S \to S'$$

instead of our $f: S \to S'$. In practice, this notation is too clumsy, so that one omits the indices S, S'. However, the reader should keep in mind the distinction between the maps

$$f^{\mathbf{R}}_{\mathbf{R}'}: \mathbf{R} \to \mathbf{R}' \qquad \text{and} \qquad f^{\mathbf{R}}_{\mathbf{R}}: \mathbf{R} \to \mathbf{R}$$

both defined by the rule $x \mapsto x^2$. The first map is surjective whereas the second one is *not*.

Let S, S' be sets, and $f: S \to S'$ a mapping. We say that f is **bijective** if f is both injective and surjective. This means that given an element $x' \in S'$, there exists a unique element $x \in S$ such that $f(x) = x'$. (Existence because f is surjective, and uniqueness because f is injective.)

Example 4. Let J_n be the set of integers $\{1, 2, \ldots, n\}$. A bijective map $\sigma: J_n \to J_n$ is called a **permutation** of the integers from 1 to n. Thus, in particular, a permutation σ as above is a mapping $i \mapsto \sigma(i)$. We shall study permutations in greater detail later in this chapter.

Example 5. Let S be a non-empty set, and let

$$I: S \to S$$

be the map such that $I(x) = x$ for all $x \in S$. Then I is called the **identity** mapping, and also denoted by id. It is obviously bijective. Often we need to specify the set S in the notation and we write I_S or id_S for the identity mapping of S. Let T be a subset of S. The identity map $t \mapsto t$ for $t \in T$, viewed as a mapping $T \to S$ is called the **inclusion**, and is sometimes denoted by

$$T \hookrightarrow S.$$

Let S, T, U be sets, and let

$$f: S \to T \quad \text{and} \quad g: T \to U$$

be mappings. Then we can form the **composite mapping**

$$g \circ f: S \to U$$

defined by the rule

$$(g \circ f)(x) = g(f(x))$$

for all $x \in S$.

Example 6. Let $f: \mathbf{R} \to \mathbf{R}$ be the map $f(x) = x^2$, and $g: \mathbf{R} \to \mathbf{R}$ the map $g(x) = x + 1$. Then $g(f(x)) = x^2 + 1$. Note that in this case, we can form also $f(g(x)) = f(x + 1) = (x + 1)^2$, and thus that

$$f \circ g \neq g \circ f.$$

Composition of mappings is associative. This means: Let S, T, U, V be sets, and let

$$f: S \to T, \qquad g: T \to U, \qquad h: U \to V$$

be mappings. Then

$$h \circ (g \circ f) = (h \circ g) \circ f.$$

Proof. The proof is very simple. By definition, we have, for any element $x \in S$,

$$(h \circ (g \circ f))(x) = h((g \circ f)(x)) = h(g(f(x))).$$

On the other hand,

$$((h \circ g) \circ f)(x) = (h \circ g)(f(x)) = h(g(f(x))).$$

By definition, this means that $(h \circ g) \circ f = h \circ (g \circ f)$.

Let S, T, U be sets, and $f: S \to T$, $g: T \to U$ mappings. If f and g are injective, then $g \circ f$ is injective. If f and g are surjective, then $g \circ f$ is surjective. If f and g are bijective, then so is $g \circ f$.

Proof. As to the first statement, assume that f, g are injective. Let $x, y \in S$ and $x \neq y$. Then $f(x) \neq f(y)$ because f is injective, and hence $g(f(x)) \neq g(f(y))$ because g is injective. By the definition of the composite map, we conclude that $g \circ f$ is injective. The second statement will be left as an exercise. The third is a consequence of the first two and the definition of bijective.

Let $f: S \to S'$ be a mapping. By an **inverse mapping** for f we mean a mapping

$$g: S' \to S$$

such that

$$g \circ f = \mathrm{id}_S \quad \text{and} \quad f \circ g = \mathrm{id}_{S'}.$$

As an exercise, prove that if an inverse mapping for f exists, then it is unique, in the sense that if g_1, g_2 are inverse mappings for f, then $g_1 = g_2$. We then denote the inverse mapping g by f^{-1}. Thus by definition, the inverse mapping f^{-1} is characterized by the property that for all $x \in S$ and $x' \in S'$ we have

$$f^{-1}(f(x)) = x \quad \text{and} \quad f(f^{-1}(x')) = x'.$$

Let $f: S \to S'$ be a mapping. Then f is bijective if and only if f has an inverse mapping.

Proof. Suppose f is bijective. We define a mapping $g: S' \to S$ by the rule: For $x' \in S'$, let

$$g(x') = \text{unique element } x \in S \text{ such that } f(x) = x'.$$

It is immediately verified that g satisfies the conditions of being *an* inverse mapping, and therefore *the* inverse mapping. We leave the other implication as Exercise 1, namely prove that if an inverse mapping for f exists, then f is bijective.

Example 7. If $f: \mathbf{R} \to \mathbf{R}$ is the map such that

$$f(x) = x + 1,$$

then $f^{-1}: \mathbf{R} \to \mathbf{R}$ is the map such that $f^{-1}(x) = x - 1$.

Example 8. Let $\mathbf{R}^+$ denote the set of positive real numbers (i.e. real numbers > 0). Let $h: \mathbf{R}^+ \to \mathbf{R}^+$ be the map $h(x) = x^2$. Then h is bijective, and its inverse mapping is the square root mapping, i.e.

$$h^{-1}(x) = \sqrt{x}$$

for all $x \in \mathbf{R}$, $x > 0$.

Let S be a set. A bijective mapping $f: S \to S$ of S with itself is called a **permutation** of S. The set of permutations of S is denoted by

$$\mathrm{Perm}(S).$$

Proposition 2.1. *The set of permutations* Perm(S) *is a group, the law of composition being composition of mappings.*

Proof. We have already seen that composition of mappings is associative. There is a unit element in Perm(S), namely the identity I_S. If f, g are permutations of S, then we have already remarked that $f \circ g$ and $g \circ f$ are bijective, and $f \circ g$, $g \circ f$ map S onto itself, so $f \circ g$ and $g \circ f$ are permutations of S. Finally, a permutation f has an inverse f^{-1} as already remarked, so all the group axioms are satisfied, and the proposition is proved.

If σ, τ are permutations of a set S, then we often write

$$\sigma\tau \quad \text{instead of} \quad \sigma \circ \tau,$$

namely we omit the small circle when we compose permutations to fit the abstract formalism of the law of composition in a group.

Example 9. Let's go back to plane geometry. A mapping $F: \mathbf{R}^2 \to \mathbf{R}^2$ is said to be an **isometry** if F preserves distances, in other words, given two points P, $Q \in \mathbf{R}^2$ we have

$$\mathrm{dist}(P, Q) = \mathrm{dist}(F(P), F(Q)).$$

In high school geometry, it should have been mentioned that rotations, reflections, and translations preserve distances, and so are isometries. It is immediate from the definition that a composite of isometries is an isometry. It is not immediately clear that an isometry has an inverse. However, there is a basic theorem:

Let F be an isometry. Then there exist reflections $R_1, \dots, R_m$ (through lines $L_1, \dots, L_m$ respectively) such that

$$F = R_1 \cdots R_m.$$

The product is of course composition of mappings. If R is a reflection through a line L, then $R \circ R = R^2 = I$ so $R = R^{-1}$. Hence if we admit the above basic theorem, then we see that every isometry has an inverse, namely

$$F^{-1} = R_m^{-1} \cdots R_1^{-1}.$$

Since the identity I is an isometry, it follows that the set of isometries is a subgroup of the group of permutations of $\mathbf{R}^2$. For a proof of the above basic theorem, see, for instance, my book with Gene Murrow: *Geometry.*

Remark. The notation f^{-1} is also used even when f is not bijective. Let X, Y be sets and let

$$f: X \to Y$$

be a mapping. Let Z be a subset of Y. We define the **inverse image**

$$f^{-1}(Z) = \text{subset of } X \text{ consisting of those elements } x \in X \text{ such that } f(x) \in Z.$$

Thus in general f^{-1} is NOT a mapping from Y into X, but is a mapping from the set of subsets of Y to the set of subsets of X. We call $f^{-1}(Z)$ the **inverse image of Z under f**. You can work out some properties of the inverse image in Exercise 6. Often the subset Z may consist of one element y. Thus, if $y \in Y$ we define $f^{-1}(y)$ to be the set of all elements $x \in X$ such that $f(x) = y$. If y is not in the image of f, then $f^{-1}(y)$ is *empty*. If y is in the image of f, then $f^{-1}(y)$ may consist of more than one element.

Example 10. Let $f: \mathbf{R} \to \mathbf{R}$ be the mapping $f(x) = x^2$. Then

$$f^{-1}(1) = \{1, -1\},$$

and $f^{-1}(-2)$ is empty.

Example 11. Suppose $f: X \to Y$ is the inclusion mapping, so X is a subset of Y. Then $f^{-1}(Z)$ is the intersection, namely

$$f^{-1}(Z) = Z \cap X.$$

Coordinate maps. Let Y_i $(i = 1,\dots,n)$ be sets. A mapping

$$f: X \to \prod Y_i = Y_1 \times \cdots \times Y_n$$

of X into the product is given by n mappings $f_i: X \to Y_i$ such that

$$f(x) = (f_1(x),\dots,f_n(x)) \qquad \text{for all} \quad x \in X.$$

The maps f_i are called the **coordinate mappings** of f.

II, §2. EXERCISES

1. Let $f: S \to S'$ be a mapping, and assume that there exists a map $g: S' \to S$ such that

$$g \circ f = I_S \qquad \text{and} \qquad f \circ g = I_{S'},$$

in other words, f has an inverse. Show that f is both injective and surjective.

2. Let $\sigma_1,\dots,\sigma_r$ be permutations of a set S. Show that

$$(\sigma_1 \cdots \sigma_r)^{-1} = \sigma_r^{-1} \cdots \sigma_1^{-1}.$$

3. Let S be a non-empty set and G a group. Let $M(S, G)$ be the set of mappings of S into G. If f, $g \in M(S, G)$, define $fg: S \to G$ to be the map written such that $(fg)(x) = f(x)g(x)$. Show that $M(S, G)$ is a group. If G is written additively, how would you write the law of composition in $M(S, G)$?

4. Give an example of two permutations of the integers $\{1, 2, 3\}$ which do not commute.

5. Let S be a set, G a group, and $f: S \to G$ a bijective mapping. For each x, $y \in S$ define the product

$$xy = f^{-1}(f(x)f(y)).$$

Show that this multiplication defines a group structure on S.

6. Let X, Y be sets and $f: X \to Y$ a mapping. Let Z be a subset of Y. Define $f^{-1}(Z)$ to be the set of all $x \in X$ such that $f(x) \in Z$. Prove that if Z, W are subsets of Y then

$$f^{-1}(Z \cup W) = f^{-1}(Z) \cup f^{-1}(W),$$

$$f^{-1}(Z \cap W) = f^{-1}(Z) \cap f^{-1}(W).$$

II, §3. HOMOMORPHISMS

Let G, G' be groups. A **homomorphism**

$$f: G \to G'$$

of G into G' is a map having the following property: For all $x, y \in G$, we have

$$f(xy) = f(x)f(y),$$

and in additive notation, $f(x + y) = f(x) + f(y)$.

Example 1. Let G be a commutative group. The map $x \mapsto x^{-1}$ of G into itself is a homomorphism. In additive notation, this map looks like $x \mapsto -x$. The verification that it has the property defining a homomorphism is immediate.

Example 2. The map

$$z \mapsto |z|$$

is a homomorphism of the multiplicative group of non-zero complex numbers into the multiplicative group of non-zero complex numbers (in fact, into the multiplicative group of positive real numbers).

Example 3. The map

$$x \mapsto e^x$$

is a homomorphism of the additive group of real numbers into the multiplicative group of positive real numbers. Its inverse map, the logarithm, is also a homomorphism.

Let, G, H be groups and suppose H is a direct product

$$H = H_1 \times \cdots \times H_n.$$

Let $f: G \to H$ be a map, and let $f_i: G \to H_i$ be its i-th coordinate map. Then f is a homomorphism if and only if each f_i is a homomorphism.

The proof is immediate, and will be left to the reader.

For the sake of brevity, we sometimes say: "Let $f: G \to G'$ be a group-homomorphism" instead of saying: "Let G, G' be groups, and let f be a homomorphism of G into G'."

Let $f: G \to G'$ be a group-homomorphism, and let e, e' be the unit elements of G, G' respectively. Then $f(e) = e'$.

Proof. We have $f(e) = f(ee) = f(e)f(e)$. Multiplying both sides by $f(e)^{-1}$ gives the desired result.

Let $f: G \to G'$ be a group-homomorphism. Let $x \in G$. Then

$$f(x^{-1}) = f(x)^{-1}.$$

Proof. We have

$$e' = f(e) = f(xx^{-1}) = f(x)f(x^{-1}).$$

Let $f: G \to G'$ and $g: G' \to G''$ be group-homomorphisms. Then the composite map $g \circ f$ is a group-homomorphism of G into G''.

Proof. We have

$$(g \circ f)(xy) = g(f(xy)) = g(f(x)f(y)) = g(f(x))g(f(y)).$$

Let $f: G \to G'$ be a group-homomorphism. The image of f is a subgroup of G'.

Proof. If $x' = f(x)$ with $x \in G$, and $y' = f(y)$ with $y \in G$, then

$$x'y' = f(x)f(y) = f(xy)$$

is also in the image. Also, $e' = f(e)$ is in the image, and $x'^{-1} = f(x^{-1})$ is in the image. Hence the image is a subgroup.

Let $f: G \to G'$ be a group-homomorphism. We define the **kernel** of f to consist of all elements $x \in G$ such that $f(x) = e'$.

The kernel of a homomorphism $f: G \to G'$ is a subgroup of G.

The proof is routine and will be left to the reader. (The kernel contains the unit element e because $f(e)$ is the unit element of G'. And so on.)

Example 4. Let G be a group and let $a \in G$. The map

$$n \mapsto a^n$$

is a homomorphism of $\mathbf{Z}$ into G. This is merely a restatement of the rules for exponents discussed in §1. The kernel of this homomorphism consists of all integers n such that $a^n = e$, and as we have seen in §1, this kernel is either 0, or is the subgroup generated by the period of a.

Let $f: G \to G'$ be a group-homomorphism. If the kernel of f consists of e alone, then f is injective.

Proof. Let $x, y \in G$ and suppose that $f(x) = f(y)$. Then

$$e' = f(x)f(y)^{-1} = f(x)f(y^{-1}) = f(xy^{-1}).$$

Hence $xy^{-1} = e$, and consequently $x = y$, thus showing that f is injective.

An injective homomorphism will be called an **embedding**. The same terminology will be used for other objects which we shall meet later, such as rings and fields. An embedding is sometimes denoted by the special arrow

$$G \hookrightarrow G'.$$

In general, let $f: X \to Y$ be a mapping of sets. Let Z be a subset of Y. Recall from §2 that we defined the **inverse image**:

$$f^{-1}(Z) = \text{subset of elements } x \in X \text{ such that } f(x) \in Z.$$

Let $f: G \to G'$ be a homomorphism and let H' be a subgroup of G'. Let $H = f^{-1}(H')$ be its inverse image, i.e. *the set of $x \in G$ such that $f(x) \in H'$. Then H is a subgroup of G.*

Prove this as Exercise 8.

In the above statement, let us take $H' = \{e'\}$. Thus H' is the trivial subgroup of G'. Then $f^{-1}(H')$ is the kernel of f by definition.

Let $f: G \to G'$ be a group-homomorphism. We shall say that f is an **isomorphism** (or more precisely a group-isomorphism) if there exists a homomorphism $g: G' \to G$ such that $f \circ g$ and $g \circ f$ are the identity mappings of G' and G respectively. We denote an isomorphism by the notation

$$G \approx G'.$$

Remark. Roughly speaking, if a group G has a property which can be defined entirely in terms of the group operation, then every group isomorphic to G also has the property. Some of these properties are: having order n, being abelian, being cyclic, and other properties which you will encounter later, such as being solvable, being simple, having a trivial center, etc. As you encounter these properties, verify the fact that they are invariant under isomorphisms.

Example 5. The function exp is an isomorphism of the additive group of the real numbers onto the mutiplicative group of positive real numbers. Its inverse is the log.

Example 6. Let G be a commutative group. The map

$$f: x \mapsto x^{-1}$$

is an isomorphism of G onto itself. What is $f \circ f$? What is f^{-1}?

A group-homomorphism $f: G \to G'$ which is injective and surjective is an isomorphism.

Proof. We let $f^{-1}: G' \to G$ be the inverse mapping. All we need to prove is that f^{-1} is a group-homomorphism. Let $x', y' \in G'$, and let $x, y \in G$ be such that $f(x) = x'$ and $f(y) = y'$. Then $f(xy) = x'y'$. Hence by definition,

$$f^{-1}(x'y') = xy = f^{-1}(x')f^{-1}(y').$$

This proves that f^{-1} is a homomorphism, as desired.

From the preceding condition for an isomorphism, we obtain the standard test for a homomorphism to be an isomorphism.

Theorem 3.1. *Let $f: G \to G'$ be a homomorphism.*

(a) *If the kernel of f is trivial, then f is an isomorphism of G with its image $f(G)$.*
(b) *If $f: G \to G'$ is surjective and the kernel of f is trivial, then f is an isomorphism.*

Proof. We have proved previously that if the kernel of f is trivial then f is injective. Since f is always surjective onto its image, the assertion of the theorem follows from the preceding condition.

By an **automorphism** of a group, one means an isomorphism of the group onto itself. The map of Example 6 is an automorphism of the commutative group G. What does it look like in additive notation? Examples of automorphisms will be given in the exercises. (Cf. Exercises 3, 4, 5.) Denote the set of automorphisms of G by $\mathrm{Aut}(G)$.

$\mathrm{Aut}(G)$ *is a subgroup of the group of permutations of* G, *where the law of composition is composition of mappings.*

Verify this assertion in detail. See Exercise 3.

We shall now see that every group is isomorphic to a group of permutations of some set.

Example 7 (Translation). Let G be a group. For each $a \in G$, let

$$T_a: G \to G$$

be the map such that $T_a(x) = ax$. We call T_a the **left translation by** a. We contend that T_a is a bijection of G onto itself, i.e. a permutation of G. If $x \neq y$ then $ax \neq ay$ (multiply on the left by a^{-1} to see this), and hence T_a is injective. It is surjective, because given $x \in G$, we have

$$x = T_a(a^{-1}x).$$

The inverse mapping of T_a is obviously $T_{a^{-1}}$. Thus the map

$$a \mapsto T_a$$

is a map from G into the group of permutations of the set G. We contend that it is a *homomorphism*. Indeed, for $a, b \in G$ we have

$$T_{ab}(x) = abx = T_a(T_b(x)),$$

so that $T_{ab} = T_a T_b$. Furthermore, one sees at once that this homomorphism is injective. Thus the map

$$a \mapsto T_a \qquad (a \in G)$$

is an isomorphism of G onto a subgroup of the group of all permutations of G. Of course, not every permutation need be given by a translation, i.e. the image of the map is not necessarily equal to the full group of permutations of G.

The terminology of Example 7 is taken from Euclidean geometry. Let $G = \mathbf{R}^2 = \mathbf{R} \times \mathbf{R}$. We visualize G as the plane. Elements of G are called 2-dimensional vectors. If $A \in \mathbf{R} \times \mathbf{R}$, then the translation

$$T_A: \mathbf{R} \times \mathbf{R} \to \mathbf{R} \times \mathbf{R}$$

such that $T_A(X) = X + A$ for all $X \in \mathbf{R} \times \mathbf{R}$ is visualized as the usual translation of X by means of the vector A.

Example 8 (Conjugation). Let G be a group and let $a \in G$. Let

$$\mathbf{c}_a: G \to G$$

be the map defined by $x \mapsto axa^{-1}$. This map is called **conjugation** by a.

In Exercises 4 and 5 you will prove:

A conjugation $\mathbf{c}_a$ *is an automorphism of* G, *called an* **inner automorphism.** *The association* $a \mapsto \mathbf{c}_a$ *is a homomorphism of* G *into* $\operatorname{Aut}(G)$, *whose law of composition is composition of mappings.*

Let A be an abelian group, written additively. Let B, C be subgroups. We let $B + C$ consist of all sums $b + c$, with $b \in B$ and $c \in C$. You can show as an exercise that $B + C$ is a subgroup, called the **sum** of B and C. You can define the **sum** $B_1 + \cdots + B_r$ of a finite number of subgroups similarly. We say that A is the **direct sum** of B and C if every element $x \in A$ can be written uniquely in the form $x = b + c$ with $b \in B$ and $c \in C$. We then write

$$A = B \oplus C.$$

Similarly we define A to be the **direct sum**

$$A = \bigoplus B_i = B_1 \oplus \cdots \oplus B_r$$

if every element $x \in A$ can be written in the form

$$x = \sum_{i=1}^{r} b_i = b_1 + \cdots + b_r$$

with elements $b_i \in B_i$ uniquely determined by x. In Exercise 14 you will prove:

Theorem 3.2. *The abelian group* A *is a direct sum of subgroups* B *and* C *if and only if* $A = B + C$ *and* $B \cap C = \{0\}$. *This is the case if and only if the map*

$$B \times C \to A \qquad \text{given by} \qquad (b, c) \mapsto b + c$$

is an isomorphism.

Example 9 (The group of homomorphisms). Let A, B, be **abelian groups**, written additively. Let $\operatorname{Hom}(A, B)$ denote the set of homomorphisms of A into B. We can make $\operatorname{Hom}(A, B)$ into a group as follows. If f, g are homomorphisms of A into B, we define $f + g \colon A \to B$ to be the map such that

$$(f + g)(x) = f(x) + g(x)$$

for all $x \in A$. It is a simple matter to verify that the three group axioms are satisfied. In fact, if $f, g, h \in \operatorname{Hom}(A, B)$, then for all $x \in A$,

$$((f + g) + h)(x) = (f + g)(x) + h(x) = f(x) + g(x) + h(x),$$

and

$$(f + (g + h))(x) = f(x) + (g + h)(x) = f(x) + g(x) + h(x).$$

Hence $f + (g + h) = (f + g) + h$. We have an additive unit element, namely the map 0 (called zero) which to each element of A assigns the zero element of B. It obviously satisfies condition **GR 2**. Furthermore, the map $-f$ such that $(-f)(x) = -f(x)$ has the property that

$$f + (-f) = 0.$$

Finally, we must of course observe that $f + g$ and $-f$ are homomorphisms. Indeed, for $x, y \in A$,

$$\begin{aligned}(f + g)(x + y) &= f(x + y) + g(x + y) = f(x) + f(y) + g(x) + g(y)\\ &= f(x) + g(x) + f(y) + g(y)\\ &= (f + g)(x) + (f + g)(y),\end{aligned}$$

so that $f + g$ is a homomorphism. Also,

$$(-f)(x + y) = -f(x + y) = -(f(x) + f(y)) = -f(x) - f(y),$$

and hence $-f$ is a homomorphism. This proves that $\mathrm{Hom}(A, B)$ is a group.

II, §3. EXERCISES

1. Let $\mathbf{R}^*$ be the multiplicative group of non-zero real numbers. Describe explicitly the kernel of the absolute value homomorphism

$$x \mapsto |x|$$

of $\mathbf{R}^*$ into itself. What is the image of this homomorphism?

2. Let $\mathbf{C}^*$ be the multiplicative group of non-zero complex numbers. What is the kernel of the homomorphism absolute value

$$z \mapsto |z|$$

of $\mathbf{C}^*$ into $\mathbf{R}^*$?

3. Let G be a group. Prove that $\mathrm{Aut}(G)$ is a subgroup of $\mathrm{Perm}(G)$.

4. Let G be a group. Let a be an element of G. Let

$$\mathbf{c}_a \colon G \to G$$

be the map such that

$$\mathbf{c}_a(x) = axa^{-1}.$$

(a) Show that $\mathbf{c}_a: G \to G$ is an automorphism of G.
(b) Show that the set of all such maps $\mathbf{c}_a$ with $a \in G$ is a subgroup of $\operatorname{Aut}(G)$.

5. Let the notation be as in Exercise 4. Show that the association $a \mapsto \mathbf{c}_a$ is a homomorphism of G into $\operatorname{Aut}(G)$. The image of this homomorphism is called the group of **inner** automorphisms of G. Thus an inner automorphism of G is one which is equal to $\mathbf{c}_a$ for some $a \in G$.

6. If G is not commutative, is $x \mapsto x^{-1}$ a homomorphism? Prove your assertion.

7. (a) Let G be a subgroup of the group of permutations of a set S. If s, t are elements of S, we define s to be equivalent to t if there exists $\sigma \in G$ such that $\sigma s = t$. Show that this is an equivalence relation.
   (b) Let $s \in S$, and let G_s be the set of all $\sigma \in G$ such that $\sigma s = s$. Show that G_s is a subgroup of G.
   (c) If $\tau \in G$ is such that $\tau s = t$, show that $G_t = \tau G_s \tau^{-1}$.

8. Let $f: G \to G'$ be a group homomorphism. Let H' be a subgroup of G'. Show that $f^{-1}(H')$ is a subgroup of G.

9. Let G be a group and S a set of generators of G. Let $f: G \to G'$ and $g: G \to G'$ be homomorphisms of G into the same group G'. Suppose that $f(x) = g(x)$ for all $x \in S$. Prove that $f = g$.

10. Let $f: G \to G'$ be an isomorphism of groups. Let $a \in G$. Show that the period of a is the same as the period of $f(a)$.

11. Let G be a cyclic group, and $f: G \to G'$ a homomorphism. Show that the image of G is cyclic.

12. Let G be a commutative group, and n a positive integer. Show that the map $x \mapsto x^n$ is a homomorphism of G into itself.

13. Let A be an additive abelian group, and let B, C be subgroups. Let $B + C$ consist of all sums $b + c$, with $b \in B$ and $c \in C$. Show that $B + C$ is a subgroup, called the **sum** of B and C.

14. (a) Give the proof of Theorem 3.2 in all details.
    (b) Prove that an abelian group A is a direct sum of subgroups $B_1, \dots, B_r$ if and only if the map

$$(b_1, \dots, b_r) \mapsto b_1 + \cdots + b_r \quad \text{of} \quad \prod B_i \to A$$

    is an isomorphism.

15. Let A be an abelian group, written additively, and let n be a positive integer such that $nx = 0$ for all $x \in A$. Such an integer n is called an **exponent** for A. Assume that we can write $n = rs$, where r, s are positive relatively prime integers. Let A_r consist of all $x \in A$ such that $rx = 0$, and similarly A_s consist of all $x \in A$ such that $sx = 0$. Show that every element $a \in A$ can be written uniquely in the form $a = b + c$, with $b \in A_r$ and $c \in A_s$. Hence $A = A_r \oplus A_s$.

16. Let A be a finite abelian group of order n, and let $n = p_1^{r_1} \cdots p_s^{r_s}$ be its prime power factorization, the p_i being distinct. (a) Show that A is a direct sum $A = A_1 \oplus \cdots \oplus A_s$ where every element $a \in A_i$ satisfies $p_i^{r_i}a = 0$. (b) Prove that $\#(A_i) = p_i^{r_i}$.

17. Let G be a finite group. Suppose that the only automorphism of G is the identity. Prove that G is abelian and that every element has order 2. [After you know the structure theorem for abelian groups, or after you have read the general definition of a vector space, and viewing G as vector space over $\mathbf{Z}/2\mathbf{Z}$, prove that G has at most 2 elements.]

II, §4. COSETS AND NORMAL SUBGROUPS

In what follows, we need some convenient notation. Let S, S' be subsets of a group G. We define the **product** of these subsets to be

$$SS' = \text{the set of all elements } xx' \text{ with } x \in S \text{ and } x' \in S'.$$

It is easy to verify that if S_1, S_2, S_3 are three subsets of G, then

$$(S_1S_2)S_3 = S_1(S_2S_3).$$

This product simply consists of all elements xyz, with $x \in S_1$, $y \in S_2$ and $z \in S_3$. Thus the product of subsets is associative.

Example 1. Show that if H is a subgroup of G, then $HH = H$. Also if S is a non-empty subset of H, then $SH = H$. Check for yourself other properties, for instance

$$S_1(S_2 \cup S_3) = S_1S_2 \cup S_1S_3.$$

Let G be a group, and H a subgroup. Let a be an element of G. The set of all elements ax with $x \in H$ is called a **coset** of H in G. We denote it by aH, following the above notation.

In additive notation, a coset of H would be written $a + H$.

Since a group G may not be commutative, we shall in fact call aH a **left** coset of H. Similarly, we could define **right** cosets, but in the sequel, unless otherwise specified, **coset** will mean **left coset**.

Theorem 4.1. *Let aH and bH be cosets of H in the group G. Either these cosets are equal, or they have no element in common.*

Proof. Suppose that aH and bH have one element in common. We shall prove that they are equal. Let x, y be elements of H such that

$ax = by$. Since $xH = H = yH$ (see Example 1) we get

$$aH = axH = byH = bH$$

as was to be proved.

Suppose that G is a finite group. Every element $x \in G$ lies in some coset of H, namely $x \in xH$. Hence G is the union of all cosets of H. By Theorem 4.1, we can write G as the union of distinct cosets, so

$$G = \bigcup_{i=1}^{r} a_i H,$$

where the cosets $a_1H, \ldots, a_rH$ are distinct. When we write G in this manner, we say that $G = \bigcup a_i H$ is a **coset decomposition** of G, and that any element ah with $h \in H$ is a **coset representative** of the coset aH. In a coset decomposition of G as above, the coset representatives $a_1, \ldots, a_r$ represent distinct cosets, and are of course all distinct.

If a and b are coset representatives for the same coset, then

$$aH = bH.$$

Indeed, we can write $b = ah$ for some $h \in H$, and then

$$bH = ahH = aH.$$

If G is an infinite group, we can still write G as a union of distinct cosets, but there may be infinitely many of these cosets. Then we use the notation

$$G = \bigcup_{i \in I} a_i H$$

where I is some indexing set, not necessarily finite.

Theorem 4.2. *Let G be a group, and H a finite subgroup. Then the number of elements of a coset aH is equal to the number of elements in H.*

Proof. Let x, x' be distinct elements of H. Then ax and ax' are distinct, for if $ax = ax'$, then multiplying by a^{-1} on the left shows that $x = x'$. Hence, if $x_1, \ldots, x_n$ are the distinct elements of H, then $ax_1, \ldots, ax_n$ are the distinct elements of aH, whence our assertion follows.

Let G be a group and let H be a subgroup. The set of left cosets of H is denoted by

$$G/H.$$

We'll not run across the set of right cosets, but the notation is $H\backslash G$ for the set of right cosets, in case you are interested.

The number of distinct cosets of H in G is called the **index** of H in G. This index may of course be infinite. If G is a finite group, then the index of any subgroup is finite. The index of a subgroup H is denoted by $(G:H)$. Let $\#S$ denote the number of elements of a set S. As a matter of notation, we often write $\#(G/H) = (G:H)$, and

$$\#G = (G:1),$$

in other words, the order of G is the index of the trivial subgroup in G.

Theorem 4.3. *Let G be a finite group and H a subgroup. Then*:

(1) $$\text{order of } G = (G:H) \text{ (order of } H).$$

(2) *The order of a subgroup divides the order of G.*

(3) *Let $a \in G$. Then the period of a divides the order of G.*

(4) *If $G \supset H \supset K$ are subgroups, then*

$$(G:K) = (G:H)(H:K).$$

Proof. Every element of G lies in some coset (namely, a lies in the coset aH since $a = ae$). By Theorem 4.1, every element lies in precisely one coset, and by Theorem 4.2, any two cosets have the same number of elements. Formula (1) of our corollary is therefore clear. Then we see that the order of H divides the order of G. The period of an element a is the order of the subgroup generated by a, so (3) also follows. As to the last formula, we have by (1):

$$\#G = (G:H)\#H = (G:H)(H:K)\#K$$

and also

$$\#G = (G:K)\#K.$$

From this (4) follows immediately.

Example 2. Let S_n be the group of permutations of $\{1,\dots,n\}$. Let H be the subset of S_n consisting of all permutations σ such that $\sigma(n) = n$ (i.e. all permutations leaving n fixed). It is clear that H is a subgroup, and we may view H as the permutation group S_{n-1}. (We assume $n > 1$.) We wish to describe all the cosets of H. For each integer i with $1 \leqq i \leqq n$, let τ_i be the permutation such that $\tau_i(n) = i$, $\tau_i(i) = n$, and τ_i leaves all integers other than n and i fixed. We contend that the cosets

$$\tau_1 H, \ldots, \tau_n H$$

are distinct, and constitute all distinct cosets of H in S_n.

To see this, let $\sigma \in S_n$, and suppose $\sigma(n) = i$. Then

$$\tau_i^{-1}\sigma(n) = \tau_i^{-1}(i) = n.$$

Hence $\tau_i^{-1}\sigma$ lies in H, and therefore σ lies in $\tau_i H$. We have shown that every element of G lies in some coset $\tau_i H$, and hence $\tau_1 H, \ldots, \tau_n H$ yield all the cosets. We must still show that these cosets are distinct. If $i \neq j$, then for any $\sigma \in H$, $\tau_i\sigma(n) = \tau_i(n) = i$ and $\tau_j\sigma(n) = \tau_j(n) = j$. Hence $\tau_i H$ and $\tau_j H$ cannot have any element in common, since elements of $\tau_i H$ and $\tau_j H$ have distinct effects on n. This proves what we wanted.

From Theorem 4.3, we conclude that

$$\text{order of } S_n = n \cdot \text{order of } S_{n-1}.$$

By induction, we see immediately that

$$\text{order of } S_n = n! = n(n-1)\cdots 1.$$

Theorem 4.4. *Let $f: G \to G'$ be a homomorphism of groups. Let H be its kernel, and let a' be an element of G' which is in the image of f, say $a' = f(a)$ for $a \in G$. Then the set of elements x in G such that $f(x) = a'$ is precisely the coset aH.*

Proof. Let $x \in aH$, so that $x = ah$ with some $h \in H$. Then

$$f(x) = f(a)f(h) = f(a).$$

Conversely, suppose that $x \in G$, and $f(x) = a'$. Then

$$f(a^{-1}x) = f(a)^{-1}f(x) = a'^{-1}a' = e'.$$

Hence $a^{-1}x$ lies in the kernel H, say $a^{-1}x = h$ with some $h \in H$. Then $x = ah$, as was to be shown.

Let G be a group and let H be a subgroup. We shall say that H is **normal** if H satisfies either one of the following equivalent conditions:

NOR 1. For all $x \in G$ we have $xH = Hx$, that is $xHx^{-1} = H$.

NOR 2. H is the kernel of some homomorphism of G into some group.

We shall now prove that these two conditions are equivalent. Suppose first that H is the kernel of a homomorphism f. Then

$$f(xHx^{-1}) = f(x)f(H)f(x)^{-1} = 1.$$

Hence $xHx^{-1} \subset H$ for all $x \in G$, so $x^{-1}Hx \subset H$, whence $H \subset xHx^{-1}$. Hence $xHx^{-1} = H$. This proves **NOR 2** implies **NOR 1**. The converse will be proved in Theorem 4.5 and Corollary 4.6.

Warning. The condition in **NOR 1** is not the same as saying that $xhx^{-1} = h$ for all elements $h \in H$, when G is not commutative. Observe however that a subgroup of a commutative group is always normal, and even satisfies the stronger condition than **NOR 1**, namely $xhx^{-1} = h$ for all $h \in H$.

We now prove that **NOR 1** implies **NOR 2** by showing how a subgroup satisfying **NOR 1** is the kernel of a homomorphism.

Theorem 4.5. *Let G be a group and H a subgroup having the property that $xH = Hx$ for all $x \in G$. If aH and bH are cosets of H, then the product $(aH)(bH)$ is also a coset, and the collection of cosets is a group, the product being defined as above.*

Proof. We have $(aH)(bH) = aHbH = abHH = abH$. Hence the product of two cosets is a coset. Condition **GR 1** is satisfied in view of previous remarks on multiplication of subsets of G. Condition **GR 2** is satisfied, the unit element being the coset $eH = H$ itself. (Verify this in detail.) Condition **GR 3** is satisfied, the inverse of aH being $a^{-1}H$. (Again verify this in detail.) Hence Theorem 4.5 is proved.

The group of cosets in Theorem 4.5 is called the **factor group** of G by H, or G **modulo** H. We note that it is a group of left or right cosets, there being no difference between these by assumption on H. We emphasize that it is this assumption which allowed us to define multiplication of cosets. If the condition $xH = Hx$ for all $x \in G$ is not satisfied, then we cannot define a group of cosets.

Corollary 4.6. *Let G be a group and H a subgroup having the property that $xH = Hx$ for all $x \in G$. Let G/H be the factor group, and let*

$$f: G \to G/H$$

be the map which to each $a \in G$ associates the coset $f(a) = aH$. Then f is a homomorphism, and its kernel is precisely H.

Proof. The fact that f is a homomorphism is nothing but a repetition of the properties of the product of cosets. As for its kernel, it is clear that every element of H is in the kernel. Conversely, if $x \in G$, and $f(x) = xH$ is the unit element of G/H, it is the coset H itself, so $xH = H$. This means that $xe = x$ is an element of H, so H is equal to the kernel of f, as desired.

We call the homomorphism f in Corollary 4.6 the **canonical homomorphism** of G onto the factor group G/H.

Let $f: G \to G'$ be a homomorphism, and let H be its kernel. Let $x \in G$. Then for all $h \in H$ we have

$$f(xh) = f(x)f(h) = f(x).$$

We can rewrite this property in the form

$$f(xH) = f(x).$$

Thus all the elements in a coset of H have the same image under f. This is an important fact, which we shall use in the next result, which is one of the cornerstones for arguments involving homomorphisms. You should master this result thoroughly.

Corollary 4.7. *Let $f: G \to G'$ be a homomorphism, and let H be its kernel. Then the association $xH \mapsto f(xH)$ is an isomorphism*

$$G/H \xrightarrow{\approx} \operatorname{Im} f$$

of G/H with the image of f.

Proof. By the remark preceding the corollary, we can define a map

$$\bar{f}: G/H \to G' \qquad \text{by} \qquad xH \mapsto f(xH).$$

Remember that G/H is the set of cosets of H. By Theorem 3.1, we have to verify three things:

$\bar{f}$ is a homomorphism. Indeed,

$$\bar{f}(xHyH) = \bar{f}(xyH) = f(xy) = f(x)f(y) = \bar{f}(xH)\bar{f}(yH).$$

$\bar{f}$ is injective. Indeed, the kernel of $\bar{f}$ consists of those cosets xH such that $f(xH) = e'$, so consists of H itself, which is the unit element of G/H.

The image of $\bar{f}$ is the image of f, which follows directly from the way $\bar{f}$ is defined.

This proves the corollary.

The isomorphism $\bar{f}$ of the corollary is said to be **induced** by f. Note that G/H and the image of f are isomorphic not by any random isomorphism, but by the map specified in the statement of the corollary, in other words they are isomorphic by the mapping $\bar{f}$. Whenever one asserts that two groups are isomorphic, it is best to specify what is the mapping giving the isomorphism.

Example 3. Consider the subgroup **Z** of the additive group of the real numbers **R**. The factor group **R**/**Z** is sometimes called the **circle** group. Two elements $x, y \in \mathbf{R}$ are called **congruent** mod **Z** if $x - y \in \mathbf{Z}$. This congruence is an equivalence relation, and the congruence classes are precisely the cosets of **Z** in **R**. If $x \equiv y \pmod{\mathbf{Z}}$, then $e^{2\pi i x} = e^{2\pi i y}$, and conversely. Thus the map

$$x \mapsto e^{2\pi i x}$$

defines an isomorphism of **R**/**Z** with the multiplicative group of complex numbers having absolute value 1. To prove these statements, one must of course know some facts of analysis concerning the exponential function.

Example 4. Let $\mathbf{C}^*$ be the multiplicative group of non-zero complex numbers, and $\mathbf{R}^+$ the multiplicative group of positive real numbers. Given a complex number $\alpha \neq 0$, we can write

$$\alpha = ru,$$

where $r \in \mathbf{R}^+$ and u has absolute value 1. (Let $u = \alpha/|\alpha|$.) Such an expression is uniquely determined, and the map

$$\alpha \mapsto \frac{\alpha}{|\alpha|}$$

is a homomorphism of $\mathbf{C}^*$ onto the group of complex numbers of absolute value 1. The kernel is $\mathbf{R}^+$, and it follows that $\mathbf{C}^*/\mathbf{R}^+$ is isomorphic to the group of complex numbers of absolute value 1. (Cf. Exercise 14.)

For coset representatives in Examples 3 and 4, cf. Exercises 15 and 16.

The exercises list a lot of other basic facts about normal subgroups and homomorphisms. The proofs are all easy, and it is better for you to carry them out than to clutter up the text with them. You will learn them better for that. You should know these basic results as a matter of course. Especially, you will find a very useful test for a subgroup to be normal, namely:

Example 5. Let H be a subgroup of a finite group G, and suppose that the index $(G:H)$ is equal to the smallest prime number dividing the order of G. Then H is normal. In particular, a subgroup of index 2 is normal. See Exercises 29 and 30.

We shall now give a description of some standard cases of homomorphisms and isomorphisms. The easiest case is the following.

Let $K \subset H \subset G$ be normal subgroups of a group G. Then the association

$$xK \mapsto xH \qquad \text{for } x \in G$$

is a surjective homomorphism

$$G/K \to G/H,$$

which we also call the **canonical** *homomorphism. The kernel is H/K.*

The verification is immediate and will be left to you.

Note that following Corollary 4.7, we have the formula

$$G/H \approx (G/K)/(H/K),$$

which is analogous to an elementary rule of arithmetic.

Example. Let $G = \mathbf{Z}$ be the additive group of integers. The subgroups of $\mathbf{Z}$ are simply the sets of type $n\mathbf{Z}$. Let m, n be positive integers. We have $n\mathbf{Z} \subset m\mathbf{Z}$ if and only if m divides n. (Proof?) Thus if $m|n$, we have a canonical homomorphism

$$\mathbf{Z}/n\mathbf{Z} \to \mathbf{Z}/m\mathbf{Z}.$$

If we write $n = md$, then we can also write the canonical homomorphism as

$$\mathbf{Z}/md\mathbf{Z} \to \mathbf{Z}/m\mathbf{Z}.$$

The following facts are used constantly, and you will prove them as Exercises 7 and 9, using Corollary 4.7.

Theorem 4.8. *Let $f: G \to G'$ be a surjective homomorphism. Let H' be a normal subgroup of G' and let $H = f^{-1}(H')$. Then H is normal, and the map $x \mapsto f(x)H'$ is a homomorphism of G onto G'/H' whose kernel is H. Therefore we obtain an isomorphism*

$$G/H \approx G'/H'.$$

Note that the homomorphism $x \mapsto f(x)H'$ of G into G'/H' can also be described as the composite homomorphism

$$G \xrightarrow{f} G' \xrightarrow{\text{can}} G'/H'$$

where $G' \to G'/H'$ is the canonical homomorphism. Using Theorem 4.8, you can prove:

Theorem 4.9. *Let G be a group and H a subgroup. Let N be a normal subgroup of G. Then:*

(1) *HN is a subgroup of G;*

(2) *$H \cap N$ is a normal subgroup of H;*

(3) *The association*

$$f: h \mapsto hN$$

is a homomorphism of H into G/N, whose kernel is $H \cap N$. The image of f is the subgroup HN/N of G/N, so we obtain an isomorphism

$$\bar{f}: H/(H \cap N) \xrightarrow{\approx} HN/N.$$

Remark. Once you show that $H \cap N$ is the kernel of the association in (3), then you have also proved (2).

Theorem 4.8 is important in the following context. We define a group G to be **solvable** if there exists a sequence of subgroups

$$G = H_0 \supset H_1 \supset \cdots \supset H_r = \{e\}$$

such that H_{i+1} is normal in H_i for $i = 0, \ldots, r-1$, and such that H_i/H_{i+1} is abelian. It is a problem to determine which groups are solvable, and which are not. A famous theorem of Feit–Thompson states that every group of odd order is solvable. In §6 you will see a proof that the permutation group on n elements is not solvable for $n \geqq 5$. The groups of Theorem 3.8 in Chapter VI give other examples. As an application of Theorem 4.8, we shall now prove:

Theorem 4.10. *Let G be a group and K a normal subgroup. Assume that K and G/K are solvable. Then G is solvable.*

Proof. By definition, and the assumption that K is solvable, it suffices to prove the existence of a sequence of subgroups

$$G = H_0 \supset H_1 \supset \cdots \supset H_m = K$$

such that H_{i+1} is normal in H_i and H_i/H_{i+1} is abelian, for all i. Let $\bar{G} = G/K$. By assumption, there exists a sequence of subgroups

$$\bar{G} = \bar{H}_0 \supset \bar{H}_1 \supset \cdots \supset \bar{H}_m = \{\bar{e}\}$$

such that $\bar{H}_{i+1}$ is normal in $\bar{H}_i$ and $\bar{H}_i/\bar{H}_{i+1}$ is abelian for all i. Let $f: G \to G/K$ be the canonical homomorphism and let $H_i = f^{-1}(\bar{H}_i)$. By Theorem 4.8 we have an isomorphism $H_i/H_{i+1} \approx \bar{H}_i/\bar{H}_{i+1}$ and $K = f^{-1}(\bar{H}_m)$, so we have found the sequence of subgroups of G as we wanted, proving the theorem.

In mathematics, you will meet many structures, of which groups are only one basic type. Whatever structure one meets, one asks systematically for a classification of these structures, and especially one tries to answer the following questions:

1. What are the simplest structures in the category under consideration?
2. To what extent can any structure be expressed as a product of simple ones?
3. What is the structure of the group of automorphisms of a given object?

In a sense, the objects having the "simplest" structure are the building blocks for the more complicated objects. Taking direct products is an easy way to form more complicated objects, but there are more complex ways. For instance, we define a group to be **simple** if it has no normal subgroup except the group itself and the trivial subgroup consisting of the unit element. Let G be a finite group. Then one can find a sequence of subgroups

$$G = H_0 \supset H_1 \supset H_2 \supset \cdots \supset H_r = \{e\}$$

such that H_{k+1} is normal in H_k and such that H_k/H_{k+1} is simple.

The Jordan–Holder theorem states that at least the sequence of simple factor groups H_k/H_{k+1} is uniquely determined up to a permutation and up to isomorphism. You can look up a proof in a more advanced algebra text (e.g., my *Algebra*). Such a sequence already gives information about G. To get a full knowledge of G, one would have to know how these factor groups are pieced together. Usually it is not true that G is the direct product of all these factor groups. There is also the question of classifying all simple groups. In §7 we shall see how a finite abelian group is a direct product of cyclic factors. In Chapter VI, Theorem 3.8, you will find an example of a simple group. Of course, every cyclic group of prime order is simple. As an exercise, prove that a simple abelian group is cyclic of prime order or consists only of the unit element.

II, §4. EXERCISES

1. Let G be a group and H a subgroup. If $x, y \in G$, define x to be equivalent to y if x is in the coset yH. Prove that this is an equivalence relation.

2. Let $f: G \to G'$ be a homomorphism with kernel H. Assume that G is finite.
   (a) Show that

   $$\text{order of } G = (\text{order of image of } f)(\text{order of } H).$$

   (b) Suppose that G, G' are finite groups and that the orders of G and G' are relatively prime. Prove that f is trivial, that is, $f(G) = e'$, the unit element of G'.

3. The index formula of Theorem 4.3(4) holds even if G is not finite. All that one needs to assume is that H, K are of finite index. Namely, using only that assumption, prove that

   $$(G:K) = (G:H)(H:K).$$

   In fact, suppose that

   $$G = \bigcup_{i=1}^{m} a_i H \quad \text{and} \quad H = \bigcup_{j=1}^{r} b_j K$$

   are coset decompositions of G with respect to H, and H with respect to K. Prove that

   $$G = \bigcup_{i,j} a_i b_j K$$

   is a coset decomposition of G with respect to K. Thus you have to prove that G is the union of the cosets $a_i b_j K$ $(i = 1, \ldots, m; j = 1, \ldots, r)$ and that these cosets are all distinct.

4. Let p be a prime number and let G be a group with a subgroup H of index p in G. Let S be a subgroup of G such that $G \supset S \supset H$. Prove that $S = H$ or $S = G$. (This theorem applies in particular if H is of index 2.)

5. Show that the group of inner automorphisms is normal in the group of all automorphisms of a group G.

6. Let H_1, H_2 be two normal subgroups of G. Show that $H_1 \cap H_2$ is normal.

7. Let $f: G \to G'$ be a homomorphism and let H' be a normal subgroup of G'. Let $H = f^{-1}(H')$.
   (a) Prove that H is normal in G.
   (b) Prove Theorem 4.8.

8. Let $f: G \to G'$ be a surjective homomorphism. Let H be a normal subgroup of G. Show that $f(H)$ is a normal subgroup of G'.

9. Prove Theorem 4.9.

10. Let G be a group. Define the **center** of G to be the subset of all elements a in G such that $ax = xa$ for all $x \in G$. Show that the center is a subgroup, and that it is a normal subgroup. Show that it is the kernel of the conjugation homomorphism $x \mapsto \gamma_x$ in Exercise 5, §3.

11. Let G be a group and H a subgroup. Let N_H be the set of all $x \in G$ such that $xHx^{-1} = H$. Show that N_H is a group containing H, and H is normal in N_H. The group N_H is called the **normalizer** of H.

12. (a) Let G be the set of all maps of $\mathbf{R}$ into itself of type $x \mapsto ax + b$, where $a \in \mathbf{R}$, $a \neq 0$ and $b \in \mathbf{R}$. Show that G is a group under composition. We denote such a map by $\sigma_{a,b}$. Thus $\sigma_{a,b}(x) = ax + b$.
   (b) To each map $\sigma_{a,b}$ we associate the number a. Show that the association

$$\sigma_{a,b} \mapsto a$$

   is a homomorphism of G into $\mathbf{R}^*$. Describe the kernel.

13. View $\mathbf{Z}$ as a subgroup of the additive group of rational numbers $\mathbf{Q}$. Show that given an element $\bar{x} \in \mathbf{Q}/\mathbf{Z}$ there exists an integer $n \geqq 1$ such that $n\bar{x} = 0$.

14. Let D be the subgroup of $\mathbf{R}$ generated by 2π. Let $\mathbf{R}^+$ be the multiplicative group of positive real numbers, and $\mathbf{C}^*$ the multiplicative group of non-zero complex numbers. Show that $\mathbf{C}^*$ is isomorphic to $\mathbf{R}^+ \times \mathbf{R}/D$ under the map

$$(r, \theta) \mapsto re^{i\theta}.$$

   (Of course, you must use properties of the complex exponential map.)

15. Show that every coset of $\mathbf{Z}$ in $\mathbf{R}$ has a unique coset representative x such that $0 \leqq x < 1$. [*Hint*: For each real number y, let n be the integer such that $n \leqq y < n + 1$.]

16. Show that every coset of $\mathbf{R}^+$ in $\mathbf{C}^*$ has a unique representative complex number of absolute value 1.

17. Let G be a group and let $x_1,\dots,x_r$ be a set of generators. Let H be a subgroup.
    (a) Assume that $x_iHx_i^{-1} = H$ for $i = 1,\dots,r$. Show that H is normal in G.
    (b) Suppose G is finite. Assume that $x_iHx_i^{-1} \subset H$ for $i = 1,\dots,r$. Show that H is normal in G.
    (c) Suppose that H is generated by elements $y_1,\dots,y_m$. Assume that $x_iy_jx_i^{-1} \in H$ for all i, j. Assume again that G is finite. Show that H is normal.

18. Let G be the group of Exercise 8 of §1. Let H be the subgroup generated by x, so $H = \{e, x, x^2, x^3\}$. Prove that H is normal.

19. Let G be the group of Exercise 9, §1, that is, G is the quaternion group. Let H be the subgroup generated by i, so $H = \{e, i, i^2, i^3\}$. Prove that H is normal.

20. Let G be the group of Exercise 10, §1. Let H be the subgroup generated by x, so $H = \{e, x,\dots,x^5\}$. Prove that H is normal.

Commutators and solvable groups

21. (a) Let G be a commutative group, and H a subgroup. Show that G/H is commutative.
    (b) Let G be a group and H a normal subgroup. Show that G/H is commutative if and only if H contains all elements $xyx^{-1}y^{-1}$ for $x, y \in G$.

Define the **commutator subgroup** G^c to be the subgroup generated by all elements

$$xyx^{-1}y^{-1} \quad \text{with} \quad x, y \in G.$$

Such elements are called **commutators**.

22. (a) Show that the commutator subgroup is a normal subgroup.
    (b) Show that G/G^c is abelian.

23. Let G be a group, H a subgroup, and N a normal subgroup. Prove that if G/N is abelian, then $H/(H \cap N)$ is abelian.

24. (a) Let G be a solvable group, and H a subgroup. Prove that H is solvable.
    (b) Let G be a solvable group, and $f: G \to G'$ a surjective homomorphism. Show that G' is solvable.

25. (a) Prove that a simple finite abelian group is cyclic of prime order.
    (b) Let G be a finite abelian group. Prove that there exists a sequence of subgroups

    $$G \supset H_1 \supset \cdots \supset H_r = \{e\}$$

    such that H_i/H_{i+1} is cyclic of prime order for all i.

Conjugation

26. Let G be a group. Let S be the set of subgroups of G. If H, K are subgroups of G, define H to be **conjugate** to K if there exists an element $x \in G$ such that $xHx^{-1} = K$. Prove that conjugacy is an equivalence relation in S.

27. Let G be a group and S the set of subgroups of G. For each $x \in G$, let $\mathbf{c}_x: S \to S$ be the map such that

$$\mathbf{c}_x(H) = xHx^{-1}.$$

Show that $\mathbf{c}_x$ is a permutation of S, and that the map $x \mapsto \mathbf{c}_x$ is a homomorphism

$$\mathbf{c}: G \to \mathrm{Perm}(S).$$

Translations

28. Let G be a group and H a subgroup of G. Let S be the set of cosets of H in G. For each $x \in G$, let $T_x: S \to S$ be the map which to each coset yH associates the coset xyH. Prove that T_x is a permutation of S, and that the map $x \mapsto T_x$ is a homomorphism

$$T: G \to \mathrm{Perm}(S).$$

29. Let H be a subgroup of G, of finite index n. Let S be the set of cosets of H. Let

$$T: x \mapsto T_x$$

be the homomorphism of $G \to \mathrm{Perm}(S)$ described in the preceding exercise. Let K be the kernel of this homomorphism. Prove:
(a) K is contained in H.
(b) $\#(G/K)$ divides $n!$.
(c) If H is of index 2, then H is normal, and in fact, $H = K$.
[*Hint*: use the index formula and prove that $(H:K) = 1$. If you get into trouble, look at Proposition 8.3.]

30. Let G be a finite group and let p be the smallest prime number dividing the order of G. Let H be a subgroup of index p. Prove that H is normal in G. [*Hint*: Let S be the set of cosets of H. Consider the homomorphism

$$x \mapsto T_x \quad \text{of} \quad G \to \mathrm{Perm}(S).$$

Let K be the kernel of this homomorphism. Use the preceding exercise and the index formula as well as Corollary 4.7.]

31. Let G be a group and let H_1, H_2 be subgroups of finite index. Prove that $H_1 \cap H_2$ has finite index. In fact, prove something stronger. Prove that there exists a normal subgroup N of finite index such that $N \subset H_1 \cap H_2$. [*Hint*: Let S_1 be the set of cosets of H_1 and let S_2 be the set of cosets of H_2. Let $S = S_1 \times S_2$. Define a map

$$T: G \to \mathrm{Perm}(S) = \mathrm{Perm}(S_1 \times S_2)$$

by the formula

$$T_x(aH_1, bH_2) = (xaH_1, xbH_2), \quad \text{for } x, a, b \in G.$$

Show that this map $x \mapsto T_x$ is a homomorphism. What is the order of S? Show that the kernel of this homomorphism is contained in $H_1 \cap H_2$. Use Corollary 4.7.]

II, §5. APPLICATION TO CYCLIC GROUPS

Let G be a cyclic group and let a be a generator. Let $d\mathbf{Z}$ be the kernel of the homomorphism

$$\mathbf{Z} \to G \quad \text{such that} \quad n \mapsto a^n.$$

In case the kernel of this homomorphism is 0, we get an isomorphism of $\mathbf{Z}$ with G. In the second case, as an application of Corollary 4.7, when the kernel is not 0, we get an isomorphism

$$\mathbf{Z}/d\mathbf{Z} \xrightarrow{\approx} G.$$

Theorem 5.1. *Any two cyclic groups of order d are isomorphic. If a is a generator of G of period d, then there is a unique isomorphism*

$$f\colon \mathbf{Z}/d\mathbf{Z} \to G$$

such that $f(1) = a$. If G_1, G_2 are cyclic of order d, and a_1, a_2 are generators of G_1, G_2 respectively, then there is a unique isomorphism

$$g\colon G_1 \to G_2$$

such that $g(a_1) = a_2$.

Proof. By the remarks before the theorem, let

$$f_1\colon \mathbf{Z}/d\mathbf{Z} \to G_1 \quad \text{and} \quad f_2\colon \mathbf{Z}/d\mathbf{Z} \to G_2$$

be isomorphisms such that $f_1(n) = a_1^n$ and $f_2(n) = a_2^n$ for all $n \in \mathbf{Z}$. Then

$$h = f_2 \circ f_1^{-1}\colon G_1 \to G_2$$

is an isomorphism such that $h(a_1) = a_2$. Conversely, let $g: G_1 \to G_2$ be an isomorphism such that $g(a_1) = a_2$. Then

$$g(a_1^n) = g(a_1)^n = a_2^n$$

and therefore $g = h$, thereby proving the uniqueness.

Remark. The uniqueness part of the preceding result is a special case of a very general principle:

Theorem 5.2. *A homomorphism is uniquely determined by its values on a set of generators.*

We expand on what this means. Let G, G' be groups. Suppose that G is generated by a subset of elements S. In other words, every element of G can be written as a product

$$x = x_1 \cdots x_r \qquad \text{with} \quad x_i \in S \quad \text{or} \quad x_i^{-1} \in S.$$

Let $b_1, \ldots, b_r$ be elements of G'. If there is a homomorphism

$$f: G \to G'$$

such that $f(x_i) = b_i$ for $i = 1, \ldots, r$, then such a homomorphism is **uniquely determined**. In other words, if

$$g: G \to G'$$

is a homomorphism such that $g(x_i) = b_i$ for $i = 1, \ldots, r$ then $g = f$. The proof is immediate, because for any element x written as above, $x = x_1 \cdots x_r$ with $x_i \in S$ or $x_i^{-1} \in S$, we have

$$g(x) = g(x_1) \cdots g(x_r) = f(x_1) \cdots f(x_r) = f(x).$$

Of course, given arbitrary elements $b_1, \ldots, b_r \in G'$ there does not necessarily exist a homomorphism $f: G \to G'$ such that $f(x_i) = b_i$. Sometimes such an f exists and sometimes it does not exist. Look at Exercise 12 to see an example.

Theorem 5.3. *Let G be a cyclic group. Then a factor group of G is cyclic and a subgroup of G is cyclic.*

Proof. We leave the case of a factor group as an exercise. See Exercise 11 of §3. Let us prove that a subgroup of G is cyclic. Let a be a generator of G, so that we have a surjective homomorphism $f: \mathbf{Z} \to G$ such that $f(n) = a^n$. Let H be a subgroup of G. Then $f^{-1}(H)$ (the set of $n \in \mathbf{Z}$ such that $f(n) \in H$) is a subgroup A of $\mathbf{Z}$, and hence is cyclic. In fact, we know that there exists a unique integer $d \geqq 0$ such that $f^{-1}(H)$ consists of all integers which can be written in the form md with $m \in \mathbf{Z}$. Since f is surjective, it follows that f maps A on all of H, i.e. every element of H is of the form a^{md} with some integer m. It follows that H is cyclic, and in fact a^d is a generator. In fact we have proved:

Theorem 5.4. *Let G be cyclic of order n, and let $n = md$ be a factorization in positive integers m, d. Then G has a unique subgroup of order m. If $G = \langle a \rangle$, then this subgroup is generated by $a^d = a^{n/m}$.*

II, §5. EXERCISES

1. Show that a group of order 4 is isomorphic to one of the following groups:
   (a) The group with two distinct elements a, b such that
   $$a^2 = b^2 = e \quad \text{and} \quad ab = ba.$$
   (b) The group G having an element a such that $G = \{e, a, a^2, a^3\}$ and $a^4 = e$.

2. Prove that every group of prime order is cyclic.

3. Let G be a cyclic group of order n. For each $a \in \mathbf{Z}$ define $f_a: G \to G$ by $f_a(x) = x^a$.
   (a) Prove that f_a is a homomorphism of G into itself.
   (b) Prove that f_a is an automorphism of G if and only if a is prime to n.

4. Again assume that G is a group of order p, prime. What is the order of $\mathrm{Aut}(G)$? Proof?

5. Let G and Z be cyclic groups of order n. Show that $\mathrm{Hom}(G, Z)$ is cyclic of order n. [*Hint*: If a is a generator of G, show that for each $z \in Z$ there exists a unique homomorphism $f: G \to Z$ such that $f(a) = z$.]

6. Let G be the group of Exercise 8, §1 and let H be the subgroup generated by x. Either recall or prove that H is normal. Show that G/H is cyclic of order 2.

7. Let G be the group of Exercise 10, §1, and let H be the subgroup generated by x. Show that G/H is cyclic of order 2.

8. Let Z be the center of a group G, and suppose that G/Z is cyclic. Prove that G is abelian.

9. Let G be a finite group which contains a subgroup H, and assume that H is contained in the center of G. Assume that $(G : H) = p$ for some prime number p. Prove that G is abelian.

10. Let G be the group of order 8 of Exercise 8 of §1, so G is generated by elements x, y satisfying $x^4 = e = y^2$ and $yxy = x^3$.
(a) Prove that all subgroups of G are those shown on the following diagram.

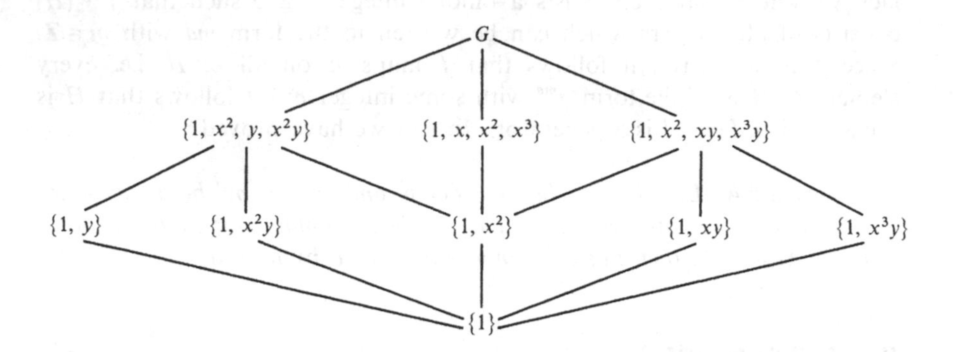

(b) Determine all the normal subgroups of G.

11. Let G be the quaternion group of Exercise 9 of §1.
(a) Make up a lattice of all subgroups of G similar to the lattice of subgroups of the preceding exercise.
(b) Prove that all subgroups of G are normal.

12. (a) Let H, N be normal subgroups of a finite group G. Assume that the orders of H, N are relatively prime. Prove that $xy = yx$ for all $x \in H$ and $y \in N$, and that $HN \approx H \times N$. [*Hint*: Show that $xyx^{-1}y^{-1} \in H \cap N$.]
(b) Let G be a finite group and let $H_1,\dots,H_r$ be normal subgroups such that the order of H_i is relatively prime to the order of H_j for $i \neq j$. Prove that

$$H_1 \cdots H_r \approx H_1 \times \cdots \times H_r.$$

13. Let G be a finite group. Let N be a normal subgroup such that N and G/N have orders relatively prime. Let H be a subgroup of G having the same order as G/N. Prove that $G = HN$.

14. Let G be a finite group. Let N be a normal subgroup such that N and G/N have orders relatively prime. Let φ be an automorphism of G. Prove that $\varphi(N) = N$.

15. Let G be a group and H an abelian normal subgroup. For $x \in G$ let $\mathbf{c}_x$ denote conjugation.
(a) Show that the association $x \mapsto \mathbf{c}_x$ induces a homomorphism

$$G/H \to \operatorname{Aut}(H).$$

(b) Suppose that G is finite, that $\#(G/H)$ is relatively prime to $\#\operatorname{Aut}(H)$, and that G/H is cyclic. Prove that G is abelian.

16. Let G be a group of order p^2. Prove that G is abelian. [*Hint*: Let $a \in G$, $a \neq e$. If a has period p^2, we are done. Otherwise a has period p (why?). Let $H = \langle a \rangle$ be the subgroup generated by a. Let $b \in G$ and $b \notin H$. Prove that

$G = \langle a, b \rangle$, namely G is generated by a and b. Using Exercise 30 of §4, conclude that H is normal. Get a homomorphism $G \to \text{Aut}(H)$, and prove that the image of this homomorphism is trivial. Hence b commutes with a. Conclude that G is abelian by using Exercise 14 of §1.]

II, §6. PERMUTATION GROUPS

In this section, we investigate more closely the permutation group S_n of n elements $\{1, \ldots, n\} = J_n$. This group is called the **symmetric group**.

If $\sigma \in S_n$, then we recall that $\sigma^{-1}: J_n \to J_n$ is the permutation such that $\sigma^{-1}(k) =$ unique integer $j \in J_n$ such that $\sigma(j) = k$. A **transposition** τ is a permutation which interchanges two numbers and leaves the others fixed, i.e. there exist integers i, $j \in J_n$, $i \neq j$ such that $\tau(i) = j$, $\tau(j) = i$, and $\tau(k) = k$ if $k \neq i$ and $k \neq j$. One sees at once that if τ is a transposition, then $\tau^{-1} = \tau$ and $\tau^2 = I$. In particular, the inverse of a transposition is a transposition. We shall prove that the transpositions generate S_n.

Theorem 6.1. *Every permutation of J_n can be expressed as a product of transpositions.*

Proof. We shall prove our assertion by induction on n. For $n = 1$, there is nothing to prove. Let $n > 1$ and assume the assertion proved for $n - 1$. Let σ be a permutation of J_n. Let $\sigma(n) = k$. Let τ be the transposition of J_n such that $\tau(k) = n$, $\tau(n) = k$. Then $\tau\sigma$ is a permutation such that

$$\tau\sigma(n) = \tau(k) = n.$$

In other words, $\tau\sigma$ leaves n fixed. We may therefore view $\tau\sigma$ as a permutation of J_{n-1}, and by induction, there exist transpositions $\tau_1, \ldots, \tau_s$ of J_{n-1}, leaving n fixed, such that

$$\tau\sigma = \tau_1 \cdots \tau_s.$$

We can now write

$$\sigma = \tau^{-1}\tau_1 \cdots \tau_s,$$

Thereby proving our proposition.

A permutation σ of $\{1, \ldots, n\}$ is sometimes denoted by

$$\begin{bmatrix} 1 & \cdots & n \\ \sigma(1) & \cdots & \sigma(n) \end{bmatrix}.$$

Thus

$$\begin{bmatrix} 1 & 2 & 3 \\ 2 & 1 & 3 \end{bmatrix}$$

denotes the permutation σ such that $\sigma(1) = 2$, $\sigma(2) = 1$, and $\sigma(3) = 3$. This permutation is in fact a transposition.

Let $i_1, \ldots, i_r$ be distinct integers in J_n. By the symbol

$$[i_1 \cdots i_r]$$

we shall mean the permutation σ such that

$$\sigma(i_1) = i_2, \quad \sigma(i_2) = i_3, \quad \ldots, \quad \sigma(i_r) = i_1,$$

and σ leaves all other integers fixed. For example

$$[132]$$

denotes the permutation σ such that $\sigma(1) = 3$, $\sigma(3) = 2$, $\sigma(2) = 1$, and σ leaves all other integers fixed. Such a permutation is called a **cycle**, or more precisely, an r-cycle.

If $\sigma = [i_1 \cdots i_r]$ is a cycle, then one verifies at once that σ^{-1} is also a cycle, and that in fact

$$\sigma^{-1} = [i_r \cdots i_1].$$

Thus if $\sigma = [132]$ then

$$\sigma^{-1} = [231].$$

Note that a 2-cycle $[ij]$ is nothing but a transposition, namely the transposition such that $i \mapsto j$ and $j \mapsto i$.

A product of cycles is easily determined. For instance,

$$[132][34] = [2134].$$

One sees this using the definition: If $\sigma = [132]$ and $\tau = [34]$, then for instance

$$\sigma(\tau(3)) = \sigma(4) = 4,$$

$$\sigma(\tau(4)) = \sigma(3) = 2,$$

$$\sigma(\tau(2)) = \sigma(2) = 1,$$

$$\sigma(\tau(1)) = \sigma(1) = 3.$$

Let G be a group. Recall that G is **solvable** if and only if there exists a sequence of subgroups

$$G = H_0 \supset H_1 \supset H_2 \supset \cdots \supset H_m = \{e\}$$

such that H_i is normal in H_{i-1} and such that the factor group H_{i-1}/H_i is abelian, for $i = 1, \ldots, m$. We shall prove that for $n \geqq 5$, the group S_n is not solvable. We need some preliminaries.

Theorem 6.2. *Let G be a group, and H a normal subgroup. Then G/H is abelian if and only if H contains all elements of the form $xyx^{-1}y^{-1}$, with $x, y \in G$.*

Proof. Let $f: G \to G/H$ be the canonical homomorphism. Assume that G/H is abelian. For any $x, y \in G$ we have

$$f(xyx^{-1}y^{-1}) = f(x)f(y)f(x)^{-1}f(y)^{-1},$$

and since G/H is abelian, the expression on the right-hand side is equal to the unit element of G/H. Hence $xyx^{-1}y^{-1} \in H$. Conversely, assume that this is the case for all $x, y \in G$. Let $\bar{x}, \bar{y}$ be elements of G/H. Since f is surjective, there exists $x, y \in G$ such that $\bar{x} = f(x)$ and $\bar{y} = f(y)$. Let $\bar{e}$ be the unit element of G/H, and e the unit element of G. Then

$$\begin{aligned} \bar{e} = f(e) = f(xyx^{-1}y^{-1}) &= f(x)f(y)f(x)^{-1}f(y)^{-1} \\ &= \bar{x}\bar{y}\bar{x}^{-1}\bar{y}^{-1}. \end{aligned}$$

Multiplying by $\bar{y}$ and $\bar{x}$ on the right, we find

$$\bar{y}\bar{x} = \bar{x}\bar{y},$$

and hence G/H is abelian.

Theorem 6.3. *If $n \geqq 5$, then S_n is not solvable.*

Proof. We shall first prove that if H, N are two subgroups of S_n such that $N \subset H$ and N is normal in H, if H contains every 3-cycle, and if H/N is abelian, then N contains every 3-cycle. To see this, let i, j, k, r, s be five distinct integers between 1 and n, and let

$$\sigma = [ijk] \quad \text{and} \quad \tau = [krs].$$

Then

$$\begin{aligned} \sigma\tau\sigma^{-1}\tau^{-1} &= [ijk][krs][kji][srk] \\ &= [rki]. \end{aligned}$$

Since the choice of i, j, k, r, s was arbitrary, we see that the cycles $[rki]$ all lie in N for all choices of distinct r, k, i thereby proving what we wanted.

Now suppose that we have a chain of subgroups

$$S_n = H_0 \supset H_1 \supset H_2 \supset \cdots \supset H_m = \{e\}$$

such that H_ν is normal in $H_{\nu-1}$ for $\nu = 1,\ldots,m$, and $H_\nu/H_{\nu-1}$ is abelian. Since S_n contains every 3-cycle, we conclude that H_1 contains every 3-cycle. By induction on ν, we conclude that $H_m = \{e\}$ contains every 3-cycle, which is impossible. Hence such a chain of subgroups cannot exist, and our theorem is proved.

The sign of a permutation

For the next theorem, we need to describe the operation of permutations on functions. Let f be a function of n real variables, so we can evaluate

$$f(x_1,\ldots,x_n) \quad \text{for} \quad x_1,\ldots,x_n \in \mathbf{R}.$$

Let σ be a permutation of J_n. Then we **define the function** σf by

$$(\sigma f)(x_1,\ldots,x_n) = f(x_{\sigma(1)},\ldots,x_{\sigma(n)}).$$

Then we have for $\sigma, \tau \in S_n$:

$$(1) \qquad (\sigma\tau)f = \sigma(\tau f).$$

Proof. We use the definition applied to the function $g = \tau f$. Then

$$\begin{aligned}\sigma(\tau f)(x_1,\ldots,x_n) &= \tau f(x_{\sigma(1)},\ldots,x_{\sigma(n)})\\ &= f(x_{\sigma\tau(1)},\ldots,x_{\sigma\tau(n)})\\ &= ((\sigma\tau)f)(x_1,\ldots,x_n)\end{aligned}$$

thus proving (1). Note that the middle step comes from applying the definition

$$\tau f(y_1,\ldots,y_n) = f(y_{\tau(1)},\ldots,y_{\tau(n)}) \quad \text{with} \quad y_i = x_{\sigma(i)}.$$

It is trivially verified that if $I: J_n \to J_n$ is the identity permutation, then

$$(2) \qquad If = f.$$

If f, g are two functions of n variables, then we may form their sum and product as usual. The sum $f + g$ is defined by the rule

$$(f+g)(x_1,\ldots,x_n) = f(x_1,\ldots,x_n) + g(x_1,\ldots,x_n)$$

and the product fg is defined by

$$(fg)(x_1,\ldots,x_n) = f(x_1,\ldots,x_n)g(x_1,\ldots,x_n).$$

We claim that

(3) $$\sigma(f+g) = \sigma f + \sigma g \quad \text{and} \quad \sigma(fg) = (\sigma f)(\sigma g).$$

To see this, we have

$$\begin{aligned}(\sigma(f+g))(x_1,\ldots,x_n) &= (f+g)(x_{\sigma(1)},\ldots,x_{\sigma(n)})\\ &= f(x_{\sigma(1)},\ldots,x_{\sigma(n)}) + g(x_{\sigma(1)},\ldots,x_{\sigma(n)})\\ &= (\sigma f)(x_1,\ldots,x_n) + (\sigma g)(x_1,\ldots,x_n),\end{aligned}$$

thereby proving (3) for the sum. The formula for the product is done the same way. As a consequence, for every number c we have

$$\sigma(cf) = c\sigma f.$$

Theorem 6.4. *To each permutation σ of J_n it is possible to assign a sign 1 or -1, denoted by $\epsilon(\sigma)$, satisfying the following conditions:*

(a) *If τ is a transposition, then $\epsilon(\tau) = -1$.*
(b) *If σ, σ' are permutations of J_n, then*

$$\epsilon(\sigma\sigma') = \epsilon(\sigma)\epsilon(\sigma').$$

Proof. Let Δ be the function

$$\Delta(x_1,\ldots,x_n) = \prod_{i<j} (x_j - x_i),$$

the product being taken for all pairs of integers i, j satisfying

$$1 \leqq i < j \leqq n.$$

Let τ be a transposition, interchanging the two integers r and s. Say $r < s$. We wish to determine

$$\tau\Delta(x_1,\ldots,x_n) = \prod_{i<j} (x_{\tau(j)} - x_{\tau(i)})$$
$$= \prod_{i<j} \tau(x_j - x_i).$$

For one factor, we have

$$\tau(x_s - x_r) = (x_r - x_s) = -(x_s - x_r).$$

If a factor does not contain x_r or x_s in it, then it remains unchanged when we apply τ. All other factors can be considered in pairs as follows:

$$(x_k - x_s)(x_k - x_r) \quad \text{if} \quad k > s,$$
$$(x_s - x_k)(x_k - x_r) \quad \text{if} \quad r < k < s,$$
$$(x_s - x_k)(x_r - x_k) \quad \text{if} \quad k < r.$$

Each one of these pairs remains unchanged when we apply τ. Hence we see that

$$\tau\Delta = -\Delta.$$

Now let σ be an arbitrary permutation. Clearly, $\sigma\Delta = \pm\Delta$. Define $\varepsilon(\sigma)$ to be the sign 1 or -1 such that

$$\sigma(\Delta) = \varepsilon(\sigma)\Delta.$$

It is now useful to use the notation $\pi(\sigma)f$ instead of σf when applying a permutation to a function f. Thus σ is in the group of permutations of $\{1,\ldots,n\}$, whereas $\pi(\sigma)$ is in the group of permutations of functions of n variables. Then

$$\pi(\sigma\sigma') = \pi(\sigma)\pi(\sigma').$$

It follows at once that $\sigma \mapsto \varepsilon(\sigma)$ is a homomorphism of S_n into $\{1, -1\}$. The theorem follows.

In particular, if σ is a product of transpositions

$$\sigma = \tau_1 \cdots \tau_m$$

then

$$\varepsilon(\sigma) = (-1)^m.$$

Hence in any such product, m is either always odd or always even, depending on σ. We restate this as a corollary.

Corollary 6.5. *If a permutation σ of J_n is expressed as a product of transpositions,*

$$\sigma = \tau_1 \cdots \tau_s,$$

then s is even or odd according as $\epsilon(\sigma) = 1$ or -1.

Corollary 6.6. *If σ is a permutation of J_n, then*

$$\epsilon(\sigma) = \epsilon(\sigma^{-1}).$$

Proof. We have

$$1 = \epsilon(\mathrm{id}) = \epsilon(\sigma\sigma^{-1}) = \epsilon(\sigma)\epsilon(\sigma^{-1}).$$

Hence either $\epsilon(\sigma)$ and $\epsilon(\sigma^{-1})$ are both equal to 1, or both equal to -1, as desired.

As a matter of terminology, a permutation is called **even** if its sign is 1, and it is called **odd** if its sign is -1. Thus every transposition is odd. From Theorem 6.4 we see that the map

$$\epsilon: S_n \to \{1, -1\}$$

is a homomorphism of S_n onto the group consisting of the two elements $1, -1$. The kernel of this homomorphism by definition consists of the even permutations, and is called the **alternating group** A_n. If τ is a transposition, then A_n and τA_n are obviously distinct cosets of A_n, and every permutation lies in A_n or τA_n. (*Proof*: If $\sigma \in S_n$ and $\sigma \notin A_n$, then $\epsilon(\sigma) = -1$, so $\epsilon(\tau\sigma) = 1$ and hence $\tau\sigma \in A_n$, whence $\sigma \in \tau^{-1}A_n = \tau A_n$.) Therefore,

$$A_n, \quad \tau A_n$$

are distinct cosets of A_n in S_n, and there is no other coset. Since A_n is the kernel of a homomorphism, it is normal in S_n. We have $\tau A_n = A_n\tau$, which can also be verified easily directly.

II, §6. EXERCISES

1. Determine the sign of the following permutations.

(a) $\begin{bmatrix} 1 & 2 & 3 \\ 2 & 3 & 1 \end{bmatrix}$ (b) $\begin{bmatrix} 1 & 2 & 3 \\ 3 & 1 & 2 \end{bmatrix}$ (c) $\begin{bmatrix} 1 & 2 & 3 \\ 3 & 2 & 1 \end{bmatrix}$

(d) $\begin{bmatrix} 1 & 2 & 3 & 4 \\ 2 & 3 & 1 & 4 \end{bmatrix}$ (e) $\begin{bmatrix} 1 & 2 & 3 & 4 \\ 2 & 1 & 4 & 3 \end{bmatrix}$ (f) $\begin{bmatrix} 1 & 2 & 3 & 4 \\ 3 & 2 & 4 & 1 \end{bmatrix}$

2. In each one of the cases of Exercise 1, write the inverse of the permutation.

3. Show that the number of odd permutations of $\{1,\dots,n\}$ for $n \geqq 2$ is equal to the number of even permutations.

4. Show that the groups S_2, S_3, S_4 are solvable. [*Hint*: For S_4, find a subgroup H of order 4 in A_4. Consider the homomorphism of A_4 into S_3 given by translation on the cosets of H. Analyze the kernel of this homomorphism.]

5. Let σ be the r-cycle $[i_1 \cdots i_r]$. Show that $\epsilon(\sigma) = (-1)^{r+1}$. *Hint*: Use induction. If $r = 2$, then σ is a transposition. If $r > 2$, then

$$[i_1 \cdots i_r] = [i_1 i_r][i_1 \cdots i_{r-1}].$$

6. Two cycles $[i_1 \cdots i_r]$ and $[j_1 \cdots j_s]$ are said to be **disjoint** if no integer i_ν is equal to any integer j_μ. Prove that a permutation is equal to a product of disjoint cycles.

7. Express the permutations of Exercise 1 as a product of disjoint cycles.

8. Show that the group of Exercise 8, §1 exists by exhibiting it as a subgroup of S_4 as follows. Let $\sigma = [1234]$ and $\tau = [24]$. Show that the subgroup generated by σ, τ has order 8, and that σ, τ satisfy the same relations as x, y in the exercise *loc. cit.*

9. Show that the group of Exercise 10, §1 exists by exhibiting it as a subgroup of S_6.

10. Let n be an even positive integer. Show that there exists a group of order $2n$, generated by two elements σ, τ such that $\sigma^n = e = \tau^2$ and $\sigma\tau = \tau\sigma^{n-1}$. [Also see Exercises 6, 7 of Chapter VI, §2.] Draw a picture of a regular n-gon, number the vertices, and use the picture as an inspiration to get σ, τ.

11. Let G be a finite group of order $2k$ for some positive integer k.
    (a) Prove that G has an element of period 2. [*Hint*: show that there exists $x \in G$, $x \neq e$, such that $x = x^{-1}$.]
    (b) Assume that k is odd. Let $a \in G$ have period 2 and let $T_a: G \to G$ be translation by a. Prove that T_a is an odd permutation.

12. Let G be a finite group of even order $2k$, and assume that k is odd. Prove that G has a normal subgroup of order k. [*Hint*: Use the previous exercise.]

II, §7. FINITE ABELIAN GROUPS

The groups referred to in the title of this section occur so frequently that it is worth while to state a theorem which describes their structure completely. Throughout this section we write our abelian groups additively.

Let A be an abelian group. An element $a \in A$ is said to be a **torsion element** if it has finite period. The subset of all torsion elements of A is a subgroup of A called the **torsion subgroup** of A. (If a has period m and b has period n then, writing the group law additively, we see that $a \pm b$ has a period dividing mn.) Let p be a prime number. A **p-group** is a finite group whose order is a power of p. Finally, a group is said to be **of exponent m** if every element has period dividing m.

If A is an abelian group and p a prime number, we denote by $A(p)$ the subgroup of all elements $x \in A$ whose period is a power of p. Then $A(p)$ is a torsion group, and is a p-group if it is finite.

Theorem 7.1. *Let A be an additive abelian group of exponent n. If $n = mm'$ is a factorization with $(m, m') = 1$, then A is the direct sum*

$$A = A_m \oplus A_{m'}.$$

The group A is the direct sum of its subgroups $A(p)$ for all primes p dividing n.

Proof. For the first statement, since m, m' are relatively prime, there exist integers r, s such that

$$rm + sm' = 1.$$

Let $x \in A$. Then

$$(*) \qquad x = 1x = rmx + sm'x.$$

But $rmx \in A_{m'}$ because $m'rmx = rm'mx = rnx = 0$. Similarly, $sm'x \in A_m$. Hence $A = A_m + A_{m'}$. This sum is direct, for suppose $x \in A_m \cap A_{m'}$. By the same formula $(*)$ we see that $x = 0$, so $A_m \cap A_{m'} = \{0\}$, whence the sum is direct. The final statement is obtained by writing n as a product of distinct prime power factors

$$n = \prod p_i^{r_i},$$

and using induction. We thus get $A = \bigoplus A(p_i)$.

Note that if A is finite, we can take n equal to the order of A.

Our next task is to describe the structure of finite abelian p-groups. Let $r_1, \ldots, r_s$ be integers $\geqq 1$. A finite p-group A is said to be of **type**

$(p^{r_1},\ldots,p^{r_s})$ if A is isomorphic to the product of cyclic groups of orders $p^{r_i}(i = 1,\ldots,s)$.

Example. A group of type (p, p) is isomorphic to a product of cyclic groups $\mathbf{Z}/p\mathbf{Z} \times \mathbf{Z}/p\mathbf{Z}$.

Theorem 7.2. *Every finite abelian p-group is isomorphic to a product of cyclic p-groups. If it is of type $(p^{r_1},\ldots,p^{r_s})$ with*

$$r_1 \geqq r_2 \geqq \cdots \geqq r_s \geqq 1,$$

then the sequence of integers $(r_1,\ldots,r_s)$ is uniquely determined.

Proof. Let A be a finite abelian p-group. We shall need the following remark: Let b be an element of A, $b \neq 0$. Let k be an integer $\geqq 0$ such that $p^k b \neq 0$, and let p^m be the period of $p^k b$. Then b has period p^{k+m}. Proof: We certainly have $p^{k+m}b = 0$, and if $p^n b = 0$ then first $n \geqq k$, and second $n \geqq k + m$, otherwise the period of $p^k b$ would be smaller than p^m.

We shall now prove the existence of the desired product by induction. Let $a_1 \in A$ be an element of maximal period. We may assume without loss of generality that A is not cyclic. Let A_1 be the cyclic subgroup generated by a_1, say of period p^{r_1}. We need a lemma.

Lemma 7.3. *Let $\bar{b}$ be an element of A/A_1, of period p^r. Then there exists a representative a of $\bar{b}$ in A which also has period p^r.*

Proof. Let b be any representative of $\bar{b}$ in A. Then $p^r b$ lies in A_1, say $p^r b = na_1$ with some integer $n \geqq 0$. If $n = 0$ we let $a = b$. Suppose $n \neq 0$. We note that the period of $\bar{b}$ is $\leqq$ the period of b. Write $n = p^k t$ where t is prime to p. Then ta_1 is also a generator of A_1, and hence has period p^{r_1}. We may assume $k \leqq r_1$. Then $p^k ta_1$ has period $p^{r_1 - k}$. By our previous remark, the element b has period

$$p^{r + r_1 - k},$$

whence by hypothesis, $r + r_1 - k \leqq r_1$ and $r \leqq k$. This proves that there exists an element $c \in A_1$ such that $p^r b = p^r c$. Let $a = b - c$. Then a is a representative for $\bar{b}$ in A and $p^r a = 0$. Since period$(a) \geqq p^r$ we conclude that a has period equal to p^r.

We return to the main proof. By induction, the factor group A/A_1 is expressible as a direct sum

$$A/A_1 = \bar{A}_2 \oplus \cdots \oplus \bar{A}_s$$

of cyclic subgroups of orders $p^{r_2},\ldots,p^{r_s}$ respectively, such that $r_2 \geqq \cdots \geqq r_s$. Let $\bar{a}_i$ be a generator for $\bar{A}_i$ $(i = 2,\ldots,s)$ and let a_i be a representative in A having the same period as $\bar{a}_i$. Let A_i be the cyclic subgroup generated by a_i. We contend that A is the direct sum of $A_1, \ldots, A_s$.

We let $\bar{A} = A/A_1$. First observe that since a_i and $\bar{a}_i$ have the same period, the canonical homomorphism $A \to \bar{A}$ induces an isomorphism $A_j \xrightarrow{\approx} \bar{A}_j$ for $j = 2, \ldots, s$.

Next we prove that $A = A_1 + \cdots + A_s$. Given $x \in A$, let $\bar{x}$ denote its image in $A/A_1 = \bar{A}$. Then there exist elements $\bar{x}_j \in \bar{A}_j$ for $j = 2, \ldots, s$ such that

$$\bar{x} = \bar{x}_2 + \cdots + \bar{x}_s.$$

Hence $x - x_2 - \cdots - x_s$ lies in A_1, so there exists an element $x_1 \in A_1$ such that

$$x = x_1 + x_2 + \cdots + x_s,$$

which proves that A is the sum of the subgroups A_i $(i = 1, \ldots, s)$.

To prove that A is the direct sum, suppose $x \in A$ and

$$x = x_1 + \cdots + x_s = y_1 + \cdots + y_s \quad \text{with} \quad x_i, y_i \in A_i.$$

Subtracting, and letting $z_i = y_i - x_i$, we find that

$$0 = z_1 + \cdots + z_s \quad \text{with} \quad z_i \in A_i.$$

Then

$$\bar{0} = \bar{z}_2 + \cdots + \bar{z}_s$$

whence $\bar{z}_j = 0$ for $j = 2, \ldots, s$ because $\bar{A}$ is the direct sum of the subgroups $\bar{A}_j$ for $j = 2, \ldots, s$. But since $A_j \approx \bar{A}_j$, it follows that $z_j = 0$ for $j = 2, \ldots, s$. Hence $0 = z_1$ also. Hence, finally, $x_i = y_i$ for $i = 1, \ldots, s$, which concludes the proof of the existence part of the theorem.

We prove uniqueness, by induction on the order of A. Suppose that A is written in two ways as a product of cyclic groups, say of type

$$(p^{r_1}, \ldots, p^{r_s}) \quad \text{and} \quad (p^{m_1}, \ldots, p^{m_k})$$

with $r_1 \geqq \cdots \geqq r_s \geqq 1$ and $m_1 \geqq \cdots \geqq m_k \geqq 1$. Then pA is also a p-group, of order strictly less than the order of A, and is of type

$$(p^{r_1 - 1}, \ldots, p^{r_s - 1}) \quad \text{and} \quad (p^{m_1 - 1}, \ldots, p^{m_k - 1}),$$

it being understood that if some exponent r_i or m_j is equal to 1, then the factor corresponding to

$$p^{r_i - 1} \quad \text{or} \quad p^{m_j - 1}$$

in pA is simply the trivial group 0. Let $i = 1,\ldots,n$ be those integers such that $r_i \geqq 2$. Since $\#(pA) < \#(A)$, by induction the subsequence

$$(r_1 - 1,\ldots,r_s - 1)$$

consisting of those integers $\geqq 1$ is uniquely determined, and is the same as the corresponding subsequence of

$$(m_1 - 1,\ldots,m_k - 1).$$

In other words, we have $r_i - 1 = m_i - 1$ for all those integers $i = 1,\ldots,n$. Hence $r_i = m_i$ for $i = 1,\ldots,n$, and the two sequences

$$(p^{r_1},\ldots,p^{r_s}) \quad \text{and} \quad (p^{m_1},\ldots,p^{m_k})$$

can differ only in their last components which are equal to p. These correspond to factors of type $(p,\ldots,p)$ occurring say ν times in the first sequence and μ times in the second sequence. Thus A is of type

$$(p^{r_1},\ldots,p^{r_n}, \underbrace{p,\ldots,p}_{\nu \text{ times}}) \quad \text{and} \quad (p^{r_1},\ldots,p^{r_n}, \underbrace{p,\ldots,p}_{\mu \text{ times}}).$$

Hence the order of A is equal to

$$p^{r_1 + \cdots + r_n} p^{\nu} = p^{r_1 + \cdots + r_n} p^{\mu},$$

whence $\nu = \mu$, and our theorem is proved.

II, §7. EXERCISES

1. Let A be a finite abelian group, B a subgroup, and $C = A/B$. Assume that the orders of B and C are relatively prime. Show that there is a subgroup C' of A isomorphic to C, such that

$$A = B \oplus C'.$$

2. Using the structure theorem for abelian groups, prove the following:
   (a) An abelian group is cyclic if and only if, for every prime p dividing the order of G, there is one and only one subgroup of order p.
   (b) Let G be a finite abelian group which is not cyclic. Then there exists a prime p such that G contains a subgroup $C_1 \times C_2$ where C_1 and C_2 are cyclic of order p.

3. Let $f: A \to B$ be a homomorphism of abelian groups. Assume that there exists a homomorphism $g: B \to A$ such that $f \circ g = \mathrm{id}_B$.
   (a) Prove that A is the direct sum

$$A = \mathrm{Ker}\, f \oplus \mathrm{Im}\, g.$$

   (b) Prove that f and g are inverse isomorphisms between $g(B)$ and B.

Note: You may wish to do the following exercises only after reading §3 *of the next chapter, especially Exercises 7 through 12 of* III, §3. *Then the following Exercises 5 and 6 are completely subsumed by Exercises 7 through 12 of Chapter* III, §3.

4. Define $(\mathbf{Z}/n\mathbf{Z})^*$ to be the set of elements in $\mathbf{Z}/n\mathbf{Z}$ having coset representatives which are integers $a \in \mathbf{Z}$ prime to n. Show that $(\mathbf{Z}/n\mathbf{Z})^*$ is a multiplicative group.

5. Let G be a multiplicative cyclic group of order N and let $\mathrm{Aut}(G)$ be its group of automorphisms. For each $a \in (\mathbf{Z}/N\mathbf{Z})^*$ let $\sigma_a: G \to G$ be the map such that $\sigma_a(w) = w^a$. Show that $\sigma_a \in \mathrm{Aut}(G)$ and that the association

$$a \mapsto \sigma_a$$

   is an isomorphism of $(\mathbf{Z}/N\mathbf{Z})^*$ with $\mathrm{Aut}(G)$.

6. (a) Let n be a positive integer, which can be written $n = n_1 n_2$ where n_1, n_2 are integers $\geqq 2$ and relatively prime. Show that there is an isomorphism

$$f\colon \mathbf{Z}/n\mathbf{Z} \to \mathbf{Z}/n_1\mathbf{Z} \times \mathbf{Z}/n_2\mathbf{Z}$$

   where the map f associates to each residue class $a \bmod n\mathbf{Z}$ the pair of classes

$$a \bmod \mathbf{Z} \mapsto (a \bmod n_1\mathbf{Z},\ a \bmod n_2\mathbf{Z}).$$

   For the surjectivity, you will need the Chinese Remainder Theorem of Chapter I, §5, Exercise 5.
   (b) Extend the result to the case when $n = n_1 n_2 \cdots n_r$ is a product of pairwise relatively prime integers $\geqq 2$.

7. Let n be a positive integer, and write n as product of prime powers

$$n = \prod_{i=1}^{r} p_i^{n_i}$$

   Show that the map f of Exercise 6 restricts to an isomorphism of multiplicative groups

$$(\mathbf{Z}/n\mathbf{Z})^* \approx \prod_{i=1}^{r} (\mathbf{Z}/p_i^{n_i}\mathbf{Z})^*$$

8. Let p be a prime number. Let $1 \leqq k \leqq m$. Let $U_k = U_k(p^m)$ be the subset of $(\mathbf{Z}/p^m\mathbf{Z})^*$ consisting of those elements which have a representative in $\mathbf{Z}$ which is $\equiv 1 \bmod p^k$. Thus

$$U_k = 1 + p^k\mathbf{Z} \bmod p^m\mathbf{Z}.$$

Show that U_k is a subgroup of $(\mathbf{Z}/p^m\mathbf{Z})^*$. Thus we have a sequence of subgroups

$$U_1 \supset U_2 \supset \cdots \supset U_{m-1} \supset U_m = \{1\}.$$

9. (a) Let p be an odd prime and let $m \geqq 2$. Show that for $k \leqq m-1$ in Exercise 8, the map

$$\mathbf{Z}/p\mathbf{Z} \to U_k/U_{k+1}$$

given by $(a \bmod p) \mapsto 1 + p^k a \bmod p^{k+1}\mathbf{Z}$ is an isomorphism.
(b) Prove that the order of U_1 is p^{m-1}.
(c) Prove that U_1 is cyclic and that $1 + p \bmod p^m\mathbf{Z}$ is a generator.
[*Hint*: If p is odd, show by induction that for every positive integer r,

$$(1 + p)^{p^r} \equiv 1 + p^{r+1} \bmod p^{r+2}.$$

Is this assertion still true if $p = 2$? State and prove the analogue of this assertion for $(1 + 4)^{2^r}$.]

10. Let p be an odd prime. Show that there is an isomorphism

$$(\mathbf{Z}/p^m\mathbf{Z})^* \approx (\mathbf{Z}/p\mathbf{Z})^* \times U_1.$$

11. (a) For the prime 2, and an integer $m \geqq 2$, show that

$$(\mathbf{Z}/2^m\mathbf{Z})^* \approx \{1, -1\} \times U_2.$$

(b) Show that the group $U_2 = 1 + 4\mathbf{Z} \bmod 2^m\mathbf{Z}$ is cyclic, and that the class of $1 + 4$ is a generator. What is the order of U_2 in this case?

12. Let φ be the Euler phi function, i.e. $\varphi(n)$ is the order of $(\mathbf{Z}/n\mathbf{Z})^*$.
(a) If p is a prime number, and r an integer $\geqq 1$, show that

$$\varphi(p^r) = (p - 1)p^{r-1}.$$

(b) Prove that if m, n are positive relatively prime integers, then

$$\varphi(mn) = \varphi(m)\varphi(n).$$

(You may (should) of course use previous exercises to do this.)

II, §8. OPERATION OF A GROUP ON A SET

In some sense, this section is a continuation of §6. We shall define a general notion which contains permutation groups as a special case, and which has already been illustrated in the exercises of §3 and §4.

Let G be a group and let S be a set. An **operation** or an **action** of G on S is a homomorphism

$$\pi\colon G \to \operatorname{Perm}(S)$$

of G into the group of permutations of S. We denote the permutation associated with an element $x \in G$ by π_x. Thus the homomorphism is denoted by

$$x \mapsto \pi_x.$$

Given $s \in S$ the image of s under the permutation π_x is $\pi_x(s)$. From such an operation, we obtain a mapping

$$G \times S \to S$$

which to each pair (x, s) with $x \in G$ and $s \in S$ associates the element $\pi_x(s)$ of S. Sometimes we abbreviate the notation and write simply xs instead of $\pi_x(s)$. With this simpler notation, we have the two properties:

OP 1. *For all $x, y \in G$ and $s \in S$ we have associativity*

$$x(ys) = (xy)s.$$

OP 2. *If e is the unit element of G, then $es = s$ for all $s \in S$.*

Note that the formula of **OP 1** is simply an abbreviation for the property

$$\pi_{xy} = \pi_x \pi_y.$$

Similarly, the formula **OP 2** is an abbreviation for the property that π_e is the identity permutation, namely

$$\pi_e(s) = s \qquad \text{for all} \quad s \in S.$$

Conversely, if we are given a mapping

$$G \times S \to S \qquad \text{denoted by} \qquad (x, s) \mapsto xs,$$

satisfying **OP 1** and **OP 2**, then for each $x \in G$ the map $s \mapsto xs$ is a permutation of S which we may denote by $\pi_x(s)$. Then $x \mapsto \pi_x$ is a homomorphism of G into $\mathrm{Perm}(S)$. (Proof?) So an operation of G on a set S could also be defined as a mapping $G \times S \to S$ satisfying properties **OP 1** and **OP 2**. Hence we often use the abbreviated notation xs instead of $\pi_x(s)$ for the operation.

We shall now give examples of operations.

1. Translations. In Example 7 of §3 we have met translations: for each $x \in G$ we defined the translation

$$T_x\colon G \to G \qquad \text{by} \qquad T_x(y) = xy.$$

Thus we got a homomorphism $x \mapsto T_x$ of G into $\mathrm{Perm}(G)$. Of course T_x is not a group homomorphism, it is only a permutation of G.

Similarly, G operates by translation on the set of subsets, for if A is a subset of G, then $T_x(A) = xA$ is a subset. If H is a subgroup of G, then $T_x(H) = xH$ is a coset of H, and G operates by translation on the set of cosets of H. You should previously have worked out several exercises concerning this operation, although we did not call it by that name.

2. Conjugation. For each $x \in G$ we let $\mathbf{c}(x)\colon G \to G$ be the map such that $\mathbf{c}(x)(y) = xyx^{-1}$. Then we have seen in Example 8 of §3 that the map

$$x \mapsto \mathbf{c}(x)$$

is a homomorphism of G into $\mathrm{Aut}(G)$, and so this map gives an operation of G on itself by conjugation. The kernel of this homomorphism is the set of all $x \in G$ such that $xyx^{-1} = y$ for all $y \in G$, so the kernel is what we called the **center** of G.

We note that G also operates by conjugation on the set of subgroups of G, because the conjugate of a subgroup is a subgroup. We have met this operation before, although we did not call it by that name, and you should have worked out exercises on this operation in §4. In the case of conjugation, we do not use the notation of **OP 1** and **OP 2**, because it would lead to confusion if we write xH for conjugation. We reserve xH for the translation of H by x. For conjugation of H by x we write $\mathbf{c}(x)(H)$.

3. Example from linear algebra. Let $\mathbf{R}^n$ be the vector space of column vectors in n-space. Let G be the set of $n \times n$ matrices which are invertible. Then G is a multiplicative group, and G operates on $\mathbf{R}^n$. For $A \in G$ and $X \in \mathbf{R}^n$ we have the linear map $L_A\colon \mathbf{R}^n \to \mathbf{R}^n$ such that

$$L_A(X) = AX.$$

The map $A \mapsto L_A$ is a homomorphism of G into the multiplicative group of invertible linear maps of $\mathbf{R}^n$ onto itself. Since we write frequently AX instead of $L_A(X)$, the notation

$$(A, X) \mapsto AX$$

is particularly useful in this context, where we see directly that the two properties **OP 1** and **OP 2** are satisfied.

For other examples, see the exercises of Chapter VI, §3.

Suppose we have an operation of G on S. Let $s \in S$. We define the **isotropy group** of $s \in S$ to be the set of elements $x \in G$ such that $\pi_x(s) = s$. We denote the isotropy group by G_s, and we leave it as an exercise to verify that indeed, the isotropy group G_s is a subgroup of G.

Examples. Let G be a group and H a subgroup. Let G operate on the set of cosets of H by translation. Then the isotropy group of H is H itself, because if $x \in G$, then $xH = H$ if and only if $x \in H$.

Suppose next that G operates on itself by conjugation. Then the isotropy group of an element $a \in G$ is called the **centralizer** of a. In this case, the centralizer consists of all elements $x \in G$ such that x commutes with a, that is

$$xax^{-1} = a \quad \text{or} \quad xa = ax.$$

If we view G as operating by conjugation on the set of subgroups, then the isotropy group of a subgroup H is what we called the **normalizer** of H.

Let G operate on a set S. We use the notation of **OP 1** and **OP 2**. Let $s \in S$. The subset of S consisting of all elements xs with $x \in G$ is called the **orbit** of s under G, and is denoted by Gs. Let us denote this orbit by O. Let $t \in O$. Then

$$O = Gs = Gt.$$

This is easily seen, because $t \in O = Gs$ means that there exists $x \in G$ such that $xs = t$, and then

$$Gt = Gxs = Gs \quad \text{because} \quad Gx = G.$$

An element $t \in O$ is called a **representative** of the orbit, and we say that t **represents** the orbit. Note that the notion of orbit is analogous to the notion of coset, and the notion of representative of an orbit is analogous to the notion of coset representative. Compare the formalism of orbits with the formalism of cosets in §4.

Example. Let G operate on itself by conjugation. Then the orbit of an element x is called a **conjugacy class**, and consists of all elements

$$yxy^{-1} \quad \text{with} \quad y \in G.$$

In general, let G operate on a set S. Let $s \in S$. If x, y are in the same coset of the subgroup G_s then $xs = ys$. Indeed, we can write $y = xh$ with $h \in G_s$, so

$$ys = xhs = xs \qquad \text{since} \quad hs = s \text{ by assumption.}$$

Therefore we can define a mapping

$$\bar{f}\colon G/G_s \to S \qquad \text{by} \qquad \bar{f}(xG_s) = xs.$$

We say that $\bar{f}$ is **induced** by f.

Proposition 8.1. *Let G operate on a set S, and let $s \in S$.*

(1) *The map $x \mapsto xs$ induces a bijection between G/G_s and the orbit Gs.*

(2) *The order of the orbit Gs is equal to the index $(G : G_s)$.*

Proof. The image of $\bar{f}$ is the orbit of s since it consists of all elements xs of S with $x \in G$. The map $\bar{f}$ is injective, because if $x, y \in G$ and $xs = ys$, then $x^{-1}ys = s$ so $x^{-1}y \in G_s$, whence $y \in xG_s$, and x, y lie in the same coset of G_s. Hence $\bar{f}(xG_s) = \bar{f}(yG_s)$ implies $xs = ys$, which implies $xG_s = yG_s$, which concludes the proof.

In particular, when G operates by conjugation on the set of subgroups, and H is a subgroup, or when G operates on itself by conjugation, we obtain from Proposition 8.1 and the definitions:

Proposition 8.2

(a) *The number of conjugate subgroups to H is equal to the index of the normalizer of H.*
(b) *Let $x \in G$. The number of elements in the conjugacy class of x is the index of the centralizer $(G : G_x)$.*

The next result gives a very good test for normality. You should have done it as an exercise before, but we now do it as an example.

Proposition 8.3. *Let G be a group and H a subgroup of index 2. Then H is normal.*

Proof. Let S be the set of cosets of H and let G operate on S by translation. For each $x \in G$ let $T_x: S \to S$ be the translation such that $T_x(aH) = xaH$. Then

$$x \mapsto T_x \quad \text{is a homomorphism of} \quad G \to \text{Perm}(S).$$

Let K be the kernel. If $x \in K$, then in particular, $T_x(H) = H$, so $xH = H$ and $x \in H$. Therefore $K \subset H$. Then G/K is embedded as a subgroup of Perm(S), and Perm(S) has order 2, because S has order 2. Hence $(G:K) = 1$ or 2. But

$$(G:K) = (G:H)(H:K),$$

and $(G:H) = 2$. Hence $(H:K) = 1$, whence $H = K$, whence H is normal because K is normal. This concludes the proof.

Proposition 8.4. *Let G operate on a set S. Then two orbits of G are either disjoint or are equal.*

Proof. Let Gs and Gt be two orbits with an element in common. This element can be written

$$xs = yt \qquad \text{with some} \quad x, y \in G.$$

Hence

$$Gs = Gxs = Gyt = Gt$$

so the two orbits are equal, thus concluding the proof.

It follows from Proposition 8.4 that S is the disjoint union of the distinct orbits, and we can write

$$S = \bigcup_{i \in I} Gs_i \quad \text{(disjoint)}$$

where I is some indexing set, and the s_i represent the distinct orbits.

Suppose that S is a finite set. Let $\#(S)$ be the number of elements of S. We call $\#(S)$ the **order** of S. Then we get a decomposition of the order of S as a sum of orders of orbits, which we call the **orbit decomposition formula**; namely by Proposition 8.1:

$$\boxed{\#(S) = \sum_{i=1}^{r} (G:G_{s_i})}$$

Example. Let G operate on itself by conjugation. An element $x \in G$ is in the center of G if and only if the orbit of x is x itself, and thus has one element. In general, the order of the orbit of x is equal to the index of the centralizer of x. Thus we obtain:

Proposition 8.5. *Let G be a finite group. Let G_x be the centralizer of x. Let Z be the center of G and let $y_1, \dots, y_m$ represent the conjugacy classes which contain more than one element. Then*

$$(G:1) = (Z:1) + \sum_{i=1}^{m} (G:G_{y_i})$$

and $(G:G_{y_i}) > 1$ for $i = 1, \dots, m$.

The formula of Proposition 8.5 is called the **class formula** or **class equation**.

II, §8. EXERCISES

1. Let a group G operate on a set S. Let s, t be elements of S, and let $x \in G$ be such that $xs = t$. Show that

$$G_t = xG_sx^{-1}.$$

2. Let G operate on a set S, and let $\pi: G \to \text{Perm}(S)$ be the corresponding homomorphism. Let K be the kernel of π. Prove that K is the intersection of all isotropy groups G_s for all $s \in S$. In particular, K is contained in every isotropy group.

3. (a) Let G be a group of order p^n where p is prime and $n > 0$. Prove that G has a non-trivial center, i.e. the center of G is larger than $\{e\}$.
(b) Prove that G is solvable.

4. Let G be a finite group operating on a finite set S.
(a) For each $s \in S$ prove that

$$\sum_{t \in Gs} \frac{1}{\#(Gt)} = 1.$$

(b) $\sum_{s \in S}(1/\#(Gs))$ = number of orbits of G in S.

5. Let G be a finite group operating on a finite set S. For each $x \in G$ define $\alpha(x)$ = number of elements $s \in S$ such that $xs = s$. Prove that the number of orbits of G in S is equal to

$$\frac{1}{\#(G)} \sum_{x \in G} \alpha(x).$$

Hint: Let T be the subset of $G \times S$ consisting of all elements (x, s) such that $xs = s$. Count the order of T in two ways.

6. Let S, T be sets and let $M(S, T)$ denote the set of all mappings of S into T. Let G be a finite group operating on S. For each map $f: S \to T$ and $x \in G$ define the map $\pi_x f: S \to T$ by

$$(\pi_x f)(s) = f(x^{-1}s).$$

   (a) Prove that $x \mapsto \pi_x$ is an operation of G on $M(S, T)$.
   (b) Assume that S, T are finite. Let $n(x)$ denote the number of orbits of the cyclic group $\langle x \rangle$ on S. Prove that the number of orbits of G in $M(S, T)$ is equal to

$$\frac{1}{\#(G)} \sum_{x \in G} (\# T)^{n(x)}.$$

II, §9. SYLOW SUBGROUPS

Let p be a prime number. By a **p-group**, we mean a finite group whose order is a power of p (i.e. p^n for some integer $n > 0$). Let G be a finite group and H a subgroup. We call H a **p-subgroup** of G if H is a p-group.

Theorem 9.1. *Let G be a p-group and not trivial,* i.e. $\neq \{e\}$. *Then:*

(1) *G has a non-trivial center.*

(2) *G is solvable.*

Proof. For (1) we use the class equation. Let Z be the center of G. Then

$$(G:1) = (Z:1) + \sum (G:G_{x_i}),$$

where the sum is taken over a finite number of elements x_i with $(G:G_{x_i}) \neq 1$. Since G is a p-group, it follows that p divides $(G:1)$ and also $(G:G_{x_i})$. Hence p divides $(Z:1)$, so the center is not trivial.

For (2), $\#(G/Z)$ divides $\#(G)$ so G/Z is a p-group, and by (1), we know that $\#(G/Z) < \#(G)$. By induction G/Z is solvable. By Theorem 4.10 it follows that G is solvable.

Let G be a finite group and let H be a subgroup. Let p be a prime number. We say that H is a **p-Sylow subgroup** if the order of H is p^n and if p^n is the highest power of p dividing the order of G. We shall prove below that such subgroups always exist. For this we need a lemma.

Lemma 9.2. *Let G be a finite abelian group of order m, let p be a prime number dividing m. Then G has a subgroup of order p.*

Proof. Write $m = p^r s$ where s is prime to p. By Theorem 7.1 we can write G as a direct product

$$G = G(p) \times G'$$

where the order of G' is prime to p. Let $a \in G(p)$, $a \neq e$. Let p^k be the period of a. Let

$$b = a^{p^{k-1}}.$$

Then $b \neq e$ but $b^p = e$, and b generates a subgroup of order p as desired.

Theorem 9.3. *Let G be a finite group and p a prime number dividing the order of G. Then there exists a p-Sylow subgroup of G.*

Proof. By induction on the order of G. If the order of G is prime, our assertion is obvious. We now assume given a finite group G, and assume the theorem proved for all groups of order smaller than that of G. If there exists a proper subgroup H of G whose index is prime to p, then a p-Sylow subgroup of H will also be one of G, and our assertion follows by induction. We may therefore assume that every proper subgroup has an index divisible by p. We now let G act on itself by conjugation. From the class formula we obtain

$$(G:1) = (Z:1) + \sum_{i=1}^{m} (G:G_{y_i}).$$

Here, Z is the center of G, and the term $(Z:1)$ is the number of orbits having one element. The sum on the right is taken over the other orbits, and each index $(G:G_{y_i})$ is then > 1, hence divisible by p. Since p divides the order of G, it follows that p divides the order of Z, hence in particular that G has a non-trivial center.

By Lemma 9.2, let a be an element of period p in Z, and let H be the cyclic group generated by a. Since H is contained in Z, H is normal. Let $f: G \to G/H$ be the canonical homomorphism. Let p^n be the highest power of p dividing $(G:1)$. Then p^{n-1} divides the order of G/H. By induction on the order of the group, there is a p-Sylow subgroup K' of G/H. Let $K = f^{-1}(K')$. Then $K \supset H$ and f maps K onto K'. Hence we have an isomorphism $K/H \approx K'$. Hence K has order $p^{n-1}p = p^n$, as desired.

Theorem 9.4. *Let G be a finite group.*

(i) *If H is a p-subgroup of G, then H is contained in some p-Sylow subgroup.*
(ii) *All p-Sylow subgroups are conjugate.*
(iii) *The number of p-Sylow subgroups of G is* $\equiv 1 \bmod p$.

Proof. All proofs are applications of the technique of the class formula. We let S be the set of p-Sylow subgroups of G. Then G operates on S by conjugation.

Proof of (i). Let P be one of the p-Sylow subgroups of G. Let S_0 be the G-orbit of P. Let G_P be the normalizer of P. Then

$$\#(S_0) = (G : G_P)$$

and G_P contains P, so $\#(S_0)$ is not divisible by p. Let H be a p-subgroup of G of order > 1. Then H operates by conjugation on S_0, and hence S_0 itself breaks up into a disjoint union of H-orbits. Since $\#(H)$ is a p-power, the index in H of any proper subgroup of H is divisible by p. But we have the orbit decomposition formula

$$\#(S_0) = \sum_{i=1}^{r} (H : H_{P_i}).$$

Since $\#(S_0)$ is prime to p, it follows that one of the H-orbits in S_0 must consist of only one element, namely a certain Sylow subgroup P'. Then H is contained in the normalizer of P', and hence HP' is a subgroup of G. Furthermore, P' is normal in HP'. Since

$$HP'/P' \approx H/(H \cap P'),$$

it follows that the order of HP'/P' is a power of p, and therefore so is the order of HP'. Since P' is a maximal p-subgroup of G, we must have $HP' = P'$, and hence $H \subset P'$, which proves (i).

Remark. We have also proved:

Let H be a p-subgroup of G and suppose that H is contained in the normalizer of a p-Sylow group P'. Then $H \subset P'$.

Proof of (ii). Let H be a p-Sylow subgroup of G. We have shown that H is contained in some conjugate P' of P, and is therefore equal to that conjugate because the orders of H and P' are equal. This proves (ii).

Proof of (iii). Finally, take $H = P$ itself. Then one orbit of H in S has exactly one element, namely P itself. Let S' be another orbit of H in S. Then S' cannot have just one element P', otherwise H is contained in the normalizer of P', and by the remark $H = P'$. Let $s' \in S'$. Then the isotropy group of s' cannot be all of H, and so the index in H of the isotropy group is > 1, and is divisible by p since H is a p-group.

Consequently the number of elements in S' is divisible by p. Hence we obtain

$$\begin{aligned}\#(S) &= 1 + \text{indices divisible by } p \\ &\equiv 1 \bmod p.\end{aligned}$$

This proves (iii).

II, §9. EXERCISES

1. Let p be a prime number. What is the order of the p-Sylow subgroup of the symmetric group on p elements?

2. Prove that a group of order 15 is abelian, and in fact cyclic.

3. Let S_3 be the symmetric group on 3 elements.
   (a) Determine the 2-Sylow subgroups of S_3.
   (b) Determine the 3-Sylow subgroups of S_3.

4. (a) Prove that a group of order 12 is solvable.
   (b) Prove that S_4 is solvable.

5. Let G be a group of order p^3 which is not abelian.
   (a) Show that the center has order p.
   (b) Let Z be the center. Show that $G/Z \approx C \times C$, where C is a cyclic group of order p.

6. (a) Let H be a subgroup of a finite group G. Let P be a Sylow subgroup of G. Prove that there exists a conjugate P' of P in G such that $P' \cap H$ is a p-Sylow subgroup of H.
   (b) Assume H normal and $(G:H)$ prime to p. Prove that H contains every p-Sylow subgroup of G.

7. Let G be a group of order 6, and assume that G is not commutative. Show that G is isomorphic to S_3. [*Hint*: Show that G contains elements a, b such that $a^2 = e$, $b^3 = e$, and $aba = b^2 = b^{-1}$.]

8. Let G be a non-commutative group of order 8. Show that G is isomorphic to the group of symmetries of the square, or to the quaternion group, in other words G is isomorphic to one of the groups of Exercises 8 and 9 of §1.

9. Let G be a finite group and assume that all the Sylow subgroups are normal. Let $P_1, \ldots, P_r$ be the Sylow subgroups. Prove that the map

   $$P_1 \times \cdots \times P_r \to G \quad \text{given by} \quad (x_1, \ldots, x_r) \mapsto x_1 \cdots x_r$$

   is an isomorphism. So G is isomorphic to the direct product of its Sylow subgroups. [*Hint*: Recall Exercise 12 of §5.]

10. Let G be a finite group of order pq, where p, q are prime and $p < q$. Suppose that $q \not\equiv 1 \bmod p$. Prove that G is abelian, and in fact cyclic.

11. Let G be a finite group. Let $N(H)$ denote the normalizer of a subgroup H. Let P be a p-Sylow subgroup of G. Prove that $N(N(P)) = N(P)$.

CHAPTER III

Rings

In this chapter, we axiomatize the notions of addition and multiplication.

III, §1. RINGS

A **ring** R is a set, whose objects can be added and multiplied (i.e. we are given associations $(x, y) \mapsto x + y$ and $(x, y) \mapsto xy$ from pairs of elements of R, into R), satisfying the following conditions:

RI 1. *Under addition, R is an additive (abelian) group.*

RI 2. *For all $x, y, z \in R$ we have*

$$x(y + z) = xy + xz \quad \text{and} \quad (y + z)x = yx + zx.$$

RI 3. *For all $x, y, z \in R$, we have associativity $(xy)z = x(yz)$.*

RI 4. *There exists an element $e \in R$ such that $ex = xe = x$ for all $x \in R$.*

Example 1. Let R be the integers $\mathbf{Z}$. Then R is a ring.

Example 2. The rational numbers, the real numbers, and the complex numbers all are rings.

Example 3. Let R be the set of continuous real-valued functions on the interval $[0, 1]$. The sum and product of two functions f, g are defined as usual, namely $(f + g)(t) = f(t) + g(t)$, and $(fg)(t) = f(t)g(t)$. Then R is a ring.

More generally, let S be a non-empty set, and let R be a ring. Let $M(S, R)$ be the set of mappings of S into R. Then $M(S, R)$ is a ring, if we define the addition and product of mappings f, g by the rules

$$(f + g)(x) = f(x) + g(x) \qquad \text{and} \qquad (fg)(x) = f(x)g(x).$$

We leave the verification as a simple exercise for the reader.

Example 4 (The ring of endomorphisms). Let A be an abelian group. Let $\mathrm{End}(A)$ denote the set of homomorphisms of A into itself. We call $\mathrm{End}(A)$ the set of **endomorphisms** of A. Thus $\mathrm{End}(A) = \mathrm{Hom}(A, A)$ in the notation of Chapter II, §3. We know that $\mathrm{End}(A)$ is an additive group.

If we let the multiplicative law of composition on $\mathrm{End}(A)$ *be ordinary composition of mappings, then* $\mathrm{End}(A)$ *is a ring.*

We prove this in detail. We already know **RI 1.** As for **RI 2**, let f, g, $h \in \mathrm{End}(A)$. Then for all $x \in A$,

$$\begin{aligned}(f \circ (g + h))(x) &= f((g + h)(x)) \\ &= f(g(x) + h(x)) = f(g(x)) + f(h(x)) \\ &= f \circ g(x) + f \circ h(x).\end{aligned}$$

Hence $f \circ (g + h) = f \circ g + f \circ h$. Similarly on the other side. We observe that **RI 3** is nothing but the associativity for composition of mappings in this case, and we already know it. The unit element of **RI 4** is the identity mapping I. Thus we have seen that $\mathrm{End}(A)$ is a ring.

A ring R is said to be **commutative** if $xy = yx$ for all x, $y \in R$. The rings of Examples 1, 2, 3 are commutative. In general, the ring of Example 4 is not commutative.

As with groups, the element e of a ring R satisfying **RI 4** is unique, and is called the **unit element** of the ring. It is often denoted by 1. Note that if $1 = 0$ in the ring R, then R consists of 0 alone, in which case it is called the zero ring.

In a ring R, a number of ordinary rules of arithmetic can be deduced from the axioms. We shall list these.

We have $0x = 0$ *for all* $x \in R$.

Proof. We have

$$0x + x = 0x + ex = (0 + e)x = ex = x.$$

Hence $0x = 0$.

We have $(-e)x = -x$ *for all* $x \in R$.

Proof.

$$(-e)x + x = (-e)x + ex = (-e + e)x = 0x = 0.$$

We have $(-e)(-e) = e$.

Proof. We multiply the equation

$$e + (-e) = 0$$

by $-e$, and find

$$-e + (-e)(-e) = 0.$$

Adding e to both sides yields $(-e)(-e) = e$, as desired.

We leave it as an exercise to prove that

$$(-x)y = -xy \quad \text{and} \quad (-x)(-y) = xy$$

for all $x, y \in R$.

From condition **RI 2**, which is called the **distributive law**, we can deduce the analogous rule with several elements, namely if $x, y_1, \ldots, y_n$ are elements of the ring R, then

$$x(y_1 + \cdots + y_n) = xy_1 + \cdots + xy_n.$$

Similarly, if $x_1, \ldots, x_m$ are elements of R, then

$$\begin{aligned}(x_1 + \cdots + x_m)(y_1 + \cdots + y_n) &= x_1 y_1 + \cdots + x_m y_n \\ &= \sum_{i=1}^{m} \sum_{j=1}^{n} x_i y_j.\end{aligned}$$

The sum on the right hand side is to be taken over all indices i and j as indicated. These more general rules can be proved by induction, and we shall omit the proofs, which are tedious.

Let R be a ring. By a **subring** R' of R one means a subset of R such that the unit element of R is in R', and if $x, y \in R'$, then $-x$, $x + y$, and xy are also in R'. It then follows obviously that R' is a ring, the operations of addition and multiplication in R' being the same as those in R.

Example 5. The integers form a subring of the rational numbers, which form a subring of the real numbers.

Example 6. The real-valued differentiable functions on **R** form a subring of the ring of continuous functions.

Let R be a ring. It may happen that there exist elements $x, y \in R$ such that $x \neq 0$ and $y \neq 0$, but $xy = 0$. Such elements are called **divisors of zero**. A commutative ring without divisors of zero, and such that $1 \neq 0$ is called an **integral ring**. A commutative ring such that the subset of nonzero elements form a group under multiplication is called a **field**. Observe that in a field, we have necessarily $1 \neq 0$, and that a field has no divisors of zero (proof?).

Example 7. The integers **Z** form an integral ring. Every field is an integral ring. We shall see later that the polynomials over a field form an integral ring.

Let R be a ring. We denote by R^* the set of elements $x \in R$ such that there exists $y \in R$ such that $xy = yx = e$. In other words, R^* is the set of elements $x \in R$ which have a multiplicative inverse. The elements of R^* are called the **units** of R. We leave it as an exercise to show that the units form a multiplicative group. As an example, the units of a field form the group of non-zero elements of the field.

Let R be a ring, and $x \in R$. If n is a positive integer, we define

$$x^n = x \cdots x,$$

the product being taken n times. Then for positive integers m, n we have

$$x^{n+m} = x^n x^m \quad \text{and} \quad (x^m)^n = x^{mn}.$$

III, §1. EXERCISES

1. Let p be a prime number. Let R be the subset of all rational numbers m/n such that $n \neq 0$ and n is not divisible by p. Show that R is a ring.
2. How would you describe the units in Exercise 1? Prove whatever assertion you make.
3. Let R be an integral ring. If $a, b, c \in R$, $a \neq 0$, and $ab = ac$, then prove that $b = c$.
4. Let R be an integral ring, and $a \in R$, $a \neq 0$. Show that the map $x \mapsto ax$ is an injective mapping of R into itself.
5. Let R be a finite integral ring. Show that R is a field. [*Hint*: Use the preceding exercise.]
6. Let R be a ring such that $x^2 = x$ for all $x \in R$. Show that R is commutative.

7. Let R be a ring and $x \in R$. We define x to be **nilpotent** if there exists a positive integer n such that $x^n = 0$. If x is nilpotent, prove that $1 + x$ is a unit, and so is $1 - x$.

8. Prove in detail that the units of a ring form a multiplicative group.

9. Let R be a ring, and Z the set of all elements $a \in R$ such that $ax = xa$ for all $x \in R$. Show that Z is a subring of R, called the **center** of R.

10. Let R be the set of numbers of type $a + b\sqrt{2}$ where a, b are rational numbers. Show that R is a ring, and in fact that R is a field.

11. Let R be the set of numbers of type $a + b\sqrt{2}$ where a, b are integers. Show that R is a ring, but not a field.

12. Let R be the set of numbers of type $a + bi$ where a, b are integers and $i = \sqrt{-1}$. Show that R is a ring. List all its units.

13. Let R be the set of numbers of type $a + bi$ where a, b are rational numbers. Show that R is a field.

14. Let S be a set, R a ring, and $f: S \to R$ a bijective mapping. For each x, $y \in S$ define

$$x + y = f^{-1}(f(x) + f(y)) \quad \text{and} \quad xy = f^{-1}(f(x)f(y)).$$

Show that these sum and product define a ring structure on S.

15. In a ring R it may happen that a product xy is equal to 0 but $x \neq 0$ and $y \neq 0$. Give an example of this in the ring of $n \times n$ matrices over a field K. Also give an example in the ring of continuous functions on the interval [0, 1]. [In this exercise we assume that you know matrices and continuous functions. For matrices, see Chapter V, §3.]

III, §2. IDEALS

Let R be a ring. A **left ideal** of R is a subset J of R having the following properties: If x, $y \in J$, then $x + y \in J$ also, the zero element is in J, and if $x \in J$ and $a \in R$, then $ax \in J$.

Using the negative $-e$, we see that if J is a left ideal, and $x \in J$, then $-x \in J$ also, because $-x = (-e)x$. Thus the elements of a left ideal form an additive subgroup of R and we may as well say that a left ideal is an additive subgroup J of R such that, if $x \in J$ and $a \in R$ then $ax \in J$.

We note that R is a left ideal, called the **unit ideal**, and so is the subset of R consisting of 0 alone. We have $J = R$ if and only if $1 \in J$.

Similarly, we can define a **right ideal** and a **two-sided ideal**. Thus a two-sided ideal J is by definition an additive subgroup of R such that, if $x \in J$ and $a \in R$, then ax and $xa \in J$.

Example 1. Let R be the ring of continuous real-valued functions on the interval $[0, 1]$. Let J be the subset of functions f such that $f(\frac{1}{2}) = 0$. Then J is an ideal (two-sided, since R is commutative).

Example 2. Let R be the ring of integers $\mathbf{Z}$. Then the even integers, i.e. the integers of type $2n$ with $n \in \mathbf{Z}$, form an ideal. Do the odd integers form an ideal?

Example 3. Let R be a ring, and a an element of R. The set of elements xa, with $x \in R$, is a left ideal, called the **principal left ideal** generated by a. (Verify in detail that it is a left ideal.) We denote it by (a). More generally, let $a_1, \ldots, a_n$ be elements of R. The set of all elements

$$x_1a_1 + \cdots + x_na_n$$

with $x_i \in R$, is a left ideal, denoted by $(a_1, \ldots, a_n)$. We call $a_1, \ldots, a_n$ **generators** for this ideal.

We shall give a complete proof for this to show how easy it is, and leave the proof of further statements in the next examples as exercises. If $y_1, \ldots, y_n, x_1, \ldots, x_n \in R$ then

$$\begin{aligned}(x_1a_1 + \cdots + x_na_n) + (y_1a_1 + \cdots + y_na_n)& \\ &= x_1a_1 + y_1a_1 + \cdots + x_na_n + y_na_n \\ &= (x_1 + y_1)a_1 + \cdots + (x_n + y_n)a_n.\end{aligned}$$

If $z \in R$, then

$$z(x_1a_1 + \cdots + x_na_n) = zx_1a_1 + \cdots + zx_na_n.$$

Finally,

$$0 = 0a_1 + \cdots + 0a_n.$$

This proves that the set of all elements $x_1a_1 + \cdots + x_na_n$ with $x_i \in R$ is a left ideal.

Example 4. Let R be a ring. Let L, M be left ideals. We denote by LM the set of all elements $x_1y_1 + \cdots + x_ny_n$ with $x_i \in L$ and $y_i \in M$. It is an easy exercise for the reader to verify that LM is also a left ideal. Verify also that if L, M, N are left ideals, then $(LM)N = L(MN)$.

Example 5. Let L, M be left ideals. We define $L + M$ to be the subset consisting of all elements $x + y$ with $x \in L$ and $y \in M$. Then $L + M$ is a left ideal. Besides verifying this in detail, also show that if L, M, N are left ideals, then

$$L(M + N) = LM + LN.$$

Also formulate and prove the analogues of Examples 4 and 5 for right, and two-sided ideals.

Example 6. Let L be a left ideal, and denote by LR the set of elements $x_1 y_1 + \cdots + x_n y_n$ with $x_i \in L$ and $y_i \in R$. Then LR is a two-sided ideal. The proof is again left as an exercise.

Example 7. In Theorem 3.1 of Chapter I we proved that every ideal of $\mathbf{Z}$ is principal.

III, §2. EXERCISES

1. Show that a field has no ideal other than the zero and unit ideal.

2. Let R be a commutative ring. If M is an ideal, abbreviate MM by M^2. Let M_1, M_2 be two ideals such that $M_1 + M_2 = R$. Show that $M_1^2 + M_2^2 = R$.

3. Let R be the ring of Exercise 1 in the preceding section. Show that the subset of elements m/n in R such that m is divisible by p is an ideal.

4. Let R be a ring and J_1, J_2 left ideals. Show that $J_1 \cap J_2$ is a left ideal, and similarly for right and two-sided ideals.

5. Let R be a ring and $a \in R$. Let J be the set of all $x \in R$ such that $xa = 0$. Show that J is a left ideal.

6. Let R be a ring and L a left ideal. Let M be the set of all $x \in R$ such that $xL = 0$ (i.e. $xy = 0$ for all $y \in L$). Show that M is a two-sided ideal.

7. Let R be a commutative ring. Let L, M be ideals.
   (a) Show that $LM \subset L \cap M$.
   (b) Given an example when $LM \neq L \cap M$.
   Note that as a result of (a), if J is an ideal of R, then we get a sequence of ideals contained in each other by taking powers of J, namely
   $$J \supset J^2 \supset J^3 \supset \cdots \supset J^n \supset \cdots.$$

8. The following example will be of interest in calculus. Let R be the ring of infinitely differentiable functions defined, say, on the open interval $-1 < t < 1$. Let J_n be the set of functions $f \in R$ such that $D^k f(0) = 0$ for all integers k with $0 \leqq k \leqq n$. Here D denotes the derivative, so J_n is the set of functions all of whose derivatives up to order n vanish at 0. Show that J_n is an ideal in R.

9. Let R be the ring of real-valued functions on the interval $[0, 1]$. Let S be a subset of this interval. Show that the set of all functions $f \in R$ such that $f(x) = 0$ for all $x \in S$ is an ideal of R.

Note: If you know about matrices and linear maps then you should do immediately the exercises of Chapter V, §3, and you should look at those of Chapter V, §4. Of course, if necessary, do them after you have read the required material. But they give examples of rings and ideals.

III, §3. HOMOMORPHISMS

Let R, R' be rings. By a **ring-homomorphism** $f: R \to R'$, we shall mean a mapping having the following properties: For all $x, y \in R$,

$$f(x + y) = f(x) + f(y), \qquad f(xy) = f(x)f(y), \qquad f(e) = e'$$

(if e, e' are the unit elements of R and R' respectively).

By the **kernel** of a ring-homomorphism $f: R \to R'$, we shall mean its kernel viewed as a homomorphism of additive groups, i.e. it is the set of all elements $x \in R$ such that $f(x) = 0$. *Exercise*: Prove that the kernel is a two-sided ideal of R.

Example 1. Let R be the ring of complex-valued functions on the interval $[0, 1]$. The map which to each function $f \in R$ associates its value $f(\frac{1}{2})$ is a ring-homomorphism of R into $\mathbf{C}$.

Example 2. Let R be the ring of real-valued functions on the interval $[0, 1]$. Let R' be the ring of real-valued functions on the interval $[0, \frac{1}{2}]$. Each function $f \in R$ can be viewed as a function on $[0, \frac{1}{2}]$, and when we so view f, we call it the **restriction** of f to $[0, \frac{1}{2}]$. More generally, let S be a set, and S' a subset. Let R be the ring of real-valued functions on S. For each $f \in R$, we denote by $f|S'$ the function on S' whose value at an element $x \in S'$ is $f(x)$. Then $f|S'$ is called the **restriction** of f to S'. Let R' be the ring of real-valued functions on S'. Then the map

$$f \mapsto f|S'$$

is a ring-homomorphism of R into R'.

Since the kernel of a ring-homomorphism is defined only in terms of the additive groups involved, we know that a ring-homomorphism whose kernel is trivial is injective.

Let $f: R \to R'$ be a ring-homomorphism. If there exists a ring-homomorphism $g: R' \to R$ such that $g \circ f$ and $f \circ g$ are the respective identity mappings, then we say that f is a **ring-isomorphism**. A ring-isomorphism of a ring with itself is called an **automorphism**. As with groups, we have the following properties.

If $f: R \to R'$ is a ring-homomorphism which is a bijection, then f is a ring-isomorphism.

Furthermore, if $f: R \to R'$ and $g: R' \to R''$ are ring-homomorphisms, then the composite $g \circ f: R \to R''$ is also a ring-homomorphism.

We leave both proofs to the reader.

Remark. We have so far met with group homomorphisms and ring homomorphisms, and we have defined the notion of isomorphism in a similar way in each one of these categories of objects. Our definitions have followed a completely general pattern, which can be applied to other objects and categories (for instance modules, which we shall meet later). In general, not to prejudice what kind of objects one deals with, one may use the word **morphism** instead of homomorphism. Then an **isomorphism** (in whatever category) is a morphism f for which there exists a morphism g satisfying

$$f \circ g = \mathrm{id} \quad \text{and} \quad g \circ f = \mathrm{id}.$$

In other words, it is a morphism which has an inverse.

The symbol id denotes the identity. For this general definition to make sense, very few properties of morphisms need be satisfied: only associativity, and the existence of an identity for each object. We don't want to go fully into this abstraction now, but watch for it as you encounter the pattern latter. An **automorphism** is then defined as an isomorphism of an object with itself. It then follows completely generally directly from the definition that *the automorphisms of an object form a group.* One of the basic topics of study of mathematics is the structure of the groups of automorphisms of various objects. For instance, in Chapter II, as an exercise you determined the group of automorphisms of a cyclic group. In Galois theory, you will determine the automorphisms of certain fields.

We shall now define a notion similar to that of factor group, but applied to rings.

Let R be a ring and M a two-sided ideal. If $x, y \in R$, define x **congruent to** y mod M to mean $x - y \in M$. We write this relation in the form

$$x \equiv y \pmod{M}.$$

It is then very simple to prove the following statements.

(a) We have $x \equiv x \pmod{M}$.
(b) If $x \equiv y$ and $y \equiv z \pmod{M}$, then $x \equiv z \pmod{M}$.
(c) If $x \equiv y$ then $y \equiv x \pmod{M}$.
(d) If $x \equiv y \pmod{M}$, and $z \in R$, then $xz \equiv yz \pmod{M}$, and also $zx \equiv zy \pmod{M}$.
(e) If $x \equiv y$ and $x' \equiv y' \pmod{M}$, then $xx' \equiv yy' \pmod{M}$. Furthermore, $x + x' \equiv y + y' \pmod{M}$.

The proofs of the preceding assertions are all trivial. As an example, we shall give the proof of the first part of (e). The hypothesis means that we can write

$$x = y + z \quad \text{and} \quad x' = y' + z'$$

with $z, z' \in M$. Then

$$xx' = (y + z)(y' + z') = yy' + zy' + yz' + zz'.$$

Since M is a two-sided ideal, each one of zy', yz', zz' lies in M, and consequently their sum lies in M. Hence $xx' \equiv yy' \pmod{M}$, as was to be shown.

Remark. The present notion of congruence generalizes the notion defined for the integers in Chapter I. Indeed, if $R = \mathbf{Z}$, then the congruence

$$x \equiv y \pmod{n}$$

in Chapter I, meaning that $x - y$ is divisible by n, is equivalent to the property that $x - y$ lies in the ideal generated by n.

If $x \in R$, we let $\bar{x}$ be the set of all elements of R which are congruent to $x \pmod{M}$. Recalling the definition of a factor group, we see that $\bar{x}$ is none other than the additive coset $x + M$ of x, relative to M. Any element of that coset (also called **congruence class** of x mod M) is called a **representative** of the coset.

We let $\bar{R}$ be the set of all congruence classes of R mod M. In other words, we let $\bar{R} = R/M$ be the additive factor group of R modulo M. Then we already know that $\bar{R}$ is an additive group. We shall now define a multiplication which will make $\bar{R}$ into a ring.

If $\bar{x}$ and $\bar{y}$ are additive cosets of M, we define their product to be the coset of xy, i.e. to be $\overline{xy}$. Using condition (e) above, we see that this coset is independent of the selected representatives x in $\bar{x}$ and y in $\bar{y}$. Thus our multiplication is well defined by the rule

$$(x + M)(y + M) = (xy + M).$$

It is now a simple matter to verify that the axioms of a ring are satisfied. **RI 1** is already known since R/M is taken as the factor group. For **RI 2**, let $\bar{x}$, $\bar{y}$, $\bar{z}$ be congruence classes, with representatives x, y, z respectively in R. Then $y + z$ is a representative of $\bar{y} + \bar{z}$ by definition, and $x(y + z)$ is a representative of $\bar{x}(\bar{y} + \bar{z})$. But $x(y + z) = xy + xz$. Furthermore, xy is a representative of $\bar{x}\bar{y}$ and xz is a representative of $\bar{x}\bar{z}$. Hence by definition,

$$\bar{x}(\bar{y} + \bar{z}) = \bar{x}\bar{y} + \bar{x}\bar{z}.$$

Similarly, one proves **RI 3**. As for **RI 4**, if e denotes the unit element of R, then $\bar{e}$ is a unit element in $\bar{R}$, because $ex = x$ is a representative of $\bar{e}\bar{x}$. This proves all the axioms.

We call $\bar{R} = R/M$ the **factor ring** of R modulo M.

We observe that the map $f: R \to R/M$ such that $f(x) = \bar{x}$ is a ring-homomorphism of R onto R/M, whose kernel is M. The verification is immediate, and essentially amounts to the definition of the addition and multiplication of cosets of M.

Theorem 3.1. *Let $f: R \to S$ be a ring-homomorphism and let M be its kernel. For each coset C of M the image $f(C)$ is an element of S, and the association*

$$\bar{f}: C \mapsto f(C)$$

is a ring-isomorphism of R/M onto the image of f.

Proof. The fact that the image of f is a subring of S is left as an exercise (Exercise 1). Each coset C consists of all elements $x + z$ with some x and all $z \in M$. Thus

$$f(x + z) = f(x) + f(z) = f(x)$$

implies that $f(C)$ consists of one element. Thus we get a map

$$\bar{f}: C \mapsto f(C)$$

as asserted. If x, y represent cosets of M, then the relations

$$f(xy) = f(x)f(y),$$
$$f(x + y) = f(x) + f(y),$$
$$f(e_R) = e_S$$

show that $\bar{f}$ is a homomorphism of R/M into S. If $\bar{x} \in R/M$ is such that $\bar{f}(\bar{x}) = 0$, this means that for any representative x of $\bar{x}$ we have $f(x) = 0$, whence $x \in M$ and $\bar{x} = 0$ (in R/M). Thus $\bar{f}$ is injective. This proves what we wanted.

Example 3. If $R = \mathbf{Z}$, and n is a non-zero integer, then $R/(n) = \mathbf{Z}/(n)$ is called the ring of **integers modulo** n. We note that this is a finite ring, having exactly n elements. (Proof?) We also write $\mathbf{Z}/n\mathbf{Z}$ instead of $\mathbf{Z}/(n)$.

Example 4. Let R be any ring, with unit element e. Let $a \in R$. Since R is also an additive abelian group, we know how to define na for every integer n. If n is positive, then

$$na = a + a + \cdots + a,$$

the sum being taken n times. If $n = -k$ where k is positive, so n is negative, then

$$na = -(ka).$$

In particular, we can take $a = e$, so we get a map

$$f\colon \mathbf{Z} \to R \quad \text{such that} \quad n \mapsto ne.$$

As in Example 4 of Chapter II, §3 we know that this map f is a homomorphism of additive abelian groups. But f is also a ring homomorphism. Indeed, first note the property that for every positive integer n,

$$(ne)a = (e + \cdots + e)a = ea + \cdots + ea = n(ea) = \underbrace{a + \cdots + a}_{n \text{ times}} = na.$$

If m, n are positive integers, then

$$m(na) = \underbrace{na + \cdots + na}_{m \text{ times}} = \underbrace{\underbrace{a + \cdots + a}_{n \text{ times}} + \cdots + \underbrace{a + \cdots + a}_{n \text{ times}}}_{m \text{ times}} = \underbrace{a + \cdots + a}_{mn \text{ times}}$$

$$= (mn)a.$$

Hence putting $a = e$, we get

$$f(mn) = (mn)e = m(ne) = (me)(ne) = f(m)f(n).$$

We leave it to the reader to verify the verify the similar property when m or n is negative. The proof uses the case when m, n are positive, together with the property of homomorphisms that $f(-n) = -f(n)$.

Let $f\colon \mathbf{Z} \to R$ be a ring homomorphism. By definition we must have $f(1) = e$. Hence necessarily for every positive integer n we must have

$$f(n) = f(1 + \cdots + 1) = f(1) + \cdots + f(1) = ne,$$

and for a negative integer $m = -k$,

$$f(-k) = -f(k) = -(ke).$$

Thus there is one and only one ring homomorphism of $\mathbf{Z}$ into a ring R, which is the one we defined above.

Assume $R \neq \{0\}$. Let $f\colon \mathbf{Z} \to R$ be *the* ring homomorphism. Then the kernel of f is not all of $\mathbf{Z}$ and hence is an ideal $n\mathbf{Z}$ for some integer

$n \geqq 0$. It follows from Theorem 3.1 that $\mathbf{Z}/n\mathbf{Z}$ is isomorphic to the image of f. In practice, we do not make any distinction between $\mathbf{Z}/n\mathbf{Z}$ and its image in R, and we agree to say that R contains $\mathbf{Z}/n\mathbf{Z}$ as a subring. *Suppose that $n \neq 0$. Then we have relation*

$$na = 0 \quad \textit{for all} \quad a \in R.$$

Indeed, $na = (ne)a = 0a = 0$. Sometimes one says that R has **characteristic** n. Thus if n is the characteristic of R, then $na = 0$ for all $a \in R$.

Theorem 3.2. *Suppose that R is an integral ring, so has no divisors of 0. Then the integer n such that $\mathbf{Z}/n\mathbf{Z}$ is contained in R must be 0 or a prime number.*

Proof. Suppose n is not 0 and is not prime. Then $n = mk$ with integers $m, k \geqq 2$, and neither m, k are in the kernel of the homomorphism $f: \mathbf{Z} \to R$. Hence $me \neq 0$ and $ke \neq 0$. But $(me)(ke) = mke = 0$, contradicting the hypothesis that R has no divisors of 0. Hence n is prime.

Let K be a field and let $f: \mathbf{Z} \to K$ be the homomorphism of the integers into K. If the kernel of f is $\{0\}$, then K contains $\mathbf{Z}$ as a subring, and we say that K has **characteristic** 0. If the kernel of f is generated by a prime number p, then we say that K has **characteristic** p. The field $\mathbf{Z}/p\mathbf{Z}$ is sometimes denoted by $\mathbf{F}_p$, and is called the **prime field**, of characteristic p. This prime field $\mathbf{F}_p$ is contained in every field of characteristic p.

Let R be a ring. Recall that a **unit** in R is an element $u \in R$ which has a multiplicative inverse, that is there exists an element $v \in R$ such that $uv = e$. The set of units is denoted by R^*. This set of units is a group. Indeed, if u_1, u_2 are units, then the product $u_1 u_2$ is a unit, because it has the inverse $u_2^{-1} u_1^{-1}$. The rest of the group axioms are immediately verified from the ring axioms concerning multiplication.

Example. Let n be an integer $\geqq 2$ and let $R = \mathbf{Z}/n\mathbf{Z}$. Then the units of R consist of those elements of R which have a representative $a \in \mathbf{Z}$ which is prime to n. (Do Exercise 3.) This group of units is especially important, and we shall now describe how it occurs as a group of automorphisms.

Theorem 3.3. *Let G be a cyclic group of order N, written multiplicatively. Let $m \in \mathbf{Z}$ be relatively prime to N, and let*

$$\sigma_m: G \to G$$

be the map such that $\sigma_m(x) = x^m$. *Then* σ_m *is an automorphism of* G, *and the association*

$$m \mapsto \sigma_m$$

induces an isomorphism $(\mathbf{Z}/N\mathbf{Z})^* \xrightarrow{\approx} \mathrm{Aut}(G)$.

Proof. Since $\sigma_m(xy) = (xy)^m = x^m y^m$ (because G is commutative), it follows that σ_m is a homomorphism of G into itself. Since $(m, N) = 1$, we conclude that $x^m = e \Rightarrow x = e$. Hence $\ker(\sigma_m)$ is trivial, and since G is finite, it follows that σ_m is bijective, so σ_m is an automorphism. If $m \equiv n \bmod N$ then $\sigma_m = \sigma_n$ so σ_m depends only on the coset of $m \bmod N\mathbf{Z}$. We have

$$\sigma_{mn}(x) = x^{mn} = (x^n)^m = \sigma_n \sigma_m(x)$$

so $m \mapsto \sigma_m$ induces a homomorphism of $(\mathbf{Z}/N\mathbf{Z})^*$ into $\mathrm{Aut}(G)$. Let a be a generator of G. If $\sigma_m = \mathrm{id}$, then $a^m = a$, whence $a^{m-1} = e$ and $N|(m-1)$, so $m \equiv 1 \bmod N$. Hence the kernel of $m \mapsto \sigma_m$ in $(\mathbf{Z}/N\mathbf{Z})^*$ is trivial. Finally, let $f: G \to G$ be an automorphism. Then $f(a) = a^k$ for some $k \in \mathbf{Z}$ because a is a generator. Since f is an automorphism, we must have $(k, N) = 1$, otherwise a^k is not a generator of G. But then for all $x \in G$, $x = a^i$ (i depends on x), we get

$$f(a^i) = f(a)^i = a^{ki} = (a^i)^k,$$

so $f = \sigma_k$. Hence the injective homomorphism $(\mathbf{Z}/N\mathbf{Z})^* \to \mathrm{Aut}(G)$ given by $m \mapsto \sigma_m$ is surjective, and is therefore an isomorphism. QED.

Let R be a commutative ring. Let P be an ideal. We define P to be a **prime ideal** if $P \neq R$ and whenever $a, b \in R$ and $ab \in P$ then $a \in P$ or $b \in P$. In Exercise 17 you will prove that an ideal of $\mathbf{Z}$ is prime if and only if this ideal is 0, or is generated by a prime number.

Let R be a commutative ring. Let M be an ideal. We define M to be **maximal ideal** if $M \neq R$ and if there is no ideal J such that $R \supset J \supset M$ and $R \neq J$, $J \neq M$.

You should do Exercises 17, 18, and 19 to get acquainted with prime and maximal ideals. These exercises will prove:

Theorem 3.4. *Let* R *be a commutative ring.*

(a) *A maximal ideal is prime.*

(b) *An ideal* P *is prime if and only if* R/P *is integral.*

(c) *An ideal* M *is maximal if and only if* R/M *is a field.*

To do these exercises, you may want to use the following fact:

Let M be a maximal ideal and let $x \in R$, $x \notin M$. Then

$$M + Rx = R.$$

Indeed, $M + Rx$ is an ideal $\neq M$, so $M + Rx$ must be R since M is assumed maximal.

III, §3. EXERCISES

1. Let $f : R \to R'$ be a ring-homomorphism. Show that the image of f is a subring of R'.

2. Show that a ring-homomorphism of a field K into a ring $R \neq \{0\}$ is an isomorphism of K onto its image.

3. (a) Let n be a positive integer, and let $Z_n = \mathbf{Z}/n\mathbf{Z}$ be the factor ring of $\mathbf{Z}$ modulo n. Show that the units of Z_n are precisely those residue classes $\bar{x}$ having a representative integer $x \neq 0$ and relatively prime to n. (For the definition of unit, see the end of §1.)
   (b) Let x be an integer relatively prime to n. Let φ be the Euler function. Show that $x^{\varphi(n)} \equiv 1 \pmod{n}$.

4. (a) Let n be an integer $\geqq 2$. Show that $\mathbf{Z}/n\mathbf{Z}$ is an integral ring if and only if n is prime.
   (b) Let p be a prime number. Show that in the ring $\mathbf{Z}/(p)$, every non-zero element has a multiplicative inverse, and that the non-zero elements form a multiplicative group.
   (c) If a is an integer, $a \not\equiv 0 \pmod{p}$, show that $a^{p-1} \equiv 1 \pmod{p}$.

5. (a) Let R be a ring, and let $x, y \in R$ be such that $xy = yx$. What is $(x + y)^n$? (Cf. Exercise 2 of Chapter I, §2.)
   (b) Recall that an element x is called **nilpotent** if there exists a positive integer n such that $x^n = 0$. If R is commutative and x, y are nilpotent, show that $x + y$ is nilpotent.

6. Let F be a finite field having q elements. Prove that $x^{q-1} = 1$ for every nonzero element $x \in F$. Show that $x^q = x$ for every element x of F.

7. **Chinese Remainder Theorem.** Let R be a commutative ring and let J_1, J_2 be ideals. They are called **relatively prime** if

$$J_1 + J_2 = R.$$

Suppose J_1 and J_2 are relatively prime. Given elements $a, b \in R$ show that there exists $x \in R$ such that

$$x \equiv a \pmod{J_1} \quad \text{and} \quad x \equiv b \pmod{J_2}.$$

[This result applies in particular when $R = \mathbf{Z}$, $J_1 = (m_1)$ and $J_2 = (m_2)$ with relatively prime integers m_1, m_2.]

8. If J_1, J_2 are relatively prime, show that for every positive integer n, J_1^n and J_2^n are relatively prime.

9. Let R be a ring, and M, M' two-sided ideals. Assume that M contains M'. If $x \in R$, denote its residue class mod M by $x(M)$. Show that there is a (unique) ring-homomorphism $R/M' \to R/M$ which maps $x(M')$ on $x(M)$.

   Example. If n, m are integers $\neq 0$, such that n divides m, apply Exercise 9 to get a ring-homomorphism $\mathbf{Z}/(m) \to \mathbf{Z}/(n)$.

10. Let R, R' be rings. Let $R \times R'$ be the set of all pairs (x, x') with $x \in R$ and $x' \in R'$. Show how one can make $R \times R'$ into a ring, by defining addition and multiplication componentwise. In particular, what is the unit element of $R \times R'$?

11. Let $R, R_1, \ldots, R_n$ be rings and let $f: R \to R_1 \times \cdots \times R_n$. Show that f is a ring homomorphism if and only if each coordinate map $f_i: R \to R_i$ is a ring homomorphism.

12. (a) Let J_1, J_2 be relatively prime ideals in a commutative ring R. Show that the map $a \bmod J_1 \cap J_2 \mapsto (a \bmod J_1, a \bmod J_2)$ induces an isomorphism

$$f: R/(J_1 \cap J_2) \to R/J_1 \times R/J_2.$$

   (b) Again, if J_1, J_2 are relatively prime, show that $J_1 \cap J_2 = J_1 J_2$.
   Example. If m, n are relatively prime integers, then $(m) \cap (n) = (mn)$.
   (c) If J_1, J_2 are not relatively prime, give an example to show that one does not necessarily have $J_1 \cap J_2 = J_1 J_2$.
   (d) In (a), show that f induces an isomorphism of the unit groups

$$(R/J_1 J_2)^* \xrightarrow{\approx} (R/J_1)^* \times (R/J_2)^*.$$

   (e) Let $J_1, \ldots, J_r$ be ideals of R such that J_i is relatively prime to J_k for $i \neq k$. Show that there is a natural ring isomorphism

$$R/J_1 \cdots J_r \to \prod R/J_i.$$

13. Let P be the set of positive integers and R the set of functions defined on P, with values in a commutative ring K. Define the sum in R to be the ordinary addition of functions, and define the product by the formula

$$(f * g)(m) = \sum_{xy=m} f(x)g(y),$$

where the sum is taken over all pairs (x, y) of positive integers such that $xy = m$. This sum can also be written in the form

$$(f * g)(m) = \sum_{d \mid m} f(d)g(m/d)$$

where the sum is taken over all positive divisors of m, including 1 of course.
   (a) Show that R is a commutative ring, whose unit element is the function δ such that $\delta(1) = 1$ and $\delta(x) = 0$ if $x \neq 1$.

(b) A function f is said to be **multiplicative** if $f(mn) = f(m)f(n)$ whenever m, n are relatively prime. If f, g are multiplicative, show that $f * g$ is multiplicative.

(c) Let μ be the **Moebius function** such that $\mu(1) = 1$, $\mu(p_1 \cdots p_r) = (-1)^r$ if $p_1, \ldots, p_r$ are distinct primes, and $\mu(m) = 0$ if m is divisible by p^2 for some prime p. Show that $\mu * \varphi_1 = \delta$ (where φ_1 denotes the constant function having value 1). [*Hint*: Show first that μ is multiplicative and then prove the assertion for prime powers.] The **Möbius inversion formula** of elementary number theory is then nothing else but the relation

$$\mu * \varphi_1 * f = f.$$

In other words, if for some function g we have

$$f(n) = \sum_{d|n} g(d) = (\varphi_1 * g)(n)$$

then

$$g(n) = (\mu * f)(n) = \sum_{d|n} \mu(d) f(n/d).$$

The product $f * g$ in this exercise is called the **convolution product**. Note how formalizing this product and viewing functions as elements of a ring under the convolution product simplifies the inversion formalism with the Moebius function.

14. Let $f: R \to R'$ be a ring-homomorphism. Let J' be a two-sided ideal of R', and let J be the set of elements x of R such that $f(x)$ lies in J'. Show that J is a two-sided ideal of R.

15. Let R be a commutative ring, and N the set of elements $x \in R$ such that $x^n = 0$ for some positive integer n. Show that N is an ideal.

16. In Exercise 15, if $\bar{x}$ is an element of R/N, and if there exists an integer $n \geqq 1$ such that $\bar{x}^n = 0$, show that $\bar{x} = 0$.

17. Let R be a commutative ring. An ideal P is said to be a **prime** ideal if $P \neq R$, and whenever a, $b \in R$ and $ab \in P$ then $a \in P$ or $b \in P$. Show that a non-zero ideal of $\mathbf{Z}$ is prime if and only if it is generated by a prime number.

18. Let R be a commutative ring. An ideal M of R is said to be a **maximal** ideal if $M \neq R$, and if there is no ideal J such that $R \supset J \supset M$, and $R \neq J$, $J \neq M$. Show that every maximal ideal is prime.

19. Let R be a commutative ring.
(a) Show that an ideal P is prime if and only if R/P is integral.
(b) Show that an ideal M is maximal if and only if R/M is a field.

20. Let K be a field of characteristic p. Show that $(x + y)^p = x^p + y^p$ for all x, $y \in K$.

21. Let K be a finite field of characteristic p. Show that the map $x \mapsto x^p$ is an automorphism of K.

22. Let S be a set, X a subset, and assume neither S nor X is empty. Let R be a ring. Let $F(S, R)$ be the ring of all mappings of S into R, and let

$$\rho\colon F(S, R) \to F(X, R)$$

be the restriction, i.e. if $f \in F(S, R)$, then $\rho(f)$ is just f viewed as a map of X into R. Show that ρ is surjective. Describe the kernel of ρ.

23. Let K be a field and S a set. Let x_0 be an element of S. Let $F(S, K)$ be the ring of mappings of S into K, and let J be the set of maps $f \in F(S, K)$ such that $f(x_0) = 0$. Show that J is a maximal ideal. Show that $F(S, K)/J$ is isomorphic to K.

24. Let R be a commutative ring. A map $D\colon R \to R$ is called a **derivation** if $D(x + y) = Dx + Dy$, and $D(xy) = (Dx)y + x(Dy)$ for all $x, y \in R$. If D_1, D_2 are derivations, define the bracket product

$$[D_1, D_2] = D_1 \circ D_2 - D_2 \circ D_1.$$

Show that $[D_1, D_2]$ is a derivation.

Example. Let R be the ring of infinitely differentiable real-valued functions of, say, two real variables. Any differential operator

$$f(x, y)\frac{\partial}{\partial x} \quad \text{or} \quad g(x, y)\frac{\partial}{\partial y}$$

with coefficients f, g which are infinitely differentiable functions, is a derivation on R.

III, §4. QUOTIENT FIELDS

In the preceding sections, we have assumed that the reader is acquainted with the rational numbers, in order to give examples for more abstract concepts. We shall now study how one can define the rationals from the integers. Furthermore, in the next chapter, we shall study polynomials over a field. One is accustomed to form quotients f/g $(g \neq 0)$ of polynomials, and such quotients are called rational functions. Our discussion will apply to this situation also.

Before giving the abstract discussion, we analyze the case of the rational numbers more closely. In elementary school, what is done (or what should be done), is to give rules for determining when two quotients of rational numbers are equal. This is needed, because, for instance, $\frac{3}{4} = \frac{6}{8}$. The point is that a fraction is determined by a pair of numbers, in this special example $(3, 4)$, but also by other pairs, e.g. $(6, 8)$. If we view all pairs giving rise to the same quotient as equivalent, then we get our cue how to define the fraction, namely as a certain equivalence class of pairs. Next, one must give rules for adding fractions, and the rules we shall give in general are precisely the same as those which are (or should be) given in elementary school.

Our discussion will apply to an arbitrary integral ring R. (Recall that integral means that $1 \neq 0$, that R is commutative and without divisors of 0.)

Let (a, b) and (c, d) be pairs of elements in R, with $b \neq 0$ and $d \neq 0$. We shall say that these pairs are **equivalent** if $ad = bc$. We contend that this is an equivalence relation. Going back to the definition of Chapter I, §5, we see that **ER 1** and **ER 3** are obvious. As for **ER 2**, suppose that (a, b) is equivalent to (c, d) and (c, d) is equivalent to (e, f). By definition,

$$ad = bc \quad \text{and} \quad cf = de.$$

Multiplying the first equality by f and the second by b, we obtain

$$adf = bcf \quad \text{and} \quad bcf = bde,$$

whence $adf = bde$, and $daf - dbe = 0$. Then $d(af - be) = 0$. Since R has no divisors of 0, it follows that $af - be = 0$, i.e. $af = be$. This means that (a, b) is equivalent to (e, f), and proves **ER 2**.

We denote the equivalence class of (a, b) by a/b. We must now define how to add and multiply such classes.

If a/b and c/d are such classes, we define their sum to be

$$\frac{a}{b} + \frac{c}{d} = \frac{ad + bc}{bd}$$

and their product to be

$$\frac{a}{b}\frac{c}{d} = \frac{ac}{bd}.$$

We must show of course that in defining the sum and product as above, the result is independent of the choice of pairs (a, b) and (c, d) representing the given classes. We shall do this for the sum. Suppose that

$$a/b = a'/b' \quad \text{and} \quad c/d = c'/d'.$$

We must show that

$$\frac{ad + bc}{bd} = \frac{a'd' + b'c'}{b'd'}.$$

This is true if and only if

$$b'd'(ad + bc) = bd(a'd' + b'c'),$$

or in other words

$$b'd'ad + b'd'bc = bda'd' + bdb'c'. \tag{1}$$

But $ab' = a'b$ and $cd' = c'd$ by assumption. Using this, we see at once that (1) holds. We leave the analogous statement for the product as an exercise.

We now contend that the set of all quotients a/b with $b \neq 0$ is a ring, the operations of addition and multiplication being defined as above. Note first that there is a unit element, namely $1/1$, where 1 is the unit element of R. One must now verify all the other axioms of a ring. This is tedious, but obvious at each step. As an example, we shall prove the associativity of addition. For three quotients a/b, c/d, and e/f we have

$$\left(\frac{a}{b} + \frac{c}{d}\right) + \frac{e}{f} = \frac{ad + bc}{bd} + \frac{e}{f} = \frac{fad + fbc + bde}{bdf}.$$

On the other hand,

$$\frac{a}{b} + \left(\frac{c}{d} + \frac{e}{f}\right) = \frac{a}{b} + \frac{cf + de}{df} = \frac{adf + bcf + bde}{bdf}.$$

It is then clear that the expressions on the right-hand sides of these equations are equal, thereby proving associativity of addition. The other axioms are equally easy to prove, and we shall omit this tedious routine. We note that our ring of quotients is commutative.

Let us denote the ring of all quotients a/b by K. We contend that K is a field. To see this, all we need to do is prove that every non-zero element has a multiplicative inverse. But the zero element of K is $0/1$, and if $a/b = 0/1$ then $a = 0$. Hence any non-zero element can be written in the form a/b with $b \neq 0$ and $a \neq 0$. Its inverse is then b/a, as one sees directly from the definition of multiplication of quotients.

Finally, observe that we have a natural map of R into K, namely the map

$$a \mapsto a/1.$$

It is again routine to verify that this map is an injective ring-homomorphism. Any injective ring-homomorphism will be called an **embedding.** We see that R is embedded in K in a natural way.

We call K the **quotient field of** R. When $R = \mathbf{Z}$, then K is by definition the field of rational numbers. When R is the ring of polynomials defined in the next chapter, its quotient field is called the field of **rational functions**.

Suppose that R is a subring of a field F. The set of all elements ab^{-1} with $a, b \in R$ and $b \neq 0$ is easily seen to form a field, which is a subfield of F. We also call this field the quotient field of R in F. There can be no confusion with this terminology, because the quotient field of R as defined previously is isomorphic to this subfield, under the map

$$a/b \mapsto ab^{-1}.$$

The verification is trivial, and in view of this, the element ab^{-1} of F is also denoted by a/b.

Example. Let K be a field and as usual, $\mathbf{Q}$ the rational numbers. There does not necessarily exist an embedding of $\mathbf{Q}$ into K (for instance, K may be finite). However, if an embedding of $\mathbf{Q}$ into K exists, there is only one. This is easily seen, because any homomorphism

$$f: \mathbf{Q} \to K$$

must be such that $f(1) = e$ (unit element of K). Then for any integer $n > 0$ one sees by induction that $f(n) = ne$, and consequently

$$f(-n) = -ne.$$

Furthermore,

$$e = f(1) = f(nn^{-1}) = f(n)f(n^{-1})$$

so that $f(n^{-1}) = f(n)^{-1} = (ne)^{-1}$. Thus for any quotient $m/n = mn^{-1}$ with integers m, n and $n > 0$ we must have

$$f(m/n) = (me)(ne)^{-1}$$

thus showing that f is uniquely determined. It is then customary to identify $\mathbf{Q}$ inside K and view every rational number as an element of K.

Finally, we make some remarks on the extension of an embedding of a ring into a field.

Let R be an integral ring, and

$$f: R \to E$$

an embedding of R into some field E. Let K be the quotient field of R. Then f admits a unique extension to an embedding of K into E, that is an embedding $f^*: K \to E$ whose restriction to R is equal to f.

To see the uniqueness, observe that if f^* is an extension of f, and

$$f^*: K \to E$$

is an embedding, then for all $a, b \in R$ we must have

$$f^*(a/b) = f^*(a)/f^*(b) = f(a)/f(b),$$

so the effect of f^* on K is determined by the effect of f on R. Conversely, one can *define* f^* by the formula

$$f^*(a/b) = f(a)/f(b),$$

and it is seen at once that the value of f^* is independent of the choice of the representation of the quotient a/b, that is if $a/b = c/d$ with

$$a, b, c, d \in R \quad \text{and} \quad bd \neq 0,$$

then

$$f(a)/f(b) = f(c)/f(d).$$

One also verifies routinely that f^* so defined is a homomorphism, thereby proving the existence.

III, §4. EXERCISES

1. Put in all details in the proof of the existence of the extension f^* at the end of this section.
2. A (ring-) isomorphism of a ring onto itself is also called an **automorphism**. Let R be an integral ring, and $\sigma: R \to R$ an automorphism of R. Show that σ admits a unique extension to an automorphism of the quotient field.

CHAPTER IV

Polynomials

IV, §1. POLYNOMIALS AND POLYNOMIAL FUNCTIONS

Let K be a field. Every reader of this book will have written expressions like

$$a_n t^n + a_{n-1}t^{n-1} + \cdots + a_0,$$

where $a_0, \ldots, a_n$ are real or complex numbers. We could also take these to be elements of K. But what does "t" mean? Or powers of "t" like t, $t^2, \ldots, t^n$?

In elementary courses, when $K = \mathbf{R}$ or $\mathbf{C}$ then we speak of polynomial functions. We write

$$f(t) = a_n t^n + \cdots + a_0$$

to mean the function of K into itself such that for each element $t \in K$ the value of the function f is $f(t)$, the value given by the above expression.

But in operating with polynomials, we usually work formally, without worrying about f being a function. For instance, let $a_0, \ldots, a_n$ be elements of K and $b_0, \ldots, b_m$ also be elements of K. We just write expressions like

$$f(t) = a_n t^n + \cdots + a_0,$$

$$g(t) = b_m t^m + \cdots + b_0.$$

If, say, $n > m$ we let $b_j = 0$ if $j > m$ and we also write

$$g(t) = 0t^n + \cdots + b_m t^m + \cdots + b_0,$$

and we write the sum formally as

$$(*) \qquad (f+g)(t) = (a_n + b_n)t^n + \cdots + (a_0 + b_0).$$

If $c \in K$ then we write

$$(cf)(t) = ca_n t^n + \cdots + ca_0.$$

We can also take the product which we write as

$$(fg)(t) = (a_n b_m)t^{n+m} + \cdots + a_0 b_0.$$

In fact, if we write

$$(fg)(t) = c_{n+m}t^{n+m} + \cdots + c_0,$$

then

$$(**) \qquad c_k = \sum_{i=0}^{k} a_i b_{k-i} = a_0 b_k + a_1 b_{k-1} + \cdots + a_k b_0.$$

This expression for c_k simply comes from collecting all the terms

$$a_i t^i b_{k-i} t^{k-i} = a_i b_{k-i} t^k$$

in the product which will give rise to the term involving t^k.

All that matters is that we defined a rule for the addition and multiplication of the above expressions according to formulas (*) and (**). Furthermore, it does not matter either that the coefficients a_i, b_j are in a field. The only properties we need about such coefficients is that they satisfy the ordinary properties of arithmetic, or in other words that they lie in a commutative ring. The only thing we still have to clear up is the role of the letter "t", selected arbitrarily. So we must use some device to define polynomials, and especially a "variable" t. There are several possible devices, and one of them runs as follows.

Let R be a commutative ring. Let Pol_R be the set of infinite vectors

$$(a_0, a_1, a_2, \ldots, a_n, \ldots)$$

with $a_n \in R$ and such that all but a finite number of a_n are equal to 0. Thus a vector looks like

$$(a_0, a_1, \ldots, a_d, 0, 0, 0, \ldots)$$

with zeros all the way to the right. Elements of Pol_R are called **polynomials** over R. The elements $a_0, a_1, \ldots$ are called the **coefficients** of the polynomial. The **zero polynomial** is the polynomial $(0, 0, \ldots)$ which has $a_i = 0$ for all i. We define addition of infinite vectors componentwise, just as for finite n-tuples. Then Pol_R is an additive group. We define multiplication to mimic the multiplication we know already. If

$$f = (a_0, a_1, \ldots) \qquad \text{and} \qquad g = (b_0, b_1, \ldots)$$

are polynomials with coefficients in R, we define their product to be

$$fg = (c_0, c_1, \ldots) \qquad \text{with} \qquad c_k = \sum_{i=0}^{k} a_i b_{k-i} = \sum_{i+j=k} a_i b_j.$$

It is then a routine matter to prove that under this definition of multiplication Pol_R is a commutative ring. We shall prove associativity of multiplication and leave the other axioms as exercises.

Let

$$h = (d_0, d_1, \ldots)$$

be a polynomial. Then

$$(fg)h = (e_0, e_1, \ldots)$$

where by definition

$$e_s = \sum_{k+r=s} c_k d_r = \sum_{k+r=s} \left(\sum_{i+j=k} a_i b_j \right) d_r$$
$$= \sum_{i+j+r=s} a_i b_j d_r.$$

This last sum is taken over all triples (i, j, r) of integers $\geqq 0$ such that $i + j + r = s$. If we now compute $f(gh)$ in a similar way, we find exactly the same coefficients for $(fg)h$ as for $f(gh)$, thereby proving associativity.

We leave the proofs of the other properties to the reader.

Now pick a letter, for instance t, to **denote**

$$t = (0, 1, 0, 0, 0, \ldots).$$

Thus t has coefficient 0 in the 0-th place, coefficient 1 in the first place, and all other coefficients equal to 0. Having our ring structure, we may now take powers of t, for instance

$$t,\ t^2,\ t^3,\ \ldots, t^n.$$

By induction of whatever means you want, you will prove immediately that if n is a positive integer, then

$$t^n = (0, 0, \ldots, 0, 1, 0, 0, \ldots),$$

in other words t^n is the vector having the n-th component equal to 1, and all other components equal to 0.

The association

$$a \mapsto (a, 0, 0, \ldots) \qquad \text{for} \quad a \in R$$

is an embedding of R in the polynomial ring Pol_R, in other words it is an injective ring homomorphism. We shall identify a and the vector $(a, 0, 0, 0, \ldots)$. Then we can multiply a polynomial $f = (a_0, a_1, \ldots,)$ by an element of R componentwise, that is

$$af = (aa_0, aa_1, aa_2, \ldots).$$

This corresponds to the ordinary multiplication of a polynomial by a scalar.

Observe that we can now write

$$f = a_0 + a_1 t + \cdots + a_d t^d = (a_0, a_1, \ldots, a_d, 0, 0, 0, \ldots)$$

if $a_n = 0$ for $n > d$. This is the more usual way of writing a polynomial. So we have recovered all the basic properties concerning addition and multiplication of polynomials. The polynomial ring will be denoted by $R[t]$.

Let R be a subring of a commutative ring S. If $f \in R[t]$ is a polynomial, then we may define the associated **polynomial function**

$$f_S \colon S \to S$$

by letting for $x \in S$

$$f_S(x) = f(x) = a_0 + a_1 x + \cdots + a_d x^d.$$

Therefore f_S is a function (mapping) of S into itself, determined by the polynomial f. Given an element $c \in S$, directly from the definition of multiplication of polynomials, we find:

The association

$$\mathrm{ev}_c \colon f \mapsto f(c)$$

is a ring homomorphism of $R[t]$ *into* S.

This property simply says that

$$(f + g)(c) = f(c) + g(c) \qquad \text{and} \qquad (fg)(c) = f(c)g(c).$$

Also the polynomial 1 maps to the unit element 1 of S. This homomorphism is called the **evaluation homomorphism**, and is denoted by ev_c for obvious reasons. You are used to evaluating polynomials at numbers, and all we have done is to point out that the whole procedure of evaluating polynomials is applicable to a much more general context over commutative rings. The element t in the polynomial ring is also called a **variable** over K. The evaluation $f(c)$ of f at c is also said to be obtained by **substitution of c in the polynomial.**

Note that $f = f(t)$, according to our definition of the evaluation homomorphism.

Let R be a subring of a ring S and let x be an element of S. We denote by $R[x]$ the set of all elements $f(x)$ with $f \in R[t]$. Then it is immediately verified that $R[x]$ is a commutative subring of S, which is said to be **generated by** x over R. The evaluation homomorphism

$$R[t] \to R[x]$$

is then a ring homomorphism onto $R[x]$. If the evaluation map $f \mapsto f(x)$ gives an isomorphism of $R[t]$ with $R[x]$, then we say that x is **transcendental** over R or that x is a **variable** over R.

Example. Let $\alpha = \sqrt{2}$. Then the set of all real numbers of the form

$$a + b\alpha \quad \text{with} \quad a, b \in \mathbf{Z}$$

is a subring of the real numbers, generated by $\sqrt{2}$. This is the subring $\mathbf{Z}[\sqrt{2}]$. (Prove in detail that it is a subring.) Note that α is *not* transcendental over $\mathbf{Z}$. For instance, the polynomial $t^2 - 2$ lies in the kernel of the evaluation map $f(t) \mapsto f(\sqrt{2})$.

Example. The polynomial ring $R[t]$ is generated by the variable t over R, and t is transcendental over R.

If x, y are transcendental over R, then $R[x]$, and $R[y]$ are isomorphic, since they are both isomorphic to the polynomial ring $R[t]$. Thus our definition of polynomials was merely a concrete way of dealing with a ring generated over R by a transcendental element.

Warning. When we speak of a polynomial, we always mean a polynomial as defined above. If we mean the associated **polynomial function**, we shall say so explicitly. In some cases, it may be that two polynomials can be distinct, but give rise to the same polynomial function on a given ring. For example, let $\mathbf{F}_p = \mathbf{Z}/p\mathbf{Z}$ be the field with p elements. Then for every element $x \in \mathbf{F}_p$ we have

$$x^p = x.$$

Indeed, if $x = 0$ this is obvious, and if $x \neq 0$, since the multiplicative group of $\mathbf{F}_p$ has $p - 1$ elements, we get $x^{p-1} = 1$. It follows that $x^p = x$. Thus we see that if we let $K = \mathbf{F}_p$, and we let

$$f = t^p \qquad \text{and} \qquad g = t$$

then $f_K = g_K$ but $f \neq g$. In our original notation,

$$f = (0, 0, \ldots, 0, 1, 0, 0, \ldots) \qquad \text{and} \qquad g = (0, 1, 0, 0, 0, \ldots).$$

If K is an infinite field, then this phenomenon cannot happen, as we shall prove below. Most of our work will be over infinite fields, but finite fields are sufficiently important so that we have taken care of their possibilities from the beginning. Suppose that F is a finite subfield of an infinite field K. Let $f(t)$ and $g(t) \in F[t]$. Then it may happen that $f \neq g$, $f_K \neq g_K$, but $f_F = g_F$. For instance, the polynomials t^p and t give rise to the same function on $\mathbf{Z}/p\mathbf{Z}$, but to different functions on any infinite field K containing $\mathbf{Z}/p\mathbf{Z}$ according to what we shall prove below.

We now return to a general field K.

When we write a polynomial

$$f(t) = a_n t^n + \cdots + a_0$$

with $a_i \in K$ for $i = 0, \ldots, n$ then these elements of K are called the **coefficients** of the polynomial f. If n is the largest integer such that $a_n \neq 0$, then we say that n is the **degree** of f and write $n = \deg f$. We also say that a_n is the **leading coefficient** of f. We say that a_0 is the **constant term** of f.

Example. Let

$$f(t) = 7t^5 - 8t^3 + 4t - \sqrt{2}.$$

Then f has degree 5. The leading coefficient is 7, and the constant term is $-\sqrt{2}$.

If f is the zero polynomial then we shall use the convention that $\deg f = -\infty$. We agree to the convention that

$$-\infty + -\infty = -\infty,$$

$$-\infty + a = -\infty, \qquad -\infty < a$$

for every integer a, and *no other operation with* $-\infty$ *is defined.*

A polynomial of degree 1 is also called a **linear** polynomial.

Let α be an element of K. We shall say that α is a **root** of f if $f(\alpha) = 0$.

Theorem 1.1. *Let f be a polynomial in K, written in the form*

$$f(t) = a_n t^n + \cdots + a_0.$$

and suppose f has degree $n \geqq 0$, so $f \neq 0$ (f is not the zero polynomial). Then f has at most n roots in K.

Proof. We shall need a lemma.

Lemma 1.2. *Let f be a polynomial over K, and let $\alpha \in K$. Then there exist elements $c_0, \ldots, c_n \in K$ such that*

$$f(t) = c_0 + c_1(t - \alpha) + \cdots + c_n(t - \alpha)^n.$$

Proof. We write $t = \alpha + (t - \alpha)$, and substitute this value for t in the expression of f. For each integer k with $1 \leqq k \leqq n$, we have

$$t^k = (\alpha + (t - \alpha))^k = \alpha^k + \cdots + (t - \alpha)^k$$

(the expansion being that obtained with the binomial coefficients), and therefore

$$a_k t^k = a_k \alpha^k + \cdots + a_k(t - \alpha)^k$$

can be written as a sum of powers of $(t - \alpha)$, multiplied by elements of K. Taking the sum of $a_k t^k$ for $k = 0, \ldots, n$ we find the desired expression for f, and prove the lemma.

Observe that in the lemma, we have $f(\alpha) = c_0$. Hence, if $f(\alpha) = 0$, then $c_0 = 0$, and we can write

$$f(t) = (t - \alpha)h(t),$$

where we can write

$$h(t) = d_1 + d_2(t - \alpha) + \cdots + d_n(t - \alpha)^{n-1}$$

for some elements $d_1, d_2, \ldots, d_n$ in K. Suppose that f has more than n roots in K, and say $\alpha_1, \ldots, \alpha_{n+1}$ are $n + 1$ distinct roots in K. Let $\alpha = \alpha_1$. Then $\alpha_i - \alpha_1 \neq 0$ for $i = 2, \ldots, n + 1$. Since

$$0 = f(\alpha_i) = (\alpha_i - \alpha_1)h(\alpha_i),$$

we conclude that $h(\alpha_i) = 0$ for $i = 2,\dots,n+1$. By induction on n we now see that this is impossible, thereby proving that f has at most n roots in K.

Corollary 1.3. *Let* $f(t) = a_n t^n + \cdots + a_0$ *and* $g(t) = b_n t^n + \cdots + b_0$. *Suppose that* K *is an infinite field. If* $f(c) = g(c)$ *for all* $c \in K$ *then* $f = g$, *that is* $a_k = b_k$ *for all* $k = 0,\dots,n$.

Proof. Consider the polynomial

$$f(t) - g(t) = (a_n - b_n)t^n + \cdots + (a_0 - b_0).$$

Every element of K is a root of this polynomial. Hence by Theorem 1.1, we must have $a_i - b_i = 0$ for $i = 0,\dots,n$, in other words, $a_i = b_i$, thereby proving the corollary.

Corollary 1.3 shows that over an infinite field, a polynomial is no different from a polynomial function. For what we do in this chapter, however, most results are true working formally with polynomials. So we do not necessarily assume that the base field is infinite.

Our convention on the degree of a polynomial was also useful to make the following result true without exception.

Theorem 1.4. *Let* f, g *be polynomials with coefficients in* K. *Then*

$$\deg(fg) = \deg f + \deg g.$$

Proof. Let

$$f(t) = a_n t^n + \cdots + a_0 \quad \text{and} \quad g(t) = b_m t^m + \cdots + b_0$$

with $a_n \neq 0$ and $b_m \neq 0$. Then from the multiplication rule for fg, we see that

$$f(t)g(t) = a_n b_m t^{n+m} + \text{terms of lower degree},$$

and $a_n b_m \neq 0$. Hence $\deg fg = n + m = \deg f + \deg g$. If f or g is 0, then our convention about $-\infty$ makes our assertion also come out.

Corollary 1.5. *The ring* $K[t]$ *has no divisors of zero, and is therefore an integral ring.*

Proof. If f, g are non-zero polynomials, then $\deg f$ and $\deg g$ are $\geqq 0$, whence $\deg(fg) \geqq 0$, and $fg \neq 0$, as was to be shown.

In light of Corollary 1.5, we may form the quotient field of the polynomial ring $K[t]$. This quotient field is denoted by $K(t)$ and is called the field of **rational functions**. Its elements consist of quotients

$$f(t)/g(t)$$

where f, g are polynomials. More precisely, the elements of $K(t)$ are equivalence classes of such quotients, where

$$f/g = f_1/g_1 \quad \text{if and only if} \quad fg_1 = gf_1.$$

This relation is merely the relation of elementary school arithmetic, as we have seen in Chapter III, §4.

The next theorem is the **Euclidean algorithm** or long division, taught in elementary school. It is the analogue of the Euclidean algorithm for integers.

Theorem 1.6. *Let f, g be polynomials over the field K,* i.e. *polynomials in $K[t]$, and assume* $\deg g \geqq 0$. *Then there exist polynomials q, r in $K[t]$ such that*

$$f(t) = q(t)g(t) + r(t),$$

and $\deg r < \deg g$. *The polynomials q, r are uniquely determined by these conditions.*

Proof. Let $m = \deg g \geqq 0$. Write

$$f(t) = a_n t^n + \cdots + a_0,$$
$$g(t) = b_m t^m + \cdots + b_0,$$

with $b_m \neq 0$. If $n < m$, let $q = 0$, $r = f$. If $n \geqq m$, let

$$f_1(t) = f(t) - a_n b_m^{-1} t^{n-m} g(t).$$

(This is the first step in the process of long division.) Then

$$\deg f_1 < \deg f.$$

Continuing in this way, or more formally by induction on n, we can find polynomials q_1, r such that

$$f_1 = q_1 g + r,$$

with $\deg r < \deg g$. Then

$$\begin{aligned} f(t) &= a_n b_m^{-1} t^{n-m} g(t) + f_1(t) \\ &= a_n b_m^{-1} t^{n-m} g(t) + q_1(t) g(t) + r(t) \\ &= (a_n b_m^{-1} t^{n-m} + q_1) g(t) + r(t), \end{aligned}$$

and we have consequently expressed our polynomial in the desired form. To prove the uniqueness, suppose that

$$f = q_1 g + r_1 = q_2 g + r_2,$$

with $\deg r_1 < \deg g$ and $\deg r_2 < \deg g$. Then

$$(q_1 - q_2) g = r_2 - r_1.$$

Either the degree of the left-hand side is $\geqq \deg g$, or the left-hand side is equal to 0. Either the degree of the right-hand side is $< \deg g$, or the right-hand side is equal to 0. Hence the only possibility is that they are both 0, whence

$$q_1 = q_2 \quad \text{and} \quad r_1 = r_2,$$

as was to be shown.

From the Euclidean algorithm, we can reprove a fact already proved by other means.

Corollary 1.7. *Let f be a non-zero polynomial in $K[t]$. Let $\alpha \in K$ be such that $f(\alpha) = 0$. Then there exists a polynomial $q(t)$ in $K[t]$ such that*

$$f(t) = (t - \alpha) q(t).$$

Proof. We can write

$$f(t) = q(t)(t - \alpha) + r(t),$$

where $\deg r < \deg(t - \alpha)$. But $\deg(t - \alpha) = 1$. Hence r is constant. Since

$$0 = f(\alpha) = q(\alpha)(\alpha - \alpha) + r(\alpha) = r(\alpha),$$

it follows that $r = 0$, as desired.

Corollary 1.8. *Let K be a field such that every non-constant polynomial in $K[t]$ has a root in K. Let f be such a polynomial. Then there exist elements $\alpha_1, \dots, \alpha_n \in K$ and $c \in K$ such that*

$$f(t) = c(t - \alpha_1) \cdots (t - \alpha_n).$$

Proof. In Corollary 1.7, observe that $\deg q = \deg f - 1$. Let $\alpha = \alpha_1$ in Corollary 1.7. By assumption, if q is not constant, we can find a root α_2 of q, and thus write

$$f(t) = q_2(t)(t - \alpha_1)(t - \alpha_2).$$

Proceeding inductively, we keep on going until q_n is constant.

A field K having the property stated in Corollary 1.8, that every non-constant polynomial over K has a root in K, is called **algebraically closed**. We shall prove later in the book that the complex numbers are algebraically closed. We shall also prove later that a finite field is *not* algebraically closed. You may assume this from now on.

We now reprove Theorem 1.1 using the Euclidean algorithm.

Corollary 1.9. *Let K be a field and f a polynomial of degree $n \geqq 1$. Then f has at most n roots in K.*

Proof. Let $\alpha_1, \ldots, \alpha_r$ be distinct roots of f in K. Then by the Euclidean algorithm we know there is a factorization

$$f(t) = c(t - \alpha_1) \cdots (t - \alpha_r)g(t),$$

so $r \leqq n$, as was to be shown.

Example. Let F be a finite field, say $F = \mathbf{Z}/p\mathbf{Z}$ where p is a prime number. The polynomial

$$f(t) = t^p - 1$$

is equal to $(t - 1)^p$ and therefore has only one root, namely 1.

Suppose F has characteristic p. If $p = 2$ then the polynomial $t^2 - 1$, which is $(t - 1)^2$, has only one root, namely 1. On the other hand, if $p \neq 2$ then this polynomial has two distinct roots, 1 and -1. In case $p \neq 2$ we have $1 \neq -1$ in F, for otherwise $1 = -1$ implies $1 + 1 = 2 = 0$ in F, so F would have characteristic 2.

As an application of Corollary 1.9 we can completely determine the structure of finite subgroups of the multiplicative group in a field.

Theorem 1.10. *Let K be a field and let G be a finite subgroup of the group of non-zero elements. Then G is cyclic.*

Proof. Here we shall give a proof using the structure theorem for finite abelian groups. By that theorem, we know that

$$G = \prod_p G(p)$$

is the direct product of the subgroups $G(p)$ which consist of elements whose period is a power of p. By Exercise 18 of Chapter II, §1 it will suffice to prove that each $G(p)$ is cyclic. If $G(p)$ is not cyclic, then by the structure theorem, Theorem 7.2 of Chapter II, $G(p)$ contains a product $H_1 \times H_2$ where H_1 is cyclic of order p^r and H_2 is cyclic of order p^s, with $r, s \geqq 1$. Say $r \geqq s$. Then every element of $G(p)$ lying in the product of these two factors satisfies the equation

$$t^{p^r} - 1 = 0.$$

This equation has at most p^r roots, but there are more than p^r elements in the product of the two factors (there are in fact $p^r p^s = p^{r+s}$ elements in this product). This contradiction proves the theorem.

Remark. If you don't like the use of the structure theorem for abelian groups, then give an independent proof of just those properties which are needed here. Such a proof will be given in Chapter VIII, §3. On the other hand, you could also use Exercise 2(b) of Chapter II, §7, which can be proved directly more easily than the structure theorem.

We denote by $\boldsymbol{\mu}_n$ the group of n-th **roots of unity**. This is the set of elements ζ such that $\zeta^n = 1$. Strictly speaking, we should denote by $\boldsymbol{\mu}_n(K)$ the group of n-th roots of unity in K, but we often omit the K when the reference to the field is clear by the context. Suppose that K has characteristic p. Then

$$\boldsymbol{\mu}_p = 1.$$

Indeed suppose $\zeta^p = 1$. Then $\zeta^p - 1 = 0$. But

$$\zeta^p - 1 = (\zeta - 1)^p = 0,$$

so $\zeta - 1 = 0$ and $\zeta = 1$. We shall see in §3 that if p does not divide n, then $\boldsymbol{\mu}_n$ has order n, and is thus a cyclic group of order n. A generator for $\boldsymbol{\mu}_n$ is called a **primitive n-th root of unity**. In the complex numbers, $\boldsymbol{\mu}_n = \boldsymbol{\mu}_n(\mathbf{C})$ is the ordinary group of n-th roots of unity, generated by $e^{2\pi i/n}$. The primitive n-th roots of unity are $e^{2\pi ir/n}$ with r prime to n.

Consider the field $F = \mathbf{Z}/p\mathbf{Z}$. An integer $a \in \mathbf{Z}$ such that its image in $\mathbf{Z}/p\mathbf{Z}$ is a generator of F^* is called a **primitive root mod** p. Thus the period of $a \bmod p$ is $p - 1$. Artin conjectured that there are infinitely primes p such that, for instance, 2 is a primitive root mod p. Cf. the introduction to his collected works. The answer is still not known.

For further remarks on Theorem 1.10 in connection with finite fields, and especially $\mathbf{Z}/p\mathbf{Z}$, see Chapter VIII, §3.

IV, §1. EXERCISES

1. In each of the following cases, write $f = qg + r$ with $\deg r < \deg g$.

   (a) $f(t) = t^2 - 2t + 1,\quad g(t) = t - 1$
   (b) $f(t) = t^3 + t - 1,\quad g(t) = t^2 + 1$
   (c) $f(t) = t^3 + t,\quad g(t) = t$
   (d) $f(t) = t^3 - 1,\quad g(t) = t - 1$

2. If $f(t)$ has integer coefficients and if $g(t)$ has integer coefficients and leading coefficient 1, show that when we express $f = qg + r$ with $\deg r < \deg g$, the polynomials q and r also have integer coefficients.

3. Using the intermediate value theorem of calculus, show that every polynomial of odd degree over the real numbers has a root in the real numbers.

4. Let $f(t) = t^n + \cdots + a_0$ be a polynomial with complex coefficients, of degree n, and let α be a root. Show that $|\alpha| \leqq n \cdot \max_i |a_i|$. [*Hint*: Note $a_n = 1$. Write

$$-\alpha^n = a_{n-1}\alpha^{n-1} + \cdots + a_0.$$

   If $|\alpha| > n \cdot \max_i |a_i|$, divide by α^n and take the absolute value, together with a simple estimate, to get a contradiction.]

In Exercises 5 and 6 you may assume that the roots of a polynomial in an algebraically closed field are uniquely determined up to a permutation.

5. Let $f(t) = t^3 - 1$. Show that the three roots of f in the complex numbers are

$$1,\ e^{2\pi i/3},\ e^{-2\pi i/3}.$$

   Express these roots as $a + b\sqrt{-3}$ where a, b are rational numbers.

6. Let n be an integer $\geqq 2$. How would you describe the roots of the polynomial $f(t) = t^n - 1$ in the complex numbers?

7. Let F be a field and let $\sigma: F[t] \to F[t]$ be an automorphism of the polynomial ring such that σ restricts to the identity on F. Show that there exists elements $a \in F$, $a \neq 0$, and $b \in F$ such that $\sigma t = at + b$.

Finite fields

8. Let F be a finite field. Let c be the product of all non-zero elements of F. Show that $c = -1$.

Example. Let $F = \mathbf{Z}/p\mathbf{Z}$. Then the result of Exercise 8 can also be stated in the form

$$(p-1)! \equiv -1 \pmod{p},$$

which is known as **Wilson's theorem**.

9. Let p be a prime of the form $p = 4n + 1$ where n is a positive integer. Prove that the congruence

$$x^2 \equiv -1 \pmod{p}$$

has a solution in $\mathbf{Z}$.

10. Let K be a finite field with q elements. Prove that $x^q = x$ for all $x \in K$. Hence the polynomials t^q and t give rise to the same function on K.

11. Let K be a finite field with q elements. If f, g are polynomials over K of degrees $< q$, and if $f(x) = g(x)$ for all $x \in K$, prove that $f = g$ (as polynomials in $K[t]$).

12. Let K be a finite field with q elements. Let f be a polynomial over K. Show that there exists a polynomial f^* over K of degree $< q$ such that

$$f^*(x) = f(x)$$

for all $x \in K$.

13. Let K be a finite field with q elements. Let $a \in K$. Show that there exists a polynomial f over K such that $f(a) = 0$ and $f(x) = 1$ for $x \in K$, $x \neq a$. [*Hint*: $(t - a)^{q-1}$.]

14. Let K be a finite field with q elements. Let $a \in K$. Show that there exists a polynomial f over K such that $f(a) = 1$ and $f(x) = 0$ for all $x \in K$, $x \neq a$.

15. Let K be a finite field with q elements. Let $\varphi: K \to K$ be any function of K into itself. Show that there exists a polynomial f over K such that $\varphi(x) = f(x)$ for all $x \in K$.

[For a continuation of these ideas in connection with polynomials in several variables, see Exercise 6 of §7.]

IV, §2. GREATEST COMMON DIVISOR

Having the Euclidean algorithm, we may now develop the theory of divisibility exactly as for the integers, in Chapter I.

Theorem 2.1. *Let J be an ideal of $K[t]$. Then there exists a polynomial g which is a generator of J. If J is not the zero ideal, and g is a polynomial in J which is not 0, and is of smallest degree, then g is a generator of J.*

Proof. Suppose that J is not the zero ideal. Let g be a polynomial in J which is not 0, and is of smallest degree. We assert that g is a generator for J. Let f be any element of J. By the Euclidean algorithm, we can find polynomials q, r such that

$$f = qg + r$$

with $\deg r < \deg g$. Then $r = f - qg$, and by the definition of an ideal, it follows that r also lies in J. Since $\deg r < \deg g$, we must have $r = 0$. Hence $f = qg$, and g is a generator for J, as desired.

Remark. Let g_1 be a non-zero generator for an ideal J, and let g_2 also be a generator. Then there exists a polynomial q such that $g_1 = qg_2$. Since

$$\deg g_1 = \deg q + \deg g_2,$$

it follows that $\deg g_2 \leqq \deg g_1$. By symmetry, we must have

$$\deg g_2 = \deg g_1.$$

Hence q is constant. We can write

$$g_1 = cg_2$$

with some constant c. Write

$$g_2(t) = a_n t^n + \cdots + a_0$$

with $a_n \neq 0$. Take $b = a_n^{-1}$. Then bg_2 is also a generator of J, and its leading coefficient is equal to 1. Thus we can always find a generator for an ideal ($\neq 0$) whose leading coefficient is 1. It is furthermore clear that this generator is uniquely determined.

Let f, g be non-zero polynomials. We shall say that g **divides** f, and write $g|f$, if there exists a polynomial q such that $f = gq$. Let f_1, f_2 be polynomials $\neq 0$. By a **greatest common divisor** of f_1, f_2 we shall mean a polynomial g such that g divides f_1 and f_2, and furthermore, if h divides f_1 and f_2, then h divides g.

Theorem 2.2. *Let f_1, f_2 be non-zero polynomials in $K[t]$. Let g be a generator for the ideal generated by f_1, f_2. Then g is a greatest common divisor of f_1 and f_2.*

Proof. Since f_1 lies in the ideal generated by f_1, f_2, there exists a polynomial q_1 such that

$$f_1 = q_1 g,$$

whence g divides f_1. Similarly, g divides f_2. Let h be a polynomial dividing both f_1 and f_2. Write

$$f_1 = h_1 h \quad \text{and} \quad f_2 = h_2 h$$

with some polynomials h_1 and h_2. Since g is in the ideal generated by f_1, f_2, there are polynomials g_1, g_2 such that $g = g_1 f_1 + g_2 f_2$, whence

$$g = g_1 h_1 h + g_2 h_2 h = (g_1 h_1 + g_2 h_2)h.$$

Consequently h divides g, and our theorem is proved.

Remark 1. The greatest common divisor is determined up to a non-zero constant multiple. If we select a greatest common divisor with leading coefficient 1, then it is uniquely determined.

Remark 2. Exactly the same proof applies when we have more than two polynomials. For instance, if $f_1,\dots,f_n$ are non-zero polynomials, and if g is a generator for the ideal generated by $f_1,\dots,f_n$, then g is a greatest common divisor of $f_1,\dots,f_n$.

Polynomials $f_1,\dots,f_n$ whose greatest common divisor is 1 are said to be **relatively prime**.

IV, §2. EXERCISES

1. Show that $t^n - 1$ is divisible by $t - 1$.
2. Show that $t^4 + 4$ can be factored as a product of polynomials of degree 2 with integer coefficients. [*Hint*: try $t^2 \pm 2t + 2$.]
3. If n is odd, find the quotient of $t^n + 1$ by $t + 1$.

IV, §3. UNIQUE FACTORIZATION

A polynomial p in $K[t]$ will be said to be **irreducible** (over K) if it is of degree $\geqq 1$, and if, given a factorization $p = fg$ with $f,\ g \in K[t]$, then $\deg f$ or $\deg g = 0$ (i.e. one of f, g is constant). Thus, up to a non-zero constant factor, the only divisors of p are p itself, and 1.

Example 1. The only irreducible polynomials over the complex numbers are the polynomials of degree 1, i.e. non-zero constant multiples of polynomials of type $t - \alpha$, with $\alpha \in \mathbf{C}$.

Example 2. The polynomial $t^2 + 1$ is irreducible over $\mathbf{R}$.

Theorem 3.1. *Every polynomial in $K[t]$ of degree $\geqq 1$ can be expressed as a product $p_1 \cdots p_m$ of irreducible polynomials. In such a product, the polynomials $p_1,\dots,p_m$ are uniquely determined, up to a rearrangement, and up to non-zero constant factors.*

Proof. We first prove the existence of the factorization into a product of irreducible polynomials. Let f be in $K[t]$, of degree $\geqq 1$. If f is irreducible, we are done. Otherwise, we can write

$$f = gh,$$

where $\deg g < \deg f$ and $\deg h < \deg f$. By induction we can express g and h as products of irreducible polynomials, and hence $f = gh$ can also be expressed as such a product.

We must now prove uniqueness. We need a lemma.

Lemma 3.2. *Let p be irreducible in $K[t]$. Let f, $g \in K[t]$ be non-zero polynomials, and assume p divides fg. Then p divides f or p divides g.*

Proof. Asume that p does not divide f. Then the greatest common divisor of p and f is 1, and there exist polynomials h_1, h_2 in $K[t]$ such that

$$1 = h_1 p + h_2 f.$$

(We use Theorem 2.2.) Multiplying by g yields

$$g = gh_1 p + h_2 fg.$$

But $fg = ph_3$ for some h_3, whence

$$g = (gh_1 + h_2 h_3)p,$$

and p divides g, as was to be shown.

The lemma will be applied when p divides a product of irreducible polynomials $q_1 \cdots q_s$. In that case, p divides q_1 or p divides $q_2 \cdots q_s$. Hence there exists a constant c such that $p = cq_1$, or p divides $q_2 \cdots q_s$. In the latter case, we can proceed inductively, and we conclude that in any case, there exists some i such that p and q_i differ by a constant factor.

Suppose now that we have two products of irreducible polynomials

$$p_1 \cdots p_r = q_1 \cdots q_s.$$

After renumbering the q_i, we may assume that $p_1 = c_1 q_1$ for some constant c_1. Cancelling q_1, we obtain

$$c_1 p_2 \cdots p_r = q_2 \cdots q_s.$$

Repeating our argument inductively, we conclude that there exist constants c_i such that $p_i = c_i q_i$ for all i, after making a possible permutation of $q_1, \ldots, q_s$. This proves the desired uniqueness.

Corollary 3.3. *Let f be a polynomial in $K[t]$ of degree $\geqq 1$. Then f has a factorization $f = cp_1 \cdots p_s$, where $p_1, \ldots, p_s$ are irreducible polynomials with leading coefficient* 1, *uniquely determined up to a permutation.*

Corollary 3.4. *Let K be algebraically closed. Let f be a polynomial in $K[t]$, of degree $\geqq 1$. Then f has a factorization*

$$f(t) = c(t - \alpha_1) \cdots (t - \alpha_n),$$

with $\alpha_i \in K$ and $c \in K$. The factors $t - \alpha_i$ are uniquely determined up to a permutation.

We shall deal mostly with polynomials having leading coefficient 1. Let f be such a polynomial of degree $\geqq 1$. Let $p_1, \ldots, p_r$ be the *distinct* irreducible polynomials with leading coefficient 1 occurring in its factorization. Then we can express f as a product

$$f = p_1^{m_1} \cdots p_r^{m_r},$$

where $m_1, \ldots, m_r$ are positive integers, uniquely determined by $p_1, \ldots, p_r$. This factorization will be called a **normalized factorization** for f. In particular, over an algebraically closed field, we can write

$$f(t) = (t - \alpha_1)^{m_1} \cdots (t - \alpha_r)^{m_r}.$$

A polynomial with leading coefficient 1 is sometimes called **monic**.

If p is irreducible, and $f = p^m g$, where p does not divide g, and m is an integer $\geqq 0$, then we say that m is the **multiplicity** of p in f. (We define p^0 to be 1.) We denote this multiplicity by $\operatorname{ord}_p f$, and also call it the **order** of f at p.

If α is a root of f, and

$$f(t) = (t - \alpha)^m g(t),$$

with $g(\alpha) \neq 0$, then $t - \alpha$ does not divide $g(t)$, and m is the multiplicity of $t - \alpha$ in f. We also say that m is the **multiplicity of** α in f. A root of f is said to be **simple** if its multiplicity is 1. A root is said to be **multiple** if $m > 1$.

There is an easy test for $m > 1$ in terms of the derivative.

Let $f(t) = a_n t^n + \cdots + a_0$ be a polynomial. Define its (formal) **derivative** to be

$$Df(t) = f'(t) = na_n t^{n-1} + (n-1)a_{n-1}t^{n-2} + \cdots + a_1 = \sum_{k=1}^{n} ka_k t^{k-1}.$$

Then we have the following properties, just as in ordinary calculus.

Proposition 3.5. *Let f, g be polynomials. Let $c \in K$. Then*

$$(cf)' = cf',$$

$$(f+g)' = f' + g',$$

$$(fg)' = fg' + f'g.$$

Proof. The first two properties $(cf)' = cf'$ and $(f+g)' = f' + g'$ are immediate from the definition. As to the third, namely the rule for the derivative of a product, suppose we know this rule for the product $f_1 g$ and $f_2 g$ where f_1, f_2, g are polynomials. Then we deduce it for $(f_1 + f_2)g$ as follows:

$$\begin{aligned}((f_1 + f_2)g)' &= (f_1 g + f_2 g)' = (f_1 g)' + (f_2 g)' \\ &= f_1 g' + f_1' g + f_2 g' + f_2' g \\ &= (f_1 + f_2)g' + (f_1' + f_2')g.\end{aligned}$$

Similarly, if we know the rule for the derivative of the products fg_1 and fg_2 then this rule holds for the derivative of $f(g_1 + g_2)$. Therefore it suffices to prove the rule for the derivative of a product when $f(t)$ and $g(t)$ are monomials, that is

$$f(t) = at^n \quad \text{and} \quad g(t) = bt^m$$

with $a, b \in K$. But then

$$\begin{aligned}(at^n bt^m)' = (abt^{n+m})' &= (n+m)abt^{n+m-1} \\ &= nat^{n-1}bt^m + at^n mbt^{m-1} \\ &= f'(t)g(t) + f(t)g'(t).\end{aligned}$$

This concludes the proof.

As an exercise, prove by induction:

If $f(t) = h(t)^m$ for some integer $m \geqq 1$, then

$$f'(t) = mh(t)^{m-1}h'(t).$$

Theorem 3.6. *Let K be a field. Let f be a polynomial over K, of degree $\geqq 1$, and let α be a root of f in K. Then the multiplicity of α in f is > 1 if and only if $f'(\alpha) = 0$.*

Proof. Suppose that

$$f(t) = (t - \alpha)^m g(t)$$

with $m > 1$. Taking the derivative, we find

$$f'(t) = m(t - \alpha)^{m-1}g(t) + (t - \alpha)^m g'(t).$$

Substituting α shows that $f'(\alpha) = 0$ because $m - 1 \geqq 1$. Conversely, suppose

$$f(t) = (t - \alpha)^m g(t),$$

and $g(\alpha) \neq 0$, so that m is the multiplicity of α in f. If $m = 1$ then

$$f'(t) = g(t) + (t - \alpha)g'(t),$$

so that $f'(\alpha) = g(\alpha) \neq 0$. This proves our theorem.

Example. Let K be a field in which the polynomial $t^n - 1$ factorizes into factors of degree 1, that is

$$t^n - 1 = \prod_{i=1}^{n}(t - \zeta_i).$$

The roots of $t^n - 1$ constitute the group of n-th roots of unity $\boldsymbol{\mu}_n$. Suppose that the characteristic of K is 0, or is p and $p \nmid n$. Then we claim that these n roots $\zeta_1, \ldots, \zeta_n$ are distinct, so the group $\boldsymbol{\mu}_n$ has order n. Indeed, let $f(t) = t^n - 1$. Then

$$f'(t) = nt^{n-1},$$

and if $\zeta \in \boldsymbol{\mu}_n$ then $f'(\zeta) = n\zeta^{n-1} \neq 0$ because $n \neq 0$ in K. This proves that every root occurs with multiplicity 1, whence that the n roots are distinct. By Theorem 1.10 we know that $\boldsymbol{\mu}_n$ is cyclic, and we have now found that $\boldsymbol{\mu}_n$ is cyclic of order n when $p \nmid n$. The equation $t^n - 1 = 0$ is called the **cyclotomic equation**. We shall study it especially in Exercise 13, Chapter VII, §6 and Chapter VIII, §5.

IV, §3. EXERCISES

1. Let f be a polynomial of degree 2 over a field K. Show that either f is irreducible over K, or f has a factorization into linear factors over K.

2. (a) Let f be a polynomial of degree 3 over a field K. If f is not irreducible over K, show that f has a root in K.
   (b) Let $F = \mathbf{Z}/2\mathbf{Z}$. Show that the polynomial $t^3 + t^2 + 1$ is irreducible in $F[t]$.

3. Let $f(t)$ be an irreducible polynomial with leading coefficient 1 over the real numbers. Assume $\deg f = 2$. Show that $f(t)$ can be written in the form

$$f(t) = (t - a)^2 + b^2$$

with some $a, b \in \mathbf{R}$ and $b \neq 0$. Conversely, prove that any such polynomial is irreducible over $\mathbf{R}$.

4. Let $\sigma: K \to L$ be an isomorphism of fields. If $f(t) = \sum a_i t^i$ is a polynomial over K, define σf to be the polynomial $\sum \sigma(a_i)t^i$ in $L[t]$.
   (a) Prove that the association $f \mapsto \sigma f$ is an isomorphism of $K[t]$ onto $L[t]$.
   (b) Let $\alpha \in K$ be a root of f of multiplicity m. Prove that $\sigma\alpha$ is a root of σf also of multiplicity m. [*Hint*: Use unique factorization.]

Example for Exercise 4. Let $K = \mathbf{C}$ be the complex numbers, and let σ be conjugation, so $\sigma: \mathbf{C} \to \mathbf{C}$ maps $\alpha \mapsto \bar{\alpha}$. If f is a polynomial with complex coefficients, say

$$f(t) = \alpha_n t^n + \cdots + \alpha_0.$$

Then its complex conjugate

$$\bar{f}(t) = \bar{\alpha}_n t^n + \cdots + \bar{\alpha}_0$$

is obtained by taking the complex conjugate of each coefficient. If f, g are in $\mathbf{C}[t]$, then

$$\overline{(f + g)} = \bar{f} + \bar{g}, \qquad \overline{(fg)} = \bar{f}\bar{g},$$

and if $\beta \in \mathbf{C}$, then $\overline{(\beta f)} = \bar{\beta}\bar{f}$.

5. (a) Let $f(t)$ be a polynomial with real coefficients. Let α be a root of f, which is complex but not real. Show that $\bar{\alpha}$ is also a root of f.
   (b) Assume that the complex numbers are algebraically closed. Let $f(t) \in \mathbf{R}[t]$ be a polynomial with real coefficients, of degree $\geqq 1$. Assume that f is irreducible. Prove that $\deg f = 1$ or $\deg f = 2$. It follows that if a real polynomial is expressed as a product of irreducible factors, then these factors have degree 1 or 2.

6. Let K be a field which is a subfield of a field E. Let $f(t) \in K[t]$ be an irreducible polynomial, let $g(t) \in K[t]$ be any polynomial $\neq 0$, and let $\alpha \in E$ be an element such that $f(\alpha) = g(\alpha) = 0$. In other words, f, g have a common root in E. Prove that $f(t) | g(t)$ in $K[t]$.

7. (a) Let K be a field of characteristic 0, subfield of a field E. Let $\alpha \in E$ be a root of $f(t) \in K[t]$. Prove that α has multiplicity m if and only if

$$f^{(k)}(\alpha) = 0 \quad \text{for} \quad k = 1, \ldots, m - 1 \quad \text{but} \quad f^{(m)}(\alpha) \neq 0.$$

   (As usual, $f^{(k)}$ denotes the k-th derivative of f.)

(b) Show that the assertion of (a) is not generally true if K has characteristic p.
(c) In fact, what is the value of $f^{(m)}(\alpha)$? There is a very simple expression for it.
(d) If K has characteristic 0 and if $f(t)$ is irreducible in $K[t]$, prove that α has multiplicity 1.
(e) Suppose K has characteristic p and $p \nmid n$. Let $f(t)$ be irreducible in $K[t]$, of degree n. Let α be a root of f. Proof that α has multiplicity 1.

8. Show that the following polynomials have no multiple roots in $\mathbf{C}$:
(a) $t^4 + t$ (b) $t^5 - 5t + 1$
(c) any polynomial $t^2 + bt + c$ if b, c are numbers such that $b^2 - 4c$ is not 0.

9. (a) Let K be a subfield of a field E, and $\alpha \in E$. Let J be the set of all polynomials $f(t)$ in $K[t]$ such that $f(\alpha) = 0$. Show that J is an ideal. If J is not the zero ideal, show that the monic generator of J is irreducible.
(b) Conversely, let $p(t)$ be irreducible in $K[t]$ and let α be a root. Show that the ideal of polynomials $f(t)$ in $K[t]$ such that $f(\alpha) = 0$ is the ideal generated by $p(t)$.

10. Let f, g be two polynomials, written in the form

$$f = p_1^{i_1} \cdots p_r^{i_r}$$

and

$$g = p_1^{j_1} \cdots p_r^{j_r},$$

where i_ν, j_ν are integers $\geqq 0$, and $p_1, \ldots, p_r$ are distinct irreducible polynomials.
(a) Show that the greatest common divisor of f and g can be expressed as a product $p_1^{k_1} \cdots p_r^{k_r}$ where $k_1, \ldots, k_r$ are integers $\geqq 0$. Express k_ν in terms of i_ν and j_ν.
(b) Define the least common multiple of polynomials, and express the least common multiple of f and g as a product $p_1^{k_1} \cdots p_r^{k_r}$ with integers $k_\nu \geqq 0$. Express k_ν in terms of i_ν and j_ν.

11. Give the greatest common divisor and least common multiple of the following pairs of polynomials with complex coefficients:
(a) $(t-2)^3(t-3)^4(t-i)$ and $(t-1)(t-2)(t-3)^3$
(b) $(t^2+1)(t^2-1)$ and $(t+i)^3(t^3-1)$

12. Let K be a field, $R = K[t]$ the ring of polynomials, and F the quotient field of R, i.e. the field of rational functions. Let $\alpha \in K$. Let R_α be the set of rational functions which can be written as a quotient f/g of polynomials such that $g(\alpha) \neq 0$. Show that R_α is a ring. If φ is a rational function, and $\varphi = f/g$ such that $g(\alpha) \neq 0$, define $\varphi(\alpha) = f(\alpha)/g(\alpha)$. Show that this value $\varphi(\alpha)$ is independent of the choice of representation of φ as a quotient f/g. Show that the map $\varphi \mapsto \varphi(\alpha)$ is a ring-homomorphism of R_α into K. Show that the kernel of this ring-homomorphism consists of all rational functions f/g such that $g(\alpha) \neq 0$ and $f(\alpha) = 0$. If M_α denotes this kernel, show that M_α is a maximal ideal of R_α.

13. Let W_n be the set of primitive n-th roots of unity in $\mathbf{C}^*$. Define the n-th **cyclotomic polynomial** to be

$$\Phi_n(t) = \prod_{\zeta \in W_n} (t - \zeta).$$

(a) Prove that $t^n - 1 = \prod_{d|n} \Phi_d(t)$.
(b) $\Phi_n(t) = \prod_{d|n}(t^{n/d} - 1)^{\mu(d)}$ where μ is the Moebius function.
(c) Let p be a prime number and let k be a positive integer. Prove

$$\Phi_{p^k}(t) = \Phi_p(t^{p^{k-1}}) \quad \text{and} \quad \Phi_p(t) = t^{p-1} + \cdots + 1.$$

(d) Compute explicitly $\Phi_n(t)$ for $n \leqq 10$.

14. Let R be a rational function over the field K, and express R as a quotient of polynomials, $R = g/f$. Define the derivative

$$R' = \frac{fg' - gf'}{f^2},$$

where the prime means the formal derivative of polynomials as in the text.
(a) Show that this derivative is independent of the expression of R as a quotient of polynomials, i.e. if $R = g_1/f_1$ then

$$\frac{fg' - gf'}{f^2} = \frac{f_1 g_1' - g_1 f_1'}{f_1^2}.$$

(b) Show that the derivative of rational functions satisfies the same rules as before, namely for rational functions R_1 and R_2 we have

$$(R_1 + R_2)' = R_1' + R_2' \quad \text{and} \quad (R_1 R_2)' = R_1 R_2' + R_1' R_2.$$

(c) Let $\alpha_1, \ldots, \alpha_n$ and $a_1, \ldots, a_n$ be elements of K such that

$$\frac{1}{(t - \alpha_1) \cdots (t - \alpha_n)} = \frac{a_1}{t - \alpha_1} + \cdots + \frac{a_n}{t - \alpha_n}.$$

Let $f(t) = (t - \alpha_1) \cdots (t - \alpha_n)$ and assume that $\alpha_1, \ldots, \alpha_n$ are distinct. Show that

$$a_1 = \frac{1}{(\alpha_1 - \alpha_2) \cdots (\alpha_1 - \alpha_n)} = \frac{1}{f'(\alpha_1)}.$$

15. Show that the map $R \mapsto R'/R$ is a homomorphism from the multiplicative group of non-zero rational functions to the additive group of rational functions. We call R'/R the **logarithmic derivative** of R. If

$$R(t) = \prod_{i=1}^{n} (t - \alpha_i)^{m_i}$$

where $m_1, \ldots, m_n$ are integers, what is R'/R?

16. For any polynomial f, let $n_0(f)$ be the number of distinct roots of f when f is factored into factors of degree 1. So if

$$f(t) = c\prod_{i=1}^{r} (t - \alpha_i)^{m_i},$$

where $c \neq 0$ and $\alpha_1, \ldots, \alpha_r$ are distinct, then $n_0(f) = r$. Prove:

Mason-Stothers Theorem. *Let $K = \mathbf{C}$, or more generally an algebraically closed field of characteristic 0. Let $f, g, h \in K[t]$ be relatively prime polynomials not all constant such that $f + g = h$. Then*

$$\deg f, \deg g, \deg h \leqq n_0(fgh) - 1.$$

[*Mason's proof.* You may assume that f, g, h can be factored into products of polynomials of degree 1 in some larger field. Divide the relation $f + g = h$ by h, so get $R + S = 1$ where R, S are rational functions. Take the derivative and write the resulting relation as

$$\frac{R'}{R} R + \frac{S'}{S} S = 0.$$

Solve for S/R in terms of the logarithmic derivatives. Let

$$g(t) = c_2 \prod (t - \beta_j)^{n_j} \qquad \text{and} \qquad h(t) = c_3 \prod (t - \gamma_k)^{q_k}.$$

Let $D = \prod (t - \alpha_i) \prod (t - \beta_j) \prod (t - \gamma_k)$. Use D as a common denominator and multiply R'/R and S'/S by D. Then count degrees.]

17. Assume the preceding exercise. Let f, $g \in K[t]$ be non-constant polynomials such that $f^3 - g^2 \neq 0$, and let $h = f^3 - g^2$. Prove that

$$\deg f \leqq 2 \deg h - 2 \qquad \text{and} \qquad \deg g \leqq 3 \deg h - 3.$$

18. More generally, suppose $f^m + g^n = h \neq 0$. Assume $mn > m + n$. Prove that

$$\deg f \leqq \frac{n}{mn - (m + n)} \deg h.$$

19. Let f, g, h be polynomials over a field of characteristic 0, such that

$$f^n + g^n = h^n$$

and assume f, g relatively prime. Suppose that f, g have degree $\geqq 1$. Show that $n \leqq 2$. (This is the Fermat problem for polynomials. Use Exercise 16.)

For an alternate treatment of the above, see §9.

IV, §4. PARTIAL FRACTIONS

In the preceding section, we proved that a polynomial can be expressed as a product of powers of irreducible polynomials in a unique way (up to a permutation of the factors). The same is true of a rational function, if we allow negative exponents. Let $R = g/f$ be a rational function, expressed as a quotient of polynomials g, f with $f \neq 0$. Suppose $R \neq 0$. If g, f are not relatively prime, we may cancel their greatest common divisor and thus obtain an expression of R as a quotient of relatively prime polynomials. Factoring out their constant leading coefficients, we can write

$$R = c\frac{g_1}{f_1},$$

where f_1, g_1 have leading coefficient 1. Then f_1, g_1, and c are uniquely determined, for suppose

$$cg_1/f_1 = c_2 g_2/f_2$$

for constants c, c_2 and pairs of relatively prime polynomials f_1, g_1 and f_2, g_2 with leading coefficient 1. Then

$$cg_1 f_2 = c_2 g_2 f_1.$$

From the unique factorization of polynomials, we conclude that $g_1 = g_2$ and $f_1 = f_2$ so that $c = c_2$.

If we now factorize f_1 and g_1 into products of powers of irreducible polynomials, we obtain the unique factorization of R. This is entirely analogous to the factorization of a rational number obtained in Chapter I, §4.

We wish to decompose a rational function into a sum of rational functions, such that the denominator of each term is equal to a power of an irreducible polynomial. Such a decomposition is called a **partial fraction decomposition**. We begin by a lemma which allows us to apply induction.

Lemma 4.1. *Let f_1, f_2 be non-zero, relatively prime polynomials over a field K. Then there exist polynomials h_1, h_2 over K such that*

$$\frac{1}{f_1 f_2} = \frac{h_1}{f_1} + \frac{h_2}{f_2}.$$

Proof. Since f_1, f_2 are relatively prime, there exist polynomials h_1, h_2 such that

$$h_2 f_1 + h_1 f_2 = 1.$$

Dividing both sides by $f_1 f_2$, we obtain what we want.

Theorem 4.2. *Every rational function R can be written in the form*

$$R = \frac{h_1}{p_1^{i_1}} + \cdots + \frac{h_n}{p_n^{i_n}} + h,$$

where $p_1, \ldots, p_n$ are distinct irreducible polynomials with leading coefficient 1; $i_1, \ldots, i_n$ are integers $\geqq 0$; $h_1, \ldots, h_n$, h are polynomials, satisfying

$$\deg h_\nu < \deg p_\nu^{i_\nu} \qquad \textit{and} \qquad p_\nu \nmid h_\nu$$

for $\nu = 1, \ldots, n$. In such an expression, the integers i_ν and the polynomials h_ν, h $(\nu = 1, \ldots, n)$ are uniquely determined when all $i_\nu > 0$.

Proof. We first prove the existence of the expression described in our theorem. Let $R = g/f$ where f is a non-zero polynomial, with g, f relatively prime, and write

$$f = p_1^{i_1} \cdots p_n^{i_n},$$

where $p_1, \ldots, p_n$ are distinct irreducible polynomials, and $i_1, \ldots, i_n$ are integers $\geqq 0$. By the lemma, there exist polynomials g_1, g_1^* such that

$$\frac{1}{f} = \frac{g_1}{p_1^{i_1}} + \frac{g_1^*}{p_2^{i_2} \cdots p_n^{i_n}},$$

and by induction, there exist polynomials $g_2, \ldots, g_n$ such that

$$\frac{g_1^*}{p_2^{i_2} \cdots p_n^{i_n}} = \frac{g_2}{p_2^{i_2}} + \cdots + \frac{g_n}{p_n^{i_n}}.$$

Multiplying by g, we obtain

$$\frac{g}{f} = \frac{gg_1}{p_1^{i_1}} + \cdots + \frac{gg_n}{p_n^{i_n}}.$$

By the Euclidean algorithm, we can divide gg_ν by $p_\nu^{i_\nu}$ for $\nu = 1, \ldots, n$ letting

$$gg_\nu = q_\nu p_\nu^{i_\nu} + h_\nu, \qquad \deg h_\nu < \deg p_\nu^{i_\nu}.$$

In this way obtain the desired expression for g/f, with $h = q_1 + \cdots + q_n$.

Next we prove the uniqueness. Suppose we have expressions

$$\frac{h_1}{p_1^{i_1}} + \cdots + \frac{h_n}{p_n^{i_n}} + h = \frac{\bar{h}_1}{p_1^{j_1}} + \cdots + \frac{\bar{h}_n}{p_n^{j_n}} + \bar{h},$$

satisfying the conditions stated in the theorem. (We can assume that the irreducible polynomials $p_1,\ldots,p_n$ are the same on both sides, letting some i_ν be equal to 0 if necessary.) Then there exist polynomials φ, ψ such that $\psi \neq 0$ and $p_1 \nmid \psi$, for which we can write

$$\frac{h_1}{p_1^{i_1}} - \frac{\bar{h}_1}{p_1^{j_1}} = \frac{\varphi}{\psi}.$$

Say $i_1 \leqq j_1$. Then

$$\frac{h_1 p_1^{j_1 - i_1} - \bar{h}_1}{p_1^{j_1}} = \frac{\varphi}{\psi}.$$

Since ψ is not divisible by p_1, it follows from unique factorization that $p_1^{j_1}$ divides $h_1 p_1^{j_1 - i_1} - \bar{h}_1$. If $j_1 \neq i_1$ then $p_1 | \bar{h}_1$, contrary to the conditions stated in the theorem. Hence $j_1 = i_1$. Again since ψ is not divisible by p_1, it follows now that $p_1^{j_1}$ divides $h_1 - \bar{h}_1$. By hypothesis,

$$\deg(h_1 - \bar{h}_1) < \deg p_1^{j_1}.$$

Hence $h_1 - \bar{h}_1 = 0$, whence $h_1 = \bar{h}_1$. We therefore conclude that

$$\frac{h_2}{p_2^{i_2}} + \cdots + \frac{h_n}{p_n^{i_n}} + h = \frac{\bar{h}_2}{p_2^{j_2}} + \cdots + \frac{\bar{h}_n}{p_n^{j_n}} + \bar{h},$$

and we can conclude the proof by induction.

The expression of Theorem 4.2 is called the **partial fraction decomposition** of R.

The irreducible polynomials $p_1,\ldots,p_n$ in Theorem 4.2 can be described somewhat more precisely, and the next theorem gives additional information on them, and also on h.

Theorem 4.3. *Let the notation be as in Theorem 4.2, and let the rational function R be expressed in the form $R = g/f$ where g, f are relatively prime polynomials, $f \neq 0$. Assume that all integers $i_1,\ldots,i_n$ are > 0. Then*

$$f = p_1^{i_1} \cdots p_n^{i_n}$$

is the prime power factorization of f. Furthermore, if $\deg f > \deg g$, then $h = 0$.

Proof. If we put the partial fraction expression for R in Theorem 4.2 over a common denominator, we obtain

$$(*) \qquad R = \frac{h_1 p_2^{i_2} \cdots p_n^{i_n} + \cdots + h_n p_1^{i_1} \cdots p_{n-1}^{i_{n-1}} + h p_1^{i_1} \cdots p_n^{i_n}}{p_1^{i_1} \cdots p_n^{i_n}}.$$

Then p_ν does not divide the numerator on the right in $(*)$, for any index $\nu = 1, \ldots, n$. Indeed, p_ν divides every term in this numerator *except* the term

$$h_\nu p_1^{i_1} \cdots \widehat{p_\nu^{i_\nu}} \cdots p_n^{i_n}$$

(where the roof over $p_\nu^{i_\nu}$ means that we omit this factor). This comes from the hypothesis that p_ν does not divide h_ν. Hence the numerator and denominator on the right in $(*)$ are relatively prime, thereby proving our first assertion.

As to the second, letting g be the numerator of R and f its denominator, we have $f = p_1^{i_1} \cdots p_n^{i_n}$, and

$$g = Rf = h_1 p_2^{i_2} \cdots p_n^{i_n} + \cdots + h_n p_1^{i_1} \cdots p_{n-1}^{i_{n-1}} + h p_1^{i_1} \cdots p_n^{i_n}.$$

Assume that $\deg g < \deg f$. Then every term in the preceding sum has degree $<$ def f, except possibly the last term

$$hf = h p_1^{i_1} \cdots p_n^{i_n}.$$

If $h \neq 0$, then this last term has degree $\geqq \deg f$, and we then get

$$hf = g - h_1 p_2^{i_2} \cdots p_n^{i_n} - \cdots - h_n p_1^{i_1} \cdots p_{n-1}^{i_{n-1}},$$

where the left-hand side has degree $\geqq \deg f$ and the right-hand side has degree $< \deg f$. This is impossible. Hence $h = 0$, as was to be shown.

Remark. Given a rational function $R = g/f$ where g, f are relatively prime polynomials, we can use the Euclidean algorithm and write

$$g = g_1 f + g_2,$$

where g_1, g_2 are poloynomials and $\deg g_2 < \deg f$. Then

$$\frac{g}{f} = \frac{g_2}{f} + g_1,$$

and we can apply Theorem 4.3 to the rational function g_2/f. In studying rational functions, it is always useful to perform first this long division to reduce the study of the rational function to the case when the degree of the numerator is smaller than the degree of its denominator.

Example 1. Let $\alpha_1,\dots,\alpha_n$ be distinct elements of K. Then there exist elements $a_1,\dots,a_n \in K$ such that

$$\frac{1}{(t-\alpha_1)\cdots(t-\alpha_n)} = \frac{a_1}{t-\alpha_1} + \cdots + \frac{a_n}{t-\alpha_n}.$$

Indeed, in the present case we can apply Theorems 4.2 and 4.3, with $g = 1$, and hence $\deg g < \deg f$. In Exercise 14 of the preceding section, we showed how to determine a_i in a special way.

Each expression $h_\nu/p_\nu^{i_\nu}$ in the partial fraction decomposition can be further analyzed, by writing h_ν in a special way, which we now describe.

Theorem 4.4. *Let φ be a non-constant polynomial over the field K. Let h be any polynomial over K. Then there exist polynomials $\psi_0,\dots,\psi_m$ such that*

$$h = \psi_0 + \psi_1\varphi + \cdots + \psi_m\varphi^m$$

and $\deg \psi_i < \deg \varphi$ for all $i = 0,\dots,m$. The polynomials $\psi_0,\dots,\psi_m$ are uniquely determined by these conditions.

Proof. We prove the existence of $\psi_0,\dots,\psi_m$ by induction on the degree of h. By the Euclidean algorithm, we can write

$$h = q\varphi + \psi_0$$

with the polynomials q, ψ_0, and $\deg \psi_0 < \deg \varphi$. Then $\deg q < \deg h$, so that by induction we can write

$$q = \psi_1 + \psi_2\varphi + \cdots + \psi_m\varphi^{m-1}$$

with polynomials ψ_i such that $\deg \psi_i < \deg \varphi$. Substituting, we obtain

$$h = (\psi_1 + \psi_2\varphi + \cdots + \psi_m\varphi^{m-1})\varphi + \psi_0.$$

which yields the desired expression.

As for uniqueness, we observe first that in the expression given in the theorem, namely

$$h = \psi_0 + \psi_1\varphi + \cdots + \psi_m\varphi^m = \psi_0 + \varphi(\psi_1 + \cdots + \psi_m\varphi^{m-1})$$

the polynomial ψ_0 is necessarily the remainder of the division of h by φ, so that its uniqueness is given by the Euclidean algorithm. Then, writing $h = q\varphi + \psi_0$, we conclude that

$$q = \psi_1 + \cdots + \psi_m \varphi^{m-1},$$

and q is uniquely determined. Thus $\psi_1, \ldots, \psi_m$ are uniquely determined by induction, as was to be proved.

The expression of h in terms of powers of φ as given in Theorem 4.4 is called its **φ-adic expansion**. We can apply this to the case where φ is an irreducible polynomial p, in which case this expression is the p-adic expansion of h. Suppose that

$$h = \psi_0 + \psi_1 p + \cdots + \psi_m p^m$$

is its p-adic expansion. Then dividing by p^i for some integer $i > 0$, we obtain the following theorem.

Theorem 4.5. *Let h be a polynomial and p an irreducible polynomial over the field K. Let i be an integer > 0. Then there exists a unique expression*

$$\frac{h}{p^i} = \frac{g_{-i}}{p^i} + \frac{g_{-i+1}}{p^{i-1}} + \cdots + g_0 + g_1 p + \cdots + g_s p^s,$$

where g_μ are polynomials of degree $< \deg p$.

In Theorem 4.5, we have adjusted the numbering of $g_{-i}, g_{-i+1}, \ldots$ so that it would fit the exponent of p occurring in the denominator. Otherwise, except for this numbering, these polynomials g are nothing but $\psi_0, \psi_1, \ldots$ found in the p-adic expansion of h.

Corollary 4.6. *Let $\alpha \in K$, and let h be a polynomial over K. Then*

$$\frac{h(t)}{(t-\alpha)^i} = \frac{a_{-i}}{(t-\alpha)^i} + \frac{a_{-i+1}}{(t-\alpha)^{i-1}} + \cdots + a_0 + a_1(t-\alpha) + \cdots,$$

where a_μ are elements of K, uniquely determined.

Proof. In this case, $p(t) = t - \alpha$ has degree 1, so that the coefficients in the p-adic expansion must be constants.

Example 2. To determine the partial fraction decomposition of a given rational function, one can solve a system of linear equations. We give an example. We wish to write

$$\frac{1}{(t-1)(t-2)} = \frac{a}{t-1} + \frac{b}{t-2}$$

with constants a and b. Putting the right-hand side over a common denominator, we have

$$\frac{1}{(t-1)(t-2)} = \frac{a(t-2)+b(t-1)}{(t-1)(t-2)}.$$

Setting the numerators equal to each other, we must have

$$a + b = 0,$$
$$-2a - b = 1.$$

We then solve for a and b to get $a = -1$ and $b = 1$. The general case can be handled similarly.

IV, §4. EXERCISES

1. Determine the partial fraction decomposition of the following rational functions.

   (a) $\dfrac{t+1}{(t-1)(t+2)}$ (b) $\dfrac{1}{(t+1)(t^2+2)}$

2. Let $R = g/f$ be a rational function with $\deg g < \deg f$. Let

$$\frac{g}{f} = \frac{h_1}{p_1^{i_1}} + \cdots + \frac{h_n}{p_n^{i_n}}$$

   be its partial fraction decomposition. Let $d_\nu = \deg p_\nu$. Show that the coefficients of $h_1, \ldots, h_n$ are the solutions of a system of linear equations, such that the number of variables is equal to the number of equations, namely

$$\deg f = i_1 d_1 + \cdots + i_n d_n.$$

   Theorem 4.3 shows that this system has a unique solution.

3. Find the $(t-2)$-adic expansion of the following polynomials.
   (a) $t^2 - 1$ (b) $t^3 + t - 1$ (c) $t^3 + 3$ (d) $t^4 + 2t^3 - t + 5$

4. Find the $(t-3)$-adic expansion of the polynomials in Exercise 3.

IV, §5. POLYNOMIALS OVER RINGS AND OVER THE INTEGERS

Let R be a commutative ring. We formed the polynomial ring over R, just as we did over a field, namely $R[t]$ consists of all formal sums

$$f(t) = a_n t^n + \cdots + a_0$$

with elements $a_0, \ldots, a_n$ in R which are called the **coefficients** of f. The sum and product are defined just as over a field. The ring R could be integral, in which case R has a quotient field K, and $R[t]$ is contained in $K[t]$. In fact, $R[t]$ is a subring of $K[t]$.

The advantage of dealing with coefficients in a ring is that we can deal with more refined properties of polynomials. Polynomials with coefficients in the ring of integers $\mathbf{Z}$ form a particularly interesting ring. We shall prove some special properties of such polynomials, leading to an important criterion for irreducibility of polynomials over the rational numbers. Before we do that, we make one more general comment on polynomials over rings.

Let

$$\sigma : R \to S$$

be a homomorphism of commutative rings. If $f(t) \in R[t]$ is a polynomial over R as above, then we define the polynomial σf in $S[t]$ to be the polynomial

$$(\sigma f)(t) = \sigma(a_n)t^n + \cdots + \sigma(a_0) = \sum_{i=0}^{n} \sigma(a_i)t^i.$$

Then it is immediately verified that σ thereby is a ring homomorphism

$$R[t] \to S[t]$$

which we also denote by σ. Indeed, let

$$f(t) = \sum_{i=0}^{n} a_i t^i \qquad \text{and} \qquad g(t) = \sum_{i=0}^{n} b_i t^i.$$

Then $(f + g)(t) = \sum (a_i + b_i)t^i$ so

$$\sigma(f + g)(t) = \sum \sigma(a_i + b_i)t^i = \sum (\sigma(a_i) + \sigma(b_i))t^i = \sigma f(t) + \sigma g(t).$$

Also $fg(t) = \sum c_k t^k$ where

$$c_k = \sum_{i=0}^{k} a_i b_{k-i},$$

so

$$\sigma(fg)(t) = \sum_k \sum_{i=0}^{k} \sigma(a_i b_{k-i}) t^k = \sum_k \sum_{i=0}^{k} \sigma(a_i)\sigma(b_{i-i}) t^k$$
$$= (\sigma f)(t)(\sigma g)(t).$$

This proves that σ induces a homomorphism $R[t] \to S[t]$.

Example. Let $R = \mathbf{Z}$ and let $S = \mathbf{Z}/n\mathbf{Z}$ for some integer $n \geqq 2$. For each integer $a \in \mathbf{Z}$ let $\bar{a}$ denote its residue class mod n. Then the map

$$\sum a_i t^i \mapsto \sum \bar{a}_i t^i$$

is a ring homomorphism of $\mathbf{Z}[t]$ into $(\mathbf{Z}/n\mathbf{Z})[t]$. Note that we don't need to assume that n is prime. This ring homomorphism is called **reduction mod n**. If $n = p$ is a prime number, and we let $\mathbf{F}_p = \mathbf{Z}/p\mathbf{Z}$ so $\mathbf{F}_p$ is a field, then we get a homomorphism

$$\mathbf{Z}[t] \to \mathbf{F}_p[t]$$

where $\mathbf{F}_p[t]$ is the polynomial ring over the field $\mathbf{F}_p$.

A polynomial over $\mathbf{Z}$ will be called **primitive** if its coefficients are relatively prime, that is if there is no prime p which divides all coefficients. In particular, a primitive polynomial is not the zero polynomial.

Lemma 5.1. *Let f be a polynomial $\neq 0$ over the rational numbers. Then there exists a rational number $a \neq 0$ such that af has integer coefficients, which are relatively prime, i.e. af is primitive.*

Proof. Write

$$f(t) = a_n t^n + \cdots + a_0,$$

where $a_0, \ldots, a_n$ are rational numbers, and $a_n \neq 0$. Let d be a common denominator for $a_0, \ldots, a_n$. Then df has integral coefficients, namely $da_n, \ldots, da_0$. Let b be a greatest common divisor for $da_n, \ldots, da_0$. Then

$$\frac{d}{b} f(t) = \frac{da_n}{b} t^n + \cdots + \frac{da_0}{b}$$

has relatively prime integral coefficients, as was to be shown.

Lemma 5.2 (Gauss). *Let f, g be primitive polynomials over the integers. Then fg is primitive.*

Proof. Write

$$f(t) = a_n t^n + \cdots + a_0, \qquad a_n \neq 0,$$

$$g(t) = b_m t^m + \cdots + b_0, \qquad b_m \neq 0,$$

with relatively prime $(a_n, \ldots, a_0)$ and relatively prime $(b_m, \ldots, b_0)$. Let p be a prime. It will suffice to prove that p does not divide every coefficient of fg. Let r be the largest integer such that $0 \leqq r \leqq n$, $a_r \neq 0$, and p does not divide a_r. Similarly, let b_s be the coefficient of g farthest to the left, $b_s \neq 0$, such that p does not divide b_s. Consider the coefficient of t^{r+s} in $f(t)g(t)$. This coefficient is equal to

$$\begin{aligned} c = a_r b_s + a_{r+1} b_{s-1} + \cdots \\ + a_{r-1} b_{s+1} + \cdots \end{aligned}$$

and p does not divide $a_r b_s$. However, p divides every other non-zero term in this sum since each term will be of the form

$$a_i b_{r+s-i}$$

with a_i to the left of a_r, that is $i > r$, or of the form

$$a_{r+s-j} b_j$$

with $j > s$, that is b_j to the left of b_s. Hence p does not divide c, and our lemma is proved.

We shall now give a second proof using the idea of reduction modulo a prime. Suppose f, g are primitive polynomials in $\mathbf{Z}[t]$ but fg is not primitive. So there exists a prime p which divides all coefficients of fg. Let σ be reduction $\bmod p$. Then $\sigma f \neq 0$ and $\sigma g \neq 0$ but $\sigma(fg) = 0$. However, the polynomial ring $(\mathbf{Z}/p\mathbf{Z})[t]$ has no divisors of zero, so we get a contradiction proving the lemma.

I prefer this second proof for several reasons, but the technique of picking coefficients furthest to the left or to the right will reappear in the proof of Eisenstein's criterion, so I gave both proofs.

Let $f(t) \in \mathbf{Z}[t]$ be a polynomial of degree $\geqq 1$. Suppose that f is reducible over $\mathbf{Z}$, that is

$$f(t) = g(t)h(t),$$

where g, h have coefficients in $\mathbf{Z}$ and $\deg g$, $\deg h \geqq 1$. Then of course f is reducible over the quotient field $\mathbf{Q}$. Conversely, suppose f is reducible over **Q**. Is f reducible over **Z**? The answer is yes, and we prove it in the next theorem.

Theorem 5.3 (Gauss). *Let f be a primitive polynomial in $\mathbf{Z}[t]$, of degree $\geqq 1$. If f is reducible over* **Q**, *that is if we can write $f = gh$ with g, $h \in \mathbf{Q}[t]$, and $\deg g \geqq 1$, $\deg h \geqq 1$, then f is reducible over* **Z**. *More precisely, there exist rational numbers a, b such that, if we let $g_1 = ag$ and $h_1 = bh$, then g_1, h_1 have integer coefficients, and $f = g_1 h_1$.*

Proof. By Lemma 5.1, let a, b be non-zero rational numbers such that ag and bh have integer coefficients, relatively prime. Let $g_1 = ag$ and $h_1 = bh$. Then

$$f = \frac{1}{a} g_1 \frac{1}{b} h_1,$$

whence $abf = g_1 h_1$. By Lemma 5.2, $g_1 h_1$ has relatively prime integer coefficients. Since the coefficients of f are assumed to be relatively prime integers, it follows at once that ab itself must be an integer, and cannot be divisible by any prime. Hence $ab = \pm 1$, and dividing (say) g_1 by ab we obtain what we want.

Warning. The result of Theorem 5.3 is not generally true for every integral ring R. Some restriction on R is needed, like unique factorization. We shall see later how the notion of unique factorization can be generalized, as well as its consequences.

Theorem 5.4 (Eisenstein's criterion). *Let*

$$f(t) = a_n t^n + \cdots + a_0$$

be a polynomial of degree $n \geqq 1$ with integer coefficients. Let p be a prime, and assume

$$a_n \not\equiv 0 \pmod{p}, \qquad a_i \equiv 0 \pmod{p} \quad \text{for all} \quad i < n,$$
$$a_0 \not\equiv 0 \pmod{p^2}.$$

Then f is irreducible over the rationals.

Proof. We first divide f by the greatest common divisor of its coefficients, and we may then assume that f has relatively prime coefficients. By Theorem 5.3, we must show that f cannot be written as a product $f = gh$ with g, h having integral coefficients, and $\deg g$, $\deg h \geqq 1$. Suppose this can be done, and write

$$g(t) = b_d t^d + \cdots + b_0,$$
$$h(t) = c_m t^m + \cdots + c_0,$$

with d, $m \geqq 1$ and $b_d c_m \neq 0$. Since $b_0 c_0 = a_0$ is divisible by p but not p^2, it follows that one of the numbers b_0, c_0 is not divisible by p, say b_0. Then $p|c_0$. Since $c_m b_d = a_n$ is not divisible by p, it follows that p does not divide c_m. Let c_r be the coefficient of h farthest to the right such that $c_r \not\equiv 0 \pmod{p}$. Then $r \neq 0$ and

$$a_r = b_0 c_r + b_1 c_{r-1} + \cdots + b_r c_0.$$

Since $p \nmid b_0 c_r$ but p divides every other term in this sum, we conclude that $p \nmid a_r$, a contradiction which proves our theorem.

Example. The polynomial $t^5 - 2$ is irreducible over the rational numbers, as a direct application of Theorem 5.4.

Another criterion for irreducibility is given by the next theorem, and uses the notion of reduction mod p for some prime p.

Theorem 5.5 (Reduction criterion). *Let $f(t) \in \mathbf{Z}[t]$ be a primitive polynomial with leading coefficient a_n which is not divisible by a prime p. Let $\mathbf{Z} \to \mathbf{Z}/p\mathbf{Z} = F$ be reduction* mod p, *and denote the image of f by $\bar{f}$. If $\bar{f}$ is irreducible in $F[t]$, then f is irreducible in $\mathbf{Q}[t]$.*

Proof. By Theorem 5.3 it suffices to prove that f does not have a factorization $f = gh$ with $\deg g$ and $\deg h \geqq 1$ and g, $h \in \mathbf{Z}[t]$. Suppose there is such a factorization. Let

$$f(t) = a_n t^n + \text{lower terms},$$

$$g(t) = b_r t^r + \text{lower terms},$$

$$h(t) = c_s t^s + \text{lower terms}.$$

Then $a_n = b_r c_s$, and since a_n is not divisible by p by hypothesis, it follows that b_r, c_s are not divisible by p. Hence

$$\bar{f} = \bar{g}\bar{h}$$

and $\deg \bar{g}$, $\deg \bar{h} \geqq 1$, which contradicts the hypothesis that $\bar{f}$ is irreducible over F. This proves Theorem 5.5.

Example. The polynomial $t^3 - t - 1$ is irreducible over $\mathbf{Z}/3\mathbf{Z}$, otherwise it would have a root which must be 0, 1, or $-1 \bmod 3$. You can see that this is not the case by plugging in these three values. Hence $t^3 - t - 1$ is irreducible over $\mathbf{Q}[t]$.

In Chapter VII, §3, Exercise 1, you will prove that $t^5 - t - 1$ is irreducible over $\mathbf{Z}/5\mathbf{Z}$. It follows that $t^5 - t - 1$ is irreducible over $\mathbf{Q}[t]$. You could already try to prove here that $t^5 - t - 1$ is irreducible over $\mathbf{Z}/5\mathbf{Z}$.

IV, §5. EXERCISES

1. **Integral root theorem.** Let $f(t) = t^n + \cdots + a_0$ be a polynomial of degree $n \geqq 1$ with integer coefficients, leading coefficient 1, and $a_0 \neq 0$. Show that if f has a root in the rational numbers, then this root is in fact an integer, and that this integer divides a_0.

2. Determine which of the following polynomials are irreducible over the rational numbers:
   (a) $t^3 - t + 1$ (b) $t^3 + 2t + 10$ (c) $t^3 - t - 1$ (d) $t^3 - 2t^2 + t + 15$

3. Determine which of the following polynomials are irreducible over the rational numbers:
   (a) $t^4 + 2$ (b) $t^4 - 2$ (c) $t^4 + 4$ (d) $t^4 - t + 1$

4. Let $f(t) = a_n t^n + \cdots + a_0$ be a polynomial of degree $n \geqq 1$ with integer coefficients, assumed relatively prime, and $a_n a_0 \neq 0$. If b/c is a rational number expressed as a quotient of relatively prime integers, b, $c \neq 0$, and if $f(b/c) = 0$, show that c divides a_n and b divides a_0. (This result allows us to determine effectively all possible rational roots of f since there is only a finite number of divisors of a_n and a_0.)

   Remark. The integral root theorem of Exercise 1 is a special case of the above statement.

5. Determine all rational roots of the following polynomials:
   (a) $t^7 - 1$ (b) $t^8 - 1$ (c) $2t^2 - 3t + 4$ (d) $3t^3 + t - 5$
   (e) $2t^4 - 4t + 3$

6. Let p be a prime number. Let

$$f(t) = t^{p-1} + t^{p-2} + \cdots + 1.$$

   Prove that $f(t)$ is irreducible in $\mathbf{Z}[t]$. [*Hint*: Observe that

$$f(t) = (t^p - 1)/(t - 1).$$

   Let $u = t - 1$ so $t = u + 1$. Use Eisenstein.]

The next two exercises show how to construct an irreducible polynomial over the rational numbers, of given degree d and having precisely $d - 2$ real roots (so of course a pair of complex conjugate roots). The construction will proceed in two steps. The first has nothing to do with the rational numbers.

7. **Continuity of the roots of a polynomial.** Let d be an integer $\geqq 3$. Let

$$f_n(t) = t^d + a_{d-1}^{(n)} t^{d-1} + \cdots + a_0^{(n)}$$

be a sequence of polynomials with complex coefficients $a_i^{(n)}$. Let

$$f(t) = t^d + a_{d-1} t^{d-1} + \cdots + a_0$$

be another polynomial in $\mathbf{C}[t]$. We shall say that $f_n(t)$ **converges** to $f(t)$ as $n \to \infty$ if for each $j = 0, \ldots, d-1$ we have

$$\lim_{n \to \infty} a_j^{(n)} = a_j.$$

Thus the coefficients of f_n converge to the coefficients of f.
Factorize f_n and f into factors of degree 1:

$$f(t) = (t - \alpha_1) \cdots (t - \alpha_d) \quad \text{and} \quad f_n(t) = (t - \alpha_1^{(n)}) \cdots (t - \alpha_d^{(n)}).$$

(a) Prove the following theorem.

Assume that $f_n(t)$ converges to $f(t)$, and for simplicity assume that the roots $\alpha_1, \ldots, \alpha_d$ are distinct. Then for each n we can order the roots $\alpha_1^{(n)}, \ldots, \alpha_d^{(n)}$ in such a way that for $i = 1, \ldots, d$ we have

$$\lim_{n \to \infty} \alpha_i^{(n)} = \alpha_i.$$

This shows that if the coefficients of f_n converge to the coefficients of f, then the roots of f_n converge to the roots of f.

(b) Suppose that f and f_n have real coefficients for all n. Assume that $\alpha_3, \ldots, \alpha_d$ are real, and α_1, α_2 are complex conjugate. Prove that for all n sufficiently large, $\alpha_i^{(n)}$ is real for $i = 3, \ldots, d$; and $\alpha_1^{(n)}$, $\alpha_2^{(n)}$ are complex conjugate.

8. Let d be an integer $\geqq 3$. Prove the existence of an irreducible polynomial of degree d over the rational numbers, having precisely $d - 2$ real roots (and a pair of complex conjugate roots). Use the following construction. Let $b_1, \ldots, b_{d-2}$ be distinct integers and let a be an integer > 0. Let

$$g(t) = (t^2 + a)(t - b_1) \cdots (t - b_{d-2}) = t^d + c_{d-1} t^{d-1} + \cdots + c_0.$$

Observe that $c_i \in \mathbf{Z}$. Let p be a prime number, and let

$$g_n(t) = g(t) + \frac{p}{p^{dn}},$$

so that $g_n(t)$ converges to $g(t)$.

(a) Prove that $g_n(t)$ has precisely $d-2$ real roots for n sufficiently large.
(b) Prove that $g_n(t)$ is irreducible over $\mathbf{Q}$.

(*Note*: You might use the preceding exercise for (a), but one can also give a simple proof just looking at the graph of g, using the fact that g_n is just a slight raising of g above the horizontal axis.

Obviously the same method can be used to construct irreducible polynomials over $\mathbf{Q}$ with arbitrarily many real roots and pairs of complex conjugate roots. There is particular significance for the special case of $d-2$ real roots, when d is a prime number, as you will see in Chapter VII, §4, Exercise 15. In this special case, you will then be able to prove that the Galois group of the polynomial is the full symmetric group.)

9. Let R be a factorial ring (see the definition in the next section), and let p be a prime element in R. Let d be an integer $\geqq 2$, and let

$$f(t) = t^d + c_{d-1}t^{d-1} + \cdots + c_0$$

be a polynomial with coefficients $c_i \in R$. Let n be an integer $\geqq 1$, and let

$$g(t) = f(t) + p/p^{nd}.$$

Prove that $g(t)$ is irreducible in $K[t]$, where K is the quotient field of R.

IV, §6. PRINCIPAL RINGS AND FACTORIAL RINGS

We have seen a systematic analogy between the ring of integers $\mathbf{Z}$ and the ring of polynomials $K[t]$. Both have a Euclidean algorithm; both have unique factorization into certain elements, which are called **primes** in $\mathbf{Z}$ or **irreducible polynomials** in $K[t]$. It turns out actually that the most important property is not the Euclidean algorithm, but another property which we now axiomatize.

Let R be an integral ring. We say that R is a **principal ring** if in addition every ideal of R is principal.

Examples. If $R = \mathbf{Z}$ or $R = K[t]$, then R is principal. For $\mathbf{Z}$ this was proved in Chapter I, Theorem 3.1, and for polynomials it was proved in Theorem 2.1 of the present chapter.

Practically all the properties which we have proved for $\mathbf{Z}$ or for $K[t]$ are also valid for principal rings. We shall now make a list.

Let R be an integral ring. Let $p \in R$ and $p \neq 0$. We define p to be **prime** if p is not a unit, and given a factorization

$$p = ab \quad \text{with } a, b \in R$$

then a or b is a unit.

An element $a \in R$, $a \neq 0$ is said to have **unique factorization** into primes if there exists a unit u and there exist prime elements p_i $(i = 1, \ldots, r)$ in R (not necessarily distinct) such that

$$a = up_1 \cdots p_r;$$

and if given two factorizations into prime elements

$$a = up_1 \cdots p_r = u'q_1 \cdots q_s,$$

then $r = s$ and after a permutation of the indices i, we have $p_i = u_i q_i$, where u_i is a unit $i = 1, \ldots, r$.

We note that if p is prime and u is a unit, then up is also prime, so we must allow multiplication by units in the factorization. In the ring of integers **Z**, the ordering allows us to select a representative prime element, namely a prime number, out of two possible ones differing by a unit, namely $\pm p$, by selecting the positive one. This is, of course, impossible in more general rings. However, in the ring of polynomials over a field, we can select the prime element to be the irreducible polynomial with leading coefficient 1.

A ring is called **factorial**, or a **unique factorization ring**, if it is integral, and if every element $\neq 0$ has a unique factorization into primes.

Let R be an integral ring, and a, $b \in R$, $a \neq 0$. We say that a **divides** b and write $a|b$ if there exists $c \in R$ such that $ac = b$. We say that $d \in R$, $d \neq 0$ is a **greatest common divisor** of a and b if $d|a$, $d|b$, and if any element c of R, $c \neq 0$ divides both a and b, then c also divides d. Note that a g.c.d. is determined only up to multiplication by a unit.

Proposition 6.1. *Let R be a principal ring. Let a, $b \in R$ and $ab \neq 0$. Let $(a, b) = (c)$, that is let c be a generator of the ideal (a, b). Then c is a greatest common divisor of a and b.*

Proof. Since b lies in the ideal (c), we can write $b = xc$ for some $x \in R$, so that $c|b$. Similarly, $c|a$. Let d divide both a and b, and write $a = dy$, $b = dz$ with y, $z \in R$. Since c lies in (a, b) we can write

$$c = wa + tb$$

with some w, $t \in R$. Then $c = wdy + tdz = d(wy + tz)$, whence $d|c$, and our proposition is proved.

Theorem 6.2. *Let R be a principal ring. Then R is factorial.*

Proof. We first prove that every non-zero element of R has a factorization into irreducible elements. Given $a \in R$, $a \neq 0$. If a is prime, we are done. If not, then $a = a_1 b_1$ where neither a_1 nor b_1 is a unit. Then $(a) \subset (a_1)$. We assert that

$$(a) \neq (a_1).$$

Indeed, if $(a) = (a_1)$ then $a_1 = ax$ for some $x \in R$ and then $a = axb_1$ so $xb_1 = 1$, whence both x, b_1 are units contrary to assumption. If both a_1, b_1 are prime, we are done. Suppose that a_1 is not prime. Then $a_1 = a_2 b_2$ where neither a_2 nor b_2 are units. Then $(a_1) \subset (a_2)$, and by what we have just seen, $(a_1) \neq (a_2)$. Proceeding in this way we obtain a chain of ideals

$$(a_1) \subsetneqq (a_2) \subsetneqq (a_3) \subsetneqq \cdots \subsetneqq (a_n) \subsetneqq \cdots.$$

We claim that actually, this chain must stop for some integer n. Let

$$J = \bigcup_{n=1}^{\infty} (a_n).$$

Then J is an ideal. By assumption J is principal, so $J = (c)$ for some element $c \in R$. But c lies in the ideal (a_n) for some n, and so we have the double inclusion

$$(a_n) \subset (c) \subset (a_n),$$

whence $(c) = (a_n)$. Therefore $(a_n) = (a_{n+1}) = \ldots$, and the chain of ideals could not have proper inclusions at each step. This implies that a can be expressed as a product

$$a = p_1 \cdots p_r \qquad \text{where } p_1, \ldots, p_r \text{ are prime.}$$

Next we prove the uniqueness.

Lemma 6.3. *Let R be a principal ring. Let p be a prime element. Let a, $b \in R$. If $p \mid ab$ then $p \mid a$ or $p \mid b$.*

Proof. If $p \nmid a$ then a g.c.d. of p, a is 1, and (p, a) is the unit ideal. Hence we can write

$$1 = xp + ya$$

with some x, $y \in R$. Then $b = bxp + yab$, and since $p \mid ab$, we conclude that $p \mid b$. This proves the lemma.

Suppose finally that a has two factorizations

$$a = p_1 \cdots p_r = q_1 \cdots q_s$$

into prime elements. Since p_1 divides the product furthest to the right, it follows by the lemma that p_1 divides one of the factors, which we may assume to be q_1 after renumbering these factors. Then there exists a unit u_1 such that $q_1 = u_1 p_1$. We can now cancel p_1 from both factorizations and get

$$p_2 \cdots p_r = u_1 q_2 \cdots q_s.$$

The argument is completed by induction. This concludes the proof of Theorem 6.2.

For emphasis, we state separately the following result.

Proposition 6.4. *Let R be a factorial ring. An element $p \in R$, $p \neq 0$ is prime if and only if the ideal (p) is prime.*

Note that for any integral ring R, we have the implication

$$a \in R,\ a \neq 0 \text{ and } (a) \text{ prime} \quad \Rightarrow \quad a \text{ prime}.$$

Indeed, if we write $a = bc$ with $b, c \in R$ then $b \in (a)$ or $c \in (a)$ by definition of a prime ideal. Say we can write $b = ad$ with some $d \in R$. Then $a = acd$. Hence $cd = 1$, whence c, d are units, and therefore a is prime.

In a factorial ring, we also have the converse, because of unique factorization. In a principal ring, the key step was Lemma 6.3, which means precisely that (p) is a prime ideal.

In a factorial ring, we can make the same definitions as for the integers or polynomials. If

$$a = up_1^{m_1} \cdots p_r^{m_r}$$

is a factorization with a unit u and distinct primes $p_1, \ldots, p_r$, then we define

$$m_i = \operatorname{ord}_{p_i}(a) = \textbf{order of } a \textbf{ at } p_i.$$

If a, $b \in R$ are non-zero elements, we say that a, b are **relatively prime** if the g.c.d. of a and b is a unit. Similarly elements $a_1, \ldots, a_m$ are **relatively prime** means that no prime element p divides all of them. *If R is a principal ring*, to say that a, b are relatively prime is equivalent to saying that the ideal (a, b) is the unit ideal.

Other theorems which we proved for the integers $\mathbf{Z}$ are also valid in factorial rings. We now make a list, and comment on the proofs, which are essentially identical with the previous ones, as in §5. Thus we now study factorial rings further.

Let R be factorial. Let $f(t) \in R[t]$ be a polynomial. As in the case when $R = \mathbf{Z}$, we say that f is **primitive** if there is no prime of R which divides all coefficients of f, i.e. if the coefficients of f are relatively prime.

Lemma 6.5. *Let R be a factorial ring. Let K be its quotient field. Let $f \in K[t]$ be a polynomial $\neq 0$. Then there exists an element $a \in K$, $a \neq 0$ such that af is primitive.*

Proof. Follow step by step the proof of Lemma 5.1. No other property was used besides the fact that the ring is factorial.

Lemma 6.6 (Gauss). *Let f, g be primitive polynomials over the factorial ring R. Then fg is primitive.*

Proof. Follow step by step the proof of Lemma 5.2.

Theorem 6.7. *Let R be a factorial ring and K its quotient field. Let $f \in R[t]$ be a primitive polynomial, and* $\deg f \geqq 1$. *If f is reducible over K then f is reducible over R. More precisely, if $f = gh$ with g, $h \in K[t]$ and* $\deg g \geqq 1$, $\deg h \geqq 1$, *then there exist elements a, $b \in K$ such that, if we let $g_1 = ag$ and $h_1 = bh$, then g_1, h_1 have coefficients in R, and $f = g_1 h_1$.*

Proof. Follow step by step the proof of Theorem 5.3. Of course, at the end of the proof, we found ab equal to a unit. When $R = \mathbf{Z}$, a unit is ± 1, but in the general case all we can say is that ab is a unit, so a is a unit or b is a unit. This is the only difference in the proof.

If R is a principal ring, it is usually not true that $R[t]$ is also a principal ring. We shall discuss this systematically in a moment when we consider polynomials in several variables. However, we shall prove that $R[t]$ is factorial, and we shall prove even more.

First we make a remark which will be used in the following proofs.

Lemma 6.8. *Let R be a factorial ring, and let K be its quotient field. Let*

$$g = g(t) \in R[t], \qquad h = h(t) \in R[t]$$

be primitive polynomials in R. Let $b \in K$ be such that

$$g = bh.$$

Then b is a unit of R.

Proof. Write $b = a/d$ where a, d are relatively prime elements of R, so a is a numerator and d is a denominator for b. Then

$$dg = ah.$$

Write $h(t) = c_m t^m + \cdots + c_0$. If p is a prime dividing d then p divides ac_j for $j = 0,\ldots,m$. Since $c_0,\ldots,c_m$ are relatively prime, it follows that p divides a, which is against our assumption that a, d are relatively prime in R. Similarly, no prime can divide a. Therefore b is a unit of R as asserted.

Theorem 6.9. *Let R be a factorial ring. Then $R[t]$ is factorial. The units of $R[t]$ are the units of R. The prime elements of $R[t]$ are either the primes of R, or the primitive irreducible polynomials in $R[t]$.*

Proof. Let p be a prime of R. If $p = ab$ in $R[t]$ then

$$\deg a = \deg b = 0,$$

so $a, b \in R$, and by hypothesis a or b is a unit. Hence p is also a prime of $R[t]$.

Let $p(t) = p$ be a primitive polynomial in $R[t]$ irreducible in $K[t]$. If $p = fg$ with $f, g \in R[t]$, then from unique factorization in $K[t]$ we conclude that $\deg f = \deg p$ or $\deg g = \deg p$. Say $\deg f = \deg p$. Then $g \in R$. Since the coefficients of $p(t)$ are relatively prime, it follows that g is a unit of R. Hence $p(t)$ is a prime element of $R[t]$.

Let $f(t) \in R[t]$. We can write $f = cg$ where c is the g.c.d. of the coefficients of f, and g then has relatively prime coefficients. We know that c has unique factorization in R by hypothesis. Let

$$g = q_1 \cdots q_r$$

be a factorization of g into irreducible polynomials $q_1,\ldots,q_r$ in $K[t]$. Such a factorization exists since we know that $K[t]$ is factorial. By Lemma 6.5 there exist elements $b_1,\ldots,b_r \in K$ such that if we let $p_i = b_i q_i$ then p_i has relatively prime coefficients in R. Let their product be

$$u = b_1 \cdots b_r.$$

Then

$$ug = p_1 \cdots p_r.$$

By the Gauss Lemma 6.6, the right-hand side is a polynomial in $R[t]$ with relatively prime coefficients. Since g is assumed to have relatively prime coefficients in R, it follows that $u \in R$ and u is a unit in R. Then

$$f = cu^{-1} p_1 \cdots p_r$$

is a factorization of f in $R[t]$ into prime elements of $R[t]$ and an element of R. Thus a factorization exists.

There remains to prove uniqueness (up to factors which are units, of course). Suppose that

$$f = cp_1 \cdots p_r = dq_1 \cdots q_s,$$

where $c, d \in R$, and $p_1, \ldots, p_r, q_1, \ldots, q_s$ are irreducible polynomials in $R[t]$ with relatively prime coefficients. If we read this relation in $K[t]$, and use the fact that $K[t]$ is factorial, as well as Theorem 6.7, then we conclude that after a permutation of indices, we have $r = s$ and there are elements $b_i \in K$, $i = 1, \ldots, r$ such that

$$p_i = b_i q_i \quad \text{for} \quad i = 1, \ldots, r.$$

Since p_i, q_i have relatively prime coefficients in R, it follows that in fact b_i is a unit in R by Lemma 6.8. This proves the uniqueness.

Theorem 6.10 (Eisenstein's criterion). *Let R be a factorial ring, and let K be its quotient field. Let*

$$f(t) = a_n t^n + \cdots + a_0$$

be a polynomial of degree $n \geqq 1$ with coefficients in R. Let p be a prime, and assume:

$$a_n \not\equiv 0 \pmod{p}, \qquad a_i \equiv 0 \pmod{p} \quad \text{for all} \quad i < n,$$

$$a_0 \not\equiv 0 \pmod{p^2}.$$

Then f is irreducible over K.

Proof. Follow step by step the proof of Theorem 5.4.

Example. Let F be any field. Let $K = F(t)$ be the quotient field of the ring of polynomials, and let $R = F[t]$. Then R is a factorial ring. Note that t itself is a prime element in this ring. For any element $c \in F$ the polynomial $t - c$ is also a prime element.

Let X be a variable. Then for every positive integer n, the polynomial

$$f(X) = X^n - t$$

is irreducible in $K[X]$. This follows from Eisenstein's Criterion. In this case, we let $p = t$, and:

$$a_n = 1 \not\equiv 0 \mod t, \qquad a_0 = -t, \qquad a_i = 0 \quad \text{for} \quad 0 < i < n.$$

Thus the hypotheses in Eisenstein's Criterion are satisfied.

Similarly, the polynomial

$$X^4 - (t-1)X^3 + (t-1)^8X^2 + t(t-1)^4X - (t-1)$$

is irreducible in $K[X]$. In this case, we let $p = t - 1$.

The analogy between the ring of integers $\mathbf{Z}$ and the ring of polynomials $F[t]$ is one of the most fruitful in mathematics.

Theorem 6.11 (Reduction criterion). *Let R be a factorial ring and let K be its quotient field. Let $f(t) \in R[t]$ be a primitive polynomial with leading coefficient a_n which is not divisible by a prime p of R. Let $R \to R/pR$ be reduction* mod p, *and denote the image of f by $\bar{f}$. Let F be the quotient field of R/pR. If $\bar{f}$ is irreducible in $F[t]$ then f is irreducible in $K[t]$.*

Proof. The proof is essentially similar to the proof of Theorem 5.5, but there is a slight added technical point due to the fact that R/pR is not necessarily a field. To deal with that point, we have to use the ring $R_{(p)}$ of Exercise 3, so we assume that you have done that exercise. Thus the ring $R_{(p)}$ is principal, and $R_{(p)}/pR_{(p)}$ is a field. Furthermore, $F = R_{(p)}/pR_{(p)}$. Now exactly the same argument as in Theorem 5.5 shows that if $\bar{f}$ is irreducible in $F[t]$, then f is irreducible in $R_{(p)}[t]$ and hence f is irreducible in $K[t]$, because K is also the quotient field of $R_{(p)}$. This concludes the proof.

IV, §6. EXERCISES

1. Let p be a prime number. Let R be the ring of all rational numbers m/n, where m, n are relatively prime integers, and $p \nmid n$.
   (a) What are the units of R?
   (b) Show that R is a principal ring. What are the prime elements of R?

2. Let F be a field. Let p be an irreducible polynomial in the ring $F[t]$, Let R be the ring of all rational functions $f(t)/g(t)$ such that f, g are relatively prime polynomials in $F[t]$ and $p \nmid g$.
   (a) What are the units of R?
   (b) Show that R is a principal ring. What are the prime elements of R?

3. If you are alert, you will already have generalized the first two exercises to the following. Let R be a factorial ring. Let p be a prime elment. Let $R_{(p)}$ be the set of all quotients a/b with $a, b \in R$ and b not divisible by p. Then:
   (a) The units of $R_{(p)}$ are the quotients a/b with $a, b \in R$, $p \nmid ab$.
   (b) The ring $R_{(p)}$ has a unique maximal ideal, namely $pR_{(p)}$, consisting of all elements a/b such that $p \nmid b$ and $p|a$.

(c) The ring $R_{(p)}$ is principal, and every ideal is of the form $p^m R_{(p)}$ for some integer $m \geqq 0$.
(d) The factor ring $R_{(p)}/pR_{(p)}$ is a field, which "is" the quotient field of R/pR.
If you have not already done so, then prove the above statements.

4. Let $R = \mathbf{Z}[t]$. Let p be a prime number. Show that $t - p$ is a prime element in R. Is $t^2 - p$ a prime element in R? What about $t^3 - p$? What about $t^n - p$ where n is a positive integer?

5. Let p be a prime number. Show that the ideal (p, t) is not principal in the ring $\mathbf{Z}[t]$.

6. **Trigonometric polynomials.** Let R be the ring of all functions f of the form

$$f(x) = a_0 + \sum_{k=1}^{n} (a_k \cos kx + b_k \sin kx)$$

where a_0, a_k, b_k are real numbers. Such a function is called a **trigonometric polynomial**.
(a) Prove that R is a ring.
(b) If a_n or $b_n \neq 0$ define n to be the **(trigonometric) degree** of f. Prove that if f, g are trigonometric polynomials, then

$$\deg(fg) = \deg f + \deg g.$$

Deduce that R has no divisors of zero, so is integral.
(c) Conclude that the functions $\sin x$, $1 + \cos x$, $1 - \cos x$ are prime elements in the ring. As Hale Trotter observed (*Math. Monthly*, April 1988) the relation

$$\sin^2 x = (1 + \cos x)(1 - \cos x)$$

is an example of non-unique factorization into prime elements.

7. Let R be the subset of the polynomial ring $\mathbf{Q}[t]$ consisting of all polynomials $a_0 + a_2 t^2 + a_3 t^3 + \cdots + a_n t^n$ (so the term of degree 1 is missing), with $a_i \in \mathbf{Q}$.
(a) Show that R is a ring, and that the ideal (t^2, t^3) is not principal.
(b) Show that R is not factorial.

8. Let R be the set of numbers of the form

$$a + b\sqrt{-5} \qquad \text{with} \quad a, b \in \mathbf{Z}.$$

(a) Show that R is a ring.
(b) Show that the map $a + b\sqrt{-5} \mapsto a - b\sqrt{-5}$ of R into itself is an automorphism.
(c) Show that the only units of R are ± 1.
(d) Show that 3, $2 + \sqrt{-5}$ and $2 - \sqrt{-5}$ are prime elements, and give a non-unique factorization

$$3^2 = (2 + \sqrt{-5})(2 - \sqrt{-5}).$$

(e) Similarly, show that 2, $1+\sqrt{-5}$ and $1-\sqrt{-5}$ are prime elements, which give the non-unique factorization

$$2\cdot 3=(1+\sqrt{-5})(1-\sqrt{-5}).$$

9. Let d be a positive integer which is not a square of an integer, so in particular, $d \geqq 2$. Let R be the ring of all numbers $a+b\sqrt{-d}$ with $a, b\in \mathbf{Z}$. Let $\alpha=a+b\sqrt{-d}$ be an element of this ring, and let $\bar{\alpha}$ be its complex conjugate.
   (a) Prove that α is a unit in R if and only if $\alpha\bar{\alpha}=1$.
   (b) If $\alpha\bar{\alpha}=p$, where p is a prime number, prove that α is a prime element of R.

10. Let p be an odd prime number, and suppose $p=\alpha\bar{\alpha}$, with $\alpha\in\mathbf{Z}[\sqrt{-1}]$. Prove that $p\equiv 1 \bmod 4$. (The converse is also true, but is more difficult to prove, in other words: if $p\equiv 1 \bmod 4$ then $p=\alpha\bar{\alpha}$ with some $\alpha\in\mathbf{Z}[\sqrt{-1}]$.)

11. Determine whether $3-2\sqrt{2}$ is a square in the ring $\mathbf{Z}[\sqrt{2}]$.

IV, §7. POLYNOMIALS IN SEVERAL VARIABLES

In this section, we study the most common example of a ring which is factorial but not principal.

Let F be a field. We know that $F[t]$ is a principal ring. Let $F[t]=R$. Let $t=t_1$, and let t_2 be another variable. Since $F[t]$ is factorial, it follows from Theorem 6.9 that $R[t_2]$ is factorial. Similarly,

$$F[t_1][t_2]\dots[t_n]$$

is factorial. This ring is usually denoted by

$$F[t_1,\dots,t_n],$$

and its elements are called **polynomials in n variables**. Every element of $F[t_1,\dots,t_n]$ can be written as a sum

$$f(t_1,\dots,t_n)=\sum_{i_n=0}^{d_n}\left(\sum_{i_1,\dots,i_{n-1}} a_{i_1,\dots,i_n}t_1^{i_1}\cdots t_{n-1}^{i_{n-1}}\right)t_n^{i_n}$$

and we thus see that f can be written

$$f(t_1,\dots,t_n)=\sum_{j=0}^{d_n} f_j(t_1,\dots,t_{n-1})t_n^j,$$

where f_j are polynomials in $n-1$ variables.

As a matter of notation, it is useful to abbreviate

$$a_{i_1 \cdots i_n} = a_{(i)}$$

and write

$$f(t_1, \dots, t_n) = \sum_{(i)} a_{(i)} t_1^{i_1} \cdots t_n^{i_n}.$$

The sum is taken separately over all the indices

$$0 \leqq i_1 \leqq d_1, \quad \dots, \quad 0 \leqq i_n \leqq d_n.$$

Sometimes one also abbreviates

$$t_1^{i_1} \cdots t_n^{i_n} = t^{(i)}$$

and one writes the polynomial in the form

$$f(t) = \sum_{(i)} a_{(i)} t^{(i)}.$$

The ring $F[t_1, \dots, t_n]$ is definitely *not* a principal ring for $n \geqq 2$. For instance, let $n = 2$. The ideal

$$M = (t_1, t_2)$$

generated by t_1 and t_2 in $F[t_1 t_2]$ is not principal. (Prove this as an exercise.) This ideal (t_1, t_2) is a maximal ideal, whose residue class field is F itself. You can prove this statement as an exercise. Similarly, the ideal $(t_1, \dots, t_n)$ is maximal in $F[t_1, \dots, t_n]$ and its residue class field is F.

Let f be a polynomial in n variables, and write

$$f(t_1, \dots, t_n) = \sum c_{(i)} t_1^{i_1} \cdots t_n^{i_n}, \qquad \text{where} \quad c_{(i)} = c_{i_1 \cdots i_n}.$$

We call each term

$$c_{(i)} t_1^{i_1} \cdots t_n^{i_n}$$

a **monomial**, and if $c_{(i)} \neq 0$ we define the **degree** of this monomial to be the sum of the exponents, that is

$$\deg(t_1^{i_1} \cdots t_n^{i_n}) = i_1 + \cdots + i_n.$$

A polynomial f as above is said to be **homogeneous of degree** d if all the terms with $c_{(i)} \neq 0$ have the property that

$$i_1 + \cdots + i_n = d.$$

Example. The monomial $5t_1^3 t_2 t_3^4$ has degree 8. The polynomial

$$7t_1^8 t_2 t_3^4 - \pi t_1^6 t_2 t_3$$

is not homogeneous. The polynomial

$$7t_1^4 t_2^3 t_3 - \pi t_1^2 t_2^4 t_3^2$$

is homogeneous of degree 8.

Given a polynomial $f(t_1, \dots, t_n)$ in n variables, we can write f as a sum

$$f = f_0 + f_1 + \cdots + f_d$$

where f_k is homogeneous of degree k or is 0. By convention, we agree that the zero polynomial has degree $-\infty$, because some of the terms f_k may be 0. To write f in the above fashion, all we have to do is collect together all the monomials of the same degree, that is

$$f_k = \sum_{i_1 + \cdots + i_n = k} c_{(i)} t_1^{i_1} \cdots t_n^{i_n}.$$

If $f_d \neq 0$ is the term of highest homogeneous degree in the sum for f above, then we say that f has **total degree** d, or simply **degree** d. Just as for polynomials in one variable, we have the property:

Theorem 7.1. *Let f, g be polynomials in n variables, and $f \neq 0$, $g \neq 0$. Then*

$$\deg(fg) = \deg f + \deg g.$$

Proof. Write $f = f_0 + \cdots + f_d$ and $g = g_0 + \cdots + g_e$ with $f_d \neq 0$ and $g_e \neq 0$. Then

$$fg = f_0 g_0 + \cdots + f_d g_e$$

and since the polynomial ring is integral, $f_d g_e \neq 0$. But $f_d g_e$ is the homogeneous part of f of highest degree, so $\deg(fg) = d + e$, as was to be shown.

Remark. Do not confuse the **degree** (i.e. the total degree) and the **degree of f in each variable**. If f can be written in the form

$$f(t_1, \dots, t_n) = \sum_{k=0}^{d} f_k(t_1, \dots, t_{n-1}) t_n^k$$

and $f_d \neq 0$ as a polynomial in $t_1, \ldots, t_{n-1}$ then we say that f has **degree d in t_n**. For instance, the polynomial

$$\pi t_1^3 t_2 t_3^5 + t_1 t_2 t_3 + t_3^4$$

has total degree 9, it has degree 3 in t_1, degree 1 in t_2 and degree 5 in t_3.

From Theorem 7.1, or from the fact that the ring of polynomials in several variables is

$$K[t_1, \ldots, t_n] = K[t_1][t_2] \cdots [t_n]$$

we conclude that this ring is integral. Thus we may form its quotient field, just as in the case of one variable. We denote this quotient field by

$$K(t_1, \ldots, t_n)$$

and call it the field of **rational functions** (in several variables).

The above results go as far as we wish in our study of polynomials. We end this section with some comments on polynomial functions.

Let $K^n = K \times \cdots \times K$. Just as in the case of one variable, a polynomial $f(t) \in K[t_1, \ldots, t_n]$ may be viewed as a function

$$f_{K^n}\colon K^n \to K.$$

Indeed, let $x = (x_1, \ldots, x_n)$ be an n-tuple in K^n. If

$$f(t_1, \ldots, t_n) = f(t) = \sum a_{(i)} t_1^{i_1} \cdots t_n^{i_n}$$

then we define

$$f(x_1, \ldots, x_n) = f(x) = \sum a_{(i)} x_1^{i_1} \cdots x_n^{i_n}.$$

The map $x \mapsto f(x)$ is a function of K^n into K. But also the map

$$f \mapsto f(x)$$

is a homomorphism of $K[t_1, \ldots, t_n]$ into K, called the **evaluation at** x.

More generally, let K be a subring of a commutative ring A. Let $f(t) \in K[t_1, \ldots, t_n]$ be a polynomial. If

$$x = (x_1, \ldots, x_n) \in A^n$$

is an n-tuple in A^n then again we may define $f(x)$ by the same expression, and we obtain a function

$$f_{A^n}\colon A^n \to A \qquad \text{by} \qquad x \mapsto f(x).$$

Just as in the case of one variable, different polynomials can give rise to the same function. We gave an example of this phenomenon in §1, in the context of a finite field.

Theorem 7.2. *Let K be an infinite field. Let $f \in K[t_1,\dots,t_n]$ be a polynomial in n variables. If the corresponding function*

$$x \mapsto f(x) \qquad \textit{of} \qquad K^n \to K$$

is the zero function, then f is the zero polynomial. Furthermore, if f, g are two polynomials giving the same function of K^n into K, then $f = g$.

Proof. The second assertion is a consequence of the first. Namely, given f, g which give the same function of K^n into K, let $h = f - g$. Then h gives the zero function, so $h = 0$ and $f = g$. We now prove the first by induction. Write

$$f(t_1,\dots,t_n) = \sum_{j=0}^{d} f_j(t_1,\dots,t_{n-1})t_n^j.$$

Thus we write f as a polynomial in t_n, with coefficients which are polynomials in $t_1,\dots,t_{n-1}$. Let $(c_1,\dots,c_{n-1})$ be arbitrary elements of K. By assumption, the polynomial

$$f(c_1,\dots,c_{n-1},t) = \sum_{j=0}^{d} f_j(c_1,\dots,c_{n-1})t^j$$

vanishes when we substitute any element c of K for t. In other words, this polynomial in one variable t has infinitely many roots in K. Therefore this polynomial is identically zero. This means that

$$f_j(c_1,\dots,c_{n-1}) = 0 \qquad \text{for all} \quad j = 0,\dots,d$$

and all $(n-1)$-tuples $(c_1,\dots,c_{n-1})$ in K^{n-1}. By induction, it follows that for each j the polynomial $f_j(t_1,\dots,t_{n-1})$ is the zero polynomial, whence finally $f = 0$. This concludes the proof.

Let R be a subring of a commutative ring S. Let $x_1,\dots,x_n$ be elements of S. We denote by

$$R[x_1,\dots,x_n]$$

the ring consisting of all elements $f(x_1, \ldots, x_n)$ where f ranges over all polynomials in n variables with coefficients in R. We say that

$$R[x_1, \ldots, x_n]$$

is the ring **generated by** $x_1, \ldots, x_n$ **over** R.

Example. Let $\mathbf{R}$ be the real numbers, and let φ, ψ be the two functions

$$\varphi(x) = \sin x \quad \text{and} \quad \psi(x) = \cos x.$$

Then

$$\mathbf{R}[\varphi, \psi]$$

is the subring of all functions (even the subring of all differentiable functions) generated by φ and ψ. In fact, $\mathbf{R}[\varphi, \psi]$ is the ring of trigonometric polynomials as in Exercise 6 of §6.

Let K be a subfield of a field E. Let $x_1, \ldots, x_n$ be elements of E. As we have seen, we get an evaluation homomorphism

$$K[t_1, \ldots, t_n] \to E \quad \text{by} \quad f(t_1, \ldots, t_n) \mapsto f(x_1, \ldots, x_n) = f(x).$$

If the kernel of this homomorphism is 0, that is if the evaluation map is injective, then we say that $x_1, \ldots, x_n$ are **algebraically independent**, or that they are **independent variables** over K. The polynomial ring $K[t_1, \ldots, t_n]$ is then isomorphic to the ring $K[x_1, \ldots, x_n]$ generated by $x_1, \ldots, x_n$ over K.

Example. It can be shown that if $K = \mathbf{Q}$, there always are infinitely many algebraically independent n-tuples $(x_1, \ldots, x_n)$ in the complex, or in the real numbers.

In practice, it is quite hard to determine whether two given numbers are or are not algebraically independent over $\mathbf{Q}$. For instance, let e be the natural base of logarithms. It is not known if e and π are algebraically independent. It is not even known if e/π is irrational.

IV, §7. EXERCISES

1. Let F be a field. Show that the ideal (t_1, t_2) is not principal in the ring $F[t_1, t_2]$. Similarly, show that $(t_1, \ldots, t_n)$ is not principal in the ring $F[t_1, \ldots, t_n]$.
2. Show that the polynomial $t_1 - t_2$ is irreducible in the ring $F[t_1, t_2]$.

3. In connection with Theorem 7.2 one can ask to what extent it is necessary to have an infinite field. Or even over the real numbers, let S be a subset of $\mathbf{R}^n$. Let $f(t_1, \dots, t_n)$ be a polynomial in n variables. Suppose that $f(x) = 0$ for all $x \in S$. Can we assert that $f = 0$? Not always, not even if S is infinite and $n \geqq 2$. Prove the following statement which gives a first result in this direction.

   Let K be a field, and let $S_1, \dots, S_n$ be finite subsets of K. Let $f \in K[t_1, \dots, t_n]$ and suppose that the degree of f in the variable t_i is $\leqq d_i$. Assume that $\#(S_i) > d_i$ for $i = 1, \dots, n$. Also suppose that

   $$f(x_1, \dots, x_n) = 0 \qquad \text{for all} \quad x_i \in S_i,$$

   that is $f(x) = 0$ for all $x \in S_1 \times \cdots \times S_n$. Then $f = 0$.

The proof will follow the same pattern as the proof of Theorem 7.2.

Application to finite fields

4. Prove the following analogue of Theorem 7.2 for finite fields.

   Let K be a finite field with q elements. Let $f(t) \in K[t_1, \dots, t_n]$ be a polynomial in n variables, such that the degree of f in each variable t_i is $< q$. Assume that

   $$f(a_1, \dots, a_n) = 0 \qquad \text{for all} \quad a_1, \dots, a_n \in K.$$

   Then f is the zero polynomial.

   Let K be a finite field and let $f(t) \in K[t_1, \dots, t_n]$ be a polynomial. If t_i^q occurs in some monomial in f, replace t_i^q by t_i. After doing this a finite number of times, you obtain a polynomial $g(t_1, \dots, t_n)$ such that the degree of g in each variable t_i is $< q$. Since $x^q = x$ for all $x \in K$, it follows that f, g induce the same function of K^n into K. By a **reduced polynomial associated with** f we mean a polynomial g such that the degree of g in each variable is $< q$ and such that the functions induced by f and g on K^n are equal. The existence of a reduced polynomial was proved above.

5. Given a polynomial $f \in K[t_1, \dots, t_n]$ prove that a reduced polynomial associated with f is unique, i.e. there is only one.

6. **Chevalley's theorem.** Let K be a finite field with q elements. Let

   $$f(t_1, \dots, t_n) \in K[t_1, \dots, t_n]$$

   be a polynomial in n variables of total degree d. Suppose that the constant term of f is 0, that is $f(0, \dots, 0) = 0$. Assume that $n > d$. Then there exist elements $a_1, \dots, a_n \in K$ not all 0 such that $f(a_1, \dots, a_n) = 0$. [*Hint*: Think of the exercises of §1. Compare the polynomials

   $$1 - f^{q-1} \qquad \text{and} \qquad (1 - t_1^{q-1}) \cdots (1 - t_n^{q-1}).$$

   and their degrees.]

IV, §8. SYMMETRIC POLYNOMIALS

Let R be an integral ring and let $t_1,\dots,t_n$ be algebraically independent elements over R. Let X be a variable over $R[t_1,\dots,t_n]$. We form the polynomial

$$\begin{aligned} P(X) &= (X - t_1)\cdots(X - t_n) \\ &= X^n - s_1X^{n-1} + \cdots + (-1)^n s_n \end{aligned}$$

where each $s_i = s_i(t_1,\dots,t_n)$ is a polynomial in $t_1,\dots,t_n$. Then, for instance,

$$s_1 = t_1 + \cdots + t_n \quad \text{and} \quad s_n = t_1 \cdots t_n.$$

The polynomials $s_1,\dots,s_n$ are called the **elementary symmetric polynomials** of $t_1,\dots,t_n$.

We leave it as an easy exercise to verify that s_i **is homogeneous of degree** i **in** $t_1,\dots,t_n$.

Let σ be a permutation of the integers $(1,\dots,n)$. Given a polynomial $f(t) \in R[t] = R[t_1,\dots,t_n]$, we define σf to be

$$\sigma f(t_1,\dots,t_n) = f(t_{\sigma(1)},\dots,t_{\sigma(n)}).$$

If σ, τ are two permutations, then $(\sigma\tau)f = \sigma(\tau f)$, and hence the symmetric group G on n letters operates on the polynomial ring $R[t]$. A polynomial is called **symmetric** if $\sigma f = f$ for all $\sigma \in G$. It is clear that the set of symmetric polynomials is a subring of $R[t]$, which contains the constant polynomials (i.e. R itself) and also contains the elementary symmetric polynomials $s_1,\dots,s_n$. We shall see below that these are generators.

Let $X_1,\dots,X_n$ be variables. We define the **weight** of a monomial

$$X_1^{k_1}\cdots X_n^{k_n}$$

to be $k_1 + 2k_2 + \cdots + nk_n$. We define the **weight** of a polynomial $g(X_1,\dots,X_n)$ to be the maximum of the weights of the monomials occurring in g.

Theorem 8.1. *Let $f(t) \in R[t_1,\dots,t_n]$ be symmetric of degree d. Then there exists a polynomial $g(X_1,\dots,X_n)$ of weight $\leqq d$ such that*

$$f(t) = g(s_1,\dots,s_n).$$

Proof. By induction on n. The theorem is obvious if $n = 1$, because $s_1 = t_1$.

Assume the theorem proved for polynomials in $n - 1$ variables.

If we substitute $t_n = 0$ in the expression for $P(X)$, we find

$$(X - t_1)\cdots(X - t_{n-1})X = X^n - (s_1)_0X^{n-1} + \cdots + (-1)^{n-1}(s_{n-1})_0X$$

where $(s_i)_0$ is the expression obtained by substituting $t_n = 0$ in s_i. We see that $(s_1)_0,\ldots,(s_{n-1})_0$ are precisely the elementary symmetric polynomials in $t_1,\ldots,t_{n-1}$.

We now carry out induction on d. If $d = 0$, our assertion is trivial. Assume $d > 0$, and assume our assertion proved for polynomials of degree $< d$. Let $f(t_1,\ldots,t_n)$ have degree d. There exists a polynomial $g_1(X_1,\ldots,X_{n-1})$ of weight $\leqq d$ such that

$$f(t_1,\ldots,t_{n-1},0) = g_1\big((s_1)_0,\ldots,(s_{n-1})_0\big).$$

We note that $g_1(s_1,\ldots,s_{n-1})$ has degree $\leqq d$ in $(t_1,\ldots,t_n)$. The polynomial

$$f_1(t_1,\ldots,t_n) = f(t_1,\ldots,t_n) - g_1(s_1,\ldots,s_{n-1})$$

has degree $\leqq d$ (in $t_1,\ldots,t_n$) and is symmetric. We have

$$f_1(t_1,\ldots,t_{n-1},0) = 0.$$

Hence f_1 is divisible by t_n, i.e. contains t_n as a factor. Since f_1 is symmetric, it contains $t_1\cdots t_n$ as a factor. Hence

$$f_1 = s_nf_2(t_1,\ldots,t_n)$$

for some polynomial f_2, which must be symmetric, and whose degree is $\leqq d - n < d$. By induction, there exists a polynomial g_2 in n variables and weight $\leqq d - n$ such that

$$f_2(t_1,\ldots,t_n) = g_2(s_1,\ldots,s_n).$$

We obtain

$$f(t) = g_1(s_1,\ldots,s_{n-1}) + s_ng_2(s_1,\ldots,s_n),$$

and each term on the right has weight $\leqq d$. This proves our theorem.

Theorem 8.2. *The elementary symmetric polynomials $s_1,\ldots,s_n$ are algebraically independent over R.*

Proof. If they are not, take a polynomial $f(X_1,\ldots,X_n) \in R[X_1,\ldots,X_n]$ of least degree and not equal to 0 such that

$$f(s_1,\ldots,s_n) = 0.$$

Write f as a polynomial in X_n with coefficients in $R[X_1,\dots,X_{n-1}]$,

$$f(X_1,\dots,X_n) = f_0(X_1,\dots,X_{n-1}) + \cdots + f_d(X_1,\dots,X_{n-1})X_n^d.$$

Then $f_0 \neq 0$. Otherwise, we can write

$$f(X) = X_n h(X)$$

with some polynomial h, and hence $s_n h(s_1,\dots,s_n) = 0$. From this it follows that $h(s_1,\dots,s_n) = 0$, and h has degree smaller than the degree of f.

We substitute s_i for X_i in the above relation, and get

$$0 = f_0(s_1,\dots,s_{n-1}) + \cdots + f_d(s_1,\dots,s_{n-1})s_n^d.$$

This is a relation in $R[t_1,\dots,t_n]$, and we substitute 0 for t_n in this relation. Then all terms become 0 except the first one, which gives

$$0 = f_0((s_1)_0,\dots,(s_{n-1})_0),$$

using the same notation as in the proof of Theorem 8.1. This is a non-trivial relation between the elementary symmetric polynomials in $t_1,\dots,t_{n-1}$, a contradiction.

Example. Consider the product

$$\Delta(t_1,\dots,t_n) = \Delta(t) = \prod_{i<j} (t_i - t_j).$$

For any permutation σ of $(1,\dots,n)$, as in Chapter II, Theorem 6.4, we see that

$$\sigma\Delta(t) = \pm\Delta(t).$$

Hence $\Delta(t)^2$ is symmetric, and we call it the **discriminant**:

$$D_f(s_1,\dots,s_n) = D(s_1,\dots,s_n) = \prod_{i<j} (t_i - t_j)^2 = \Delta(t)^2.$$

We thus view the discriminant as a polynomial in the elementary symmetric functions.

Let F be a field, and let

$$P(X) = (X - \alpha_1)\cdots(X - \alpha_n) = X^n - c_1X^{n-1} + \cdots + (-1)^n c_n$$

be a polynomial in $F[X]$ with roots $\alpha_1,\dots,\alpha_n \in F$. Then there is a unique homomorphism

$$\mathbf{Z}[t_1,\dots,t_n] \to F \qquad \text{mapping} \qquad t_i \mapsto \alpha_i.$$

This homomorphism is the composite

$$\mathbf{Z}[t_1, \ldots, t_n] \to F[t_1, \ldots, t_n] \to F,$$

where the first map is induced by the unique homomorphism $\mathbf{Z} \to F$, and the second map is the evaluation homomorphism sending $t_i \mapsto \alpha_i$.

Under this homomorphism, we see that

$$\Delta(t_1, \ldots, t_n)^2 \mapsto \Delta(\alpha_1, \ldots, \alpha_n)^2$$

and so

$$D(s_1, \ldots, s_n) \mapsto D(c_1, \ldots, c_n).$$

Therefore, to get a formula for the discriminant of a polynomial, it suffices to find the formula for a polynomial over the integers $\mathbf{Z}$, with algebraically independent roots $t_1, \ldots, t_n$. We shall now find such a formula for polynomials of degree 2 and 3.

Example (Quadratic polynomials). Let

$$f(X) = X^2 + bX + c = (X - t_1)(X - t_2).$$

Then by direct computation, you will find that

$$\boxed{D_f = b^2 - 4c.}$$

Example (Cubic polynomials). Consider a cubic polynomial

$$f(X) = X^3 - s_1 X^2 + s_2 X - s_3 = (X - t_1)(X - t_2)(X - t_3).$$

We want to find a formula for the discriminant, which is more complicated than in the quadratic case. There is such a formula, but in practice, it is best to make a change of variables first and reduce the problem to the case of a cubic whose X^2 term is missing. Namely, let

$$Y = X - \tfrac{1}{3}s_1 \qquad \text{so} \qquad X = Y + \tfrac{1}{3} = Y + \tfrac{1}{3}(t_1 + t_2 + t_3).$$

Then the polynomial $f(X)$ becomes

$$f(X) = f^*(Y) = Y^3 + aY + b = (Y - u_1)(Y - u_2)(Y - u_3)$$

where $a = u_1 u_2 + u_2 u_3 + u_1 u_3$ and $b = -u_1 u_2 u_3$, while $u_1 + u_2 + u_3 = 0$. We have

$$u_i = t_i - \tfrac{1}{3}s_1 \qquad \text{for} \quad i = 1, 2, 3.$$

Note that the discriminant is unchanged because the translation by $s_1/3$ cancels out. Indeed,

$$u_i - u_j = t_i - t_j \qquad \text{for all } i \neq j.$$

If we can get a formula for the discriminant of the cubic whose square term is missing, as a function of a and b, then we can get a formula for the discriminant of the general cubic simply by substituting the values of a and b as functions of s_1, s_2, s_3. You will work this out as Exercise 1.

We now consider the cubic whose square term is missing. So we let

$$f(X) = X^3 + aX + b = (X - u_1)(X - u_2)(X - u_3)$$

where u_1, u_2 are independent variables, and $u_3 = -(u_1 + u_2)$. Then the discriminant is

$$D = (u_1 - u_2)^2(u_1 - u_3)^2(u_2 - u_3)^2.$$

As a function of the elementary symmetric functions a, b the discriminant is

$$\boxed{D = -4a^3 - 27b^2.}$$

As an exercise, try to determine this by brute force. We shall now give a proof which eliminates the brute force.

Observe first that D is homogeneous of degree 6 in u_1, u_2. Furthermore, a is homogeneous of degree 2 and b is homogeneous of degree 3. By Theorem 8.1, we know that there exists some polynomial

$$g(X_2, X_3) \qquad \text{of weight 6}$$

such that $D = g(a, b)$. The only monomials $X_2^m X_3^n$ of weight 6, i.e. such that

$$2m + 3n = 6 \qquad \text{with integers } m, n \geqq 0$$

are those for which $m = 3$, $n = 0$ or $m = 0$, $n = 2$. Hence

$$g(X_2, X_3) = vX_2^3 + wX_3^2$$

where v, w are integers which must now be determined.

Observe that the integers v, w are universal, in the sense that for any special polynomial with special values of a, b, its discriminant will be given by

$$g(a, b) = va^3 + wb^2.$$

Consider the polynomial

$$f_1(X) = X(X-1)(X+1) = X^3 - X.$$

Then $a = -1$, $b = 0$ and $D_{f_1} = va^3 = -v$. But also $D_{f_1} = 4$ by using the definition of the discriminant as the product of the differences of the roots, squared. Hence we get

$$v = -4.$$

Next consider the polynomial

$$f_2(X) = X^3 - 1.$$

Then $a = 0$, $b = -1$, and $D_{f_2} = wb^2 = w$. But the three roots of f_2 are the cube roots of unity, namely

$$1, \frac{-1+\sqrt{-3}}{2}, \frac{-1-\sqrt{-3}}{2}.$$

Using the definition of the discriminant as the product of the differences of the roots, squared, we find the value $D_{f_2} = -27$. Hence we get

$$w = -27.$$

This concludes the proof of the formula for the discriminant of the cubic.

IV, §8. EXERCISES

1. Let $f(X) = X^3 + a_1X^2 + a_2X + a_3$. Show that the discriminant of f is

$$a_1^2a_2^2 - 4a_2^3 - 4a_1^3a_3 - 27a_3^2 + 18a_1a_2a_3.$$

[Reduce the question to the case of a polynomial $Y^3 + aY + b$, and use the formula for this special case.]

2. Try to work out the formula for the discriminant of $X^3 + aX + b$ by brute force.

3. Show that the discriminant of a polynomial is 0 if and only if the polynomial has a root of multiplicity > 1. (You may assume that the polynomial has coefficients in an algebraically closed field.)

4. Let $f(X) = (X - \alpha_1)\cdots(X - \alpha_n)$. Show that

$$D_f = (-1)^{n(n-1)/2}\prod_{j=1}^{n} f'(\alpha_j).$$

IV, §9. THE MASON-STOTHERS THEOREM

Bibliographical references will occur at the end of §10.

The first part of this section presents a theorem for polynomials discovered in 1981 by Stothers [Sto 81]. He used fundamental tools from algebraic geometry, which give a lot of insight not only in the particular theorem but into the possibilities of extending it to more general situations. However, partly because of the depth of the method, few people were aware of Stothers' result at the time. An elementary proof was discovered by Mason in 1983 [Mas 83]. Then Noah Snyder gave the simplest known proof [Sny 00], and we present his proof here.

We first work over an algebraically closed field of characteristic 0, the complex numbers if you wish.

Let $f(t)$ be a non-zero polynomial, with its factorization

$$f(t) = c\prod_{i=1}^{r}(t-\alpha_i)^{m_i} = c(t-\alpha_1)^{m_1}\cdots(t-\alpha_r)^{m_r}, \tag{1}$$

with a non-zero constant c, and the distinct roots α_i $(i = 1, \ldots, r)$. As before, we call m_i the **multiplicity** of α_i. Let α be a constant. If α is not a root, that is $\alpha \neq \alpha_i$ for all i, then $f(\alpha) \neq 0$. Suppose α is a root. It is convenient to write the factorization of $f(t)$ in the form

$$f(t) = (t-\alpha)^{m(\alpha)}g(t)$$

where $g(\alpha) \neq 0$ and $m(\alpha)$ is the multiplicity of α. If $\alpha = \alpha_k$ for some index k, then

$$f(t) = c(t-\alpha_k)^{m_k}\prod_{i\neq k}(t-\alpha_i)^{m_i},$$

and $g(t) = c\prod_{i\neq k}(t-\alpha_i)^{m_i}$.

We defined α to be a **multiple root** of f if and only if $m(\alpha) \geqq 2$. If $m(\alpha) = 1$, we say that α is a **simple root** of f.

Directly from the definitions, if $f(t) = (t-\alpha)^{m(\alpha)}$ with the multiplicity $m(\alpha) \geqq 1$, then the multiplicity of α in $f'(t)$ is $m(\alpha) - 1$. We generalize this statement to an arbitrary polynomial.

Lemma 9.1. *Let $f(t)$ be a polynomial over an algebraically closed field. Let α be a root of f with multiplicity $m(\alpha)$. Then the multiplicity of α in $f'(t)$ is $m(\alpha) - 1$.*

Proof. Write $f(t) = (t-\alpha)^m g(t)$ with $g(\alpha) \neq 0$. By the rule for the derivative of a product, we get

$$f'(t) = (t-\alpha)^m g'(t) + m(t-\alpha)^{m-1} g(t)$$
$$= (t-\alpha)^{m-1}\big((t-\alpha)g'(t) + mg(t)\big) = (t-\alpha)^{m-1} h(t)$$

where $h(t) = \big((t-\alpha)g'(t) + mg(t)\big)$. Note that $h(\alpha) = mg(\alpha) = g(\alpha) \neq 0$. Hence $(X-\alpha)^{m-1}$ is the highest power of $(t-\alpha)$ dividing $f'(t)$, so $m-1$ is the multiplicity of α in $f'(t)$, as was to be shown.

Define $n_0(f)$ = number of distinct roots of f, so $n_0(f) = r$ in the factorization (1).

Proposition 9.2. *Let f be a non-constant polynomial. Suppose that $f(t)$ has the factorization* (1). *Then the g.c.d.(f, f') is*

$$g.c.d.(f, f') = c_1 \prod_{i=1}^{r} (t-\alpha_i)^{m_i - 1},$$

with some constant c_1. In particular,

$$\deg(f, f') = \deg f - n_0(f).$$

Proof. The only prime polynomials which occur in the factorization of the greatest common divisor of f, f' must be among the prime polynomials $t - \alpha_i$. Lemma 9.1 gives us the multiplicities, so the factorization is as stated. The degree comes from subtracting 1 for each i, so the formula for $\deg(f, f')$ falls out.

The results of this section so far determining the g.c.d. of a polynomial and its derivative was carried out over an algebraically closed field, when the polynomial factors into irreducible factors of degree 1. One can carry out essentially the same arguments over any field F, especially over the rational numbers themselves. In this more general case, irreducible (prime) polynomials may have a degree > 1, and this degree must be taken into account. Suppose f has the factorization

$$f(t) = c \prod_{i=1}^{r} p_i(t)^{m_i}, \tag{2}$$

with a constant $c \neq 0$ and prime polynomials in $F[t]$. We define

$$n_0^F(f) = \sum_{i=1}^{r} \deg p_i.$$

There is another notation which simplifies the use of indices, namely

$$n_0^F(f) = \sum_{p|f} \deg p.$$

The sum is taken over all the prime polynomials dividing f, and thus we take the sum of the degree of all such polynomials. The analogue of Proposition 9.2 reads as follows.

Proposition 9.3. *Let f be a non-constant polynomial in $F[t]$. Suppose $f(t)$ has the factorization* (2). *Then the g.c.d.(f, f') is*

$$g.c.d.(f, f') = c_1 \prod_{i=1}^{r} p_i(t)^{m_i - 1}$$

with some constant c_1. In particular,

$$\deg(f, f') = \deg f - n_0^F(f).$$

Proof. Write it down yourself. It'll keep you in shape.

We now return to an algebraically closed field of characteristic 0.
It is immediate that if f, g are non-zero polynomials, then

$$n_0(fg) \leqq n_0(f) + n_0(g),$$

If in addition f, g are relatively prime, then we actually have an equality

$$n_0(fg) = n_0(f) + n_0(g).$$

It is obvious that $\deg f$ can be very large, but $n_0(f)$ may be small. For instance,

$$f(t) = (t - \alpha)^{1000}$$

has degree 1000, but $n_0(f) = 1$. The Mason-Stothers gives a remarkable additive condition under which the degree cannot be large.

Theorem 9.4 (Mason-Stothers). *Let f, g, h be non-constant relatively prime polynomials satisfying $f + g = h$. Then*

$$\deg f, \deg g, \deg h \leqq n_0(fgh) - 1.$$

The theorem shows in a very precise way how the additive relation $f + g = h$ implies a bound for the degrees of f, g, h, namely the number of distinct roots of the polynomial fgh, even with -1 tacked on.

Proof. (Noah Snyder [Sny 00]) We first note the identity

$$f'g - fg' = f'h - fh'.$$

To prove it, recall the property of derivatives, which gives $f' + g' = h'$. Then

$$\begin{aligned} f'g - fg' &= f'(f+g) - f(f'+g') \quad \text{(because } f'f \text{ cancels } ff'\text{)} \\ &= f'h - fh' \end{aligned}$$

which proves the desired identity.

We have $f'g - fg' \neq 0$, otherwise $f'g = fg' \neq 0$ because f, g are assumed non-constant. Since f, g are relatively prime, this would imply $g|g'$, which is impossible. Then we notice that the g.c.d. (f, f') divides the left side of the identity, (g, g') divides the left side, and (h, h') divides the right side, which is equal to the left side. Therefore, since f, g, h are relatively prime,

$$\text{the product } (f, f')(g, g')(h, h') \text{ divides } f'g - fg'.$$

This yields an inequality between the degrees, namely,

$$(*) \qquad \begin{aligned} \deg(f, f') + \deg(g, g') + \deg(h, h') &\leqq \deg(f'g - fg') \\ &\leqq \deg f + \deg g - 1. \end{aligned}$$

We now use Proposition 2.2, namely the identity applied to f, g, h:

$$\begin{aligned} \deg(f, f') &= \deg f - n_0(f) \\ \deg(g, g') &= \deg g - n_0(g) \\ \deg(h, h') &= \deg h - n_0(h). \end{aligned}$$

We substitute these values in (*) above. We then cancel $\deg f$ and $\deg g$ from both sides, and move $n_0(f), n_0(g), n_0(h)$ to the right side, thus getting

$$\deg h \leqq n_0(f) + n_0(g) + n_0(h) - 1 = n_0(fgh) - 1$$

since f, g, h are relatively prime. Since f, g, h enter essentially symmetrically in the equation $f + g = h$, the same inequality is satisfied for $\deg f$ and $\deg g$, thus concluding the proof.

In the 17th century, Fermat claimed to have proved that the equation

$$x^n + y^n = z^n$$

has no solution in positive integers if $n \geqq 3$. He never commented again on this claim, and nobody was able to prove it until 1995 when Wiles gave a proof based on very deep and extensive mathematical theories developed during the decades 1960–1990. Fermat's claim was usually called Fermat's last theorem, but is now Wiles' theorem.

The analogue for polynomials has been known for some time, at least since the 19th century. The first proof was based on more advanced techniques of algebraic geometry. However, it is an immediate consequence of the Mason-Stothers theorem, as we now shall see.

Theorem 9.5. *Let n be an integer $\geqq 3$. There is no solution of the equation*

$$u^n + v^n = w^n$$

with non-constant relatively prime polynomials u, v, w.

Proof. Let $f = u^n$, $g = v^n$, and $h = w^n$. Then the Mason-Stothers theorem yields

$$\deg u^n \leqq n_0(u^n v^n w^n) - 1.$$

However, $\deg u^n = n \cdot \deg u$ and $n_0(u^n) = n_0(f) \leqq \deg u$. Hence

$$n \cdot \deg u \leqq \deg u + \deg v + \deg w - 1.$$

Similarly, we obtain the analogous inequality for v and w, that is

$$n \cdot \deg v \leqq \deg u + \deg v + \deg w - 1$$

$$n \cdot \deg w \leqq \deg u + \deg v + \deg w - 1.$$

Adding the three inequalities yields

$$n(\deg uvw) \leqq 3(\deg uvw) - 3 < 3(\deg uvw).$$

Cancelling $\deg uvw$ yields $n < 3$, so $n \leqq 2$ since n is an integer, thus proving the theorem.

Of course, you should know that the equation

$$x^2 + y^2 = z^2$$

has infinitely many solutions in integers, and also has a solution in polynomials. The solutions are the sides of what's called a Pythagorean triangle. For instance, $x = 3, y = 4, z = 5$ is a solution. With polynomials, we can put

$$u = 1 - t^2, \quad v = 2t, \quad w = 1 + t^2.$$

Then you can verify at once that $u^2 + v^2 = w^2$. Substituting integers for t then gives infinitely many Pythagorean triangles. Do the exercises. See [Lan 85] for a general discussion of Pythagorean triples.

IV, §9. EXERCISES

1. Let p be a prime polynomial. Show that g.c.d.$(p, p') = 1$.

2. Prove Proposition 9.3.

3. Take the rational numbers $\mathbf{Q}$ for a field F. With the n_0^F defined at the end of Chapter III, state and prove the version of the Mason-Strothers theorem for polynomials in $\mathbf{Q}[t]$.

4. Pursuing the analogy between integers and polynomials, how would you define the analogue of n_0 for a positive integer a? Suppose a has the prime factorization

$$a = p_1^{m_1} \cdots p_r^{m_r}.$$

What's a reasonable way to define $n_0(a)$? Be careful. The number of distinct roots of a polynomial was a good definition when we could factor the polynomial in terms of prime polynomials which have degree 1. However, one cannot always do this for polynomials with rational coefficients because there are plenty of irreducible polynomials of degree > 1.

5. Give at least three solutions in relatively prime integers for the equation $x^2 + y^2 = z^2$.

6. Prove Davenport's theorem [Dav 65]: Let u, v be two non-constant relatively prime polynomials such that $u^3 - v^2 \neq 0$. Show that

$$\tfrac{1}{2} \deg u \leqq \deg(u^3 - v^2) - 1$$

$$\tfrac{1}{3} \deg v \leqq \deg(u^3 - v^2) - 1.$$

Final Remark. The precise analogue of the Mason-Stothers theorem is not true for integers. Masser-Oesterle have made a conjecture which is a perturbation of the Mason-Stothers inequality. For this conjecture and its history, see the next section.

IV, §10. THE *abc* CONJECTURE

In this section we describe a great contemporary conjecture. Bibliographical references are listed at the end of the section.

In the preceding section, we met the Mason-Stothers theorem for polynomials. One of the most fruitful analogies in mathematics is that between the integers **Z** and the ring of polynomials $F[t]$. Evolving from the insights of Mason, Stothers, Frey [Fr], Szpiro, and others, Masser and Oesterle formulated the *abc* conjecture for integers as follows. Let k be a non-zero integer. Define the **radical** of k to be

$$N_0(k) = \prod_{p|k} p,$$

i.e., the product of all the primes dividing k, taken with multiplicity 1.

The *abc* conjecture. *Given* $\varepsilon > 0$ *there exists a positive number* $C(\varepsilon)$ *having the following property. For any non-zero relatively prime integers* a, b, c *such that* $a + b = c$ *we have*

$$\max(|a|, |b|, |c|) \leqq C(\varepsilon)N_0(abc)^{1+\varepsilon}.$$

Observe that the inequality says that many prime factors of a, b, c occur to the first power, and that if "small" primes occur to high powers, then they have to be compensated by "large" primes occurring to the first power. For instance, one might consider the equation

$$2^n \pm 1 = k.$$

For n large, the *abc* conjecture would state that k has to be divisible by large primes to the first power. This phenomenon can be seen in the tables of [BLSTW].

Stewart and Tijdeman [ST 86] have shown that it is necessary to have the ε in the formulation of the conjecture and they gave a lower bound for $\varepsilon(N_0)$. Subsequent examples were communicated to me by Wojtek Jastrzebowski and Dan Spielman as follows. We have to give examples such that for all $C > 0$ there exist natural numbers a, b, c relatively prime such that $a + b = c$ and $a \geqq CN_0(abc)$. But trivially,

$$2^n | (5^{2^n} - 1).$$

We consider the relations $a_n + b_n = c_n$ given by

$$(5^{2^n} - 1) + 1 = 5^{2^n}.$$

It is clear that these relations provide the desired examples. Alan Baker has conjectured that $\max(|a|, |b|, |c|) \leqq CN_0\theta(N_0)$, where $\theta(N_0)$ is the number of integers n, $0 < n \leqq N_0$, divisible only by primes dividing N_0, and C is an absolute constant [Bak 98]. Valentin Blomer then showed that $\log \theta(N) \leqq (\log 4 + o(1))(\log N)/\log \log N$.

The abc conjecture implies what we shall call the **asymptotic Fermat theorem**, that for all but a finite number of n, the equation

$$x^n + y^n = z^n$$

has no solution in relatively prime integers x, y, z. Indeed, we have by the *abc* conjecture

$$|x^n| \ll |xyz|^{1+\varepsilon}, \quad |y^n| \ll |xyz|^{1+\varepsilon}, \quad |z^n| \ll |xyz|^{1+\varepsilon},$$

where the sign $\ll$ means that the left-hand side is $\leqq C(\varepsilon)$ times the right-hand side. Taking the product yields

$$|xyz|^n \ll |xyz|^{3+\varepsilon},$$

whence for $|xyz| > 1$ we get n bounded. The extent to which the *abc* conjecture is proved with an explicit constant $C(\varepsilon)$ (or say $C(1)$ to fix ideas) yields the corresponding explicit determination of the bound for n in the application.

We shall now see how the abc conjecture implies other conjectures by Hall, Szpiro, and Lang–Waldschmidt.

Hall's original conjecture *is that if u, v are relatively prime non-zero integers such that $u^3 - v^2 \neq 0$ then*

$$|u^3 - v^2| \gg |u|^{1/2-\varepsilon}.$$

Such an inequality determines a lower bound for the amount of cancellation that can occur in a difference $u^3 - v^2$.

Note that if $|u^3 - v^2|$ is small, then $|u^3| \gg\ll |v^2|$ so $|v| \gg\ll |u|^{3/2}$. More generally, following Lang–Waldschmidt, let us fix A, B and let u, v, k, m, n be variable with $mn > m + n$. Put

$$Au^m + Bv^n = k.$$

By the *abc* conjecture, we get

$$|u|^m \ll |uvN(k)|^{1+\varepsilon} \quad \text{and} \quad |v|^n \ll |uvN(k)|^{1+\varepsilon}.$$

If, say, $|Au^m| \leqq |Bv^n|$, then

$$|v|^n \ll |v^{1+n/m} N_0(k)|^{1+\varepsilon}$$

whence

$$(1) \qquad |v| \ll N_0(k)^{\frac{m}{mn-(m+n)}(1+\varepsilon)} \quad \text{and} \quad |u| \ll N_0(k)^{\frac{n}{mn-(m+n)}(1+\varepsilon)}.$$

The situation is symmetric in u and v. Again by the *abc* conjecture, we have $|k| \ll |uvN_0(k)|^{1+\varepsilon}$, so by (1) we find

$$(2) \qquad \boxed{|k| \ll N_0(k)^{\frac{mn}{mn-(m+n)}(1+\varepsilon)}.}$$

We give a significant example.

Example. Take $m = 3$ and $n = 2$. From (1) we get the Hall conjecture, by weakening the upper bound, replacing $N_0(k)$ with k. Observe also that if we want a bound for integral relatively prime solutions of $y^2 = x^3 + b$ with integral b, then we find $|x| \ll |b|^{2+\varepsilon}$. Thus the *abc* conjecture has a direct bearing on the solutions of diophantine equations of classical type.

Again take $m = 3$, $n = 2$ and take $A = 4$, $B = -27$. In this case, we write D instead of k, and find for

$$D = 4u^3 - 27v^2$$

that

$$(3) \qquad |u| \ll N_0(D)^{2+\varepsilon}, \qquad |v| \ll N_0(D)^{3+\varepsilon}.$$

These inequalities are supposed to hold at first for u, v relatively prime. If one allows an a priori bounded common factor, then (3) should also hold in this case. We call (3) the **generalized Szpiro conjecture**.

The original Szpiro conjecture was

$$|D| \ll N_0(D)^{6+\varepsilon},$$

but the generalized conjecture actually bounds $|u|$, $|v|$ in terms of the "right" power of $N_0(D)$, not just $|D|$ itself.

The current trend of thoughts in this direction was started by Frey [Fr], who associated with each solution of $a + b = c$ the polynomial

$$f(x) = x(x - 3a)(x + 3b).$$

The discriminant of the right-hand side is the product of the differences of the roots square, and so

$$D = 3^6(abc)^2.$$

We make a translation $\xi = x + b - a$ to get rid of the x^2 term, so that our equation can be rewritten

$$\eta^2 = \xi^3 - \gamma_2\xi - \gamma_3,$$

where γ_2, γ_3 are homogeneous in a, b of appropriate weight. Then

$$D = 4\gamma_2^3 - 27\gamma_3^2.$$

The use of 3 in the Frey polynomial was made so that γ_2, γ_3 come out to be integers. You should verify that when a, b, c are relatively prime, then γ_2, γ_3 are relatively prime, or their greatest common divisor is 9. (Do Exercise 1.)

The Szpiro conjecture implies asymptotic Fermat. Indeed, suppose that

$$a = u^n, \qquad b = v^n, \qquad \text{and} \qquad c = w^n.$$

Then

$$4\gamma_2^3 - 27\gamma_3^2 = 3^6(uvw)^{2n},$$

and we get a bound on n from the Szpiro conjecture $|D| \ll N_0(D)^{6+\varepsilon}$. Of course any exponent would do, e.g. $|D| \ll N_0(D)^{100}$ for asymptotic Fermat.

We have already seen that the *abc* conjecture implies generalized Szpiro.

Conversely, generalized Szpiro implies abc. Indeed, the correspondence between

$$(a, b) \leftrightarrow (\gamma_2, \gamma_3)$$

is "invertible," and has the "right" weight. A simple algebraic manipulation shows that the generalized Szpiro estimates on γ_2, γ_3 imply the desired estimates on $|a|$, $|b|$. (Do Exercise 2.)

From this equivalence, one can use the examples given at the beginning to show that the epsilon is needed in the Szpiro conjecture.

Hall made his conjecture in 1971, actually without the epsilon. The final setting of the proofs in the simple *abc* context which we gave above had to await Mason and the *abc* conjecture a decade later.

Let us return to the polynomial case and the Mason-Stothers theorem. The proofs that the *abc* conjecture implies the other conjectures apply as well in this case, so Hall, Szpiro, and Lang–Waldschmidt are also proved in the polynomial case. Actually, it had already been conjectured in [BCHS] that if f, g are non-zero polynomials such that $f^3 - g^2 \neq 0$ then

$$\deg(f(t)^3 - g(t)^2) \geqq \tfrac{1}{2}\deg f(t) + 1.$$

This (and its analogue for higher degrees) was proved by Davenport [Dav] in 1965. As for ordinary integers, the point of the theorem is to determine a lower bound for the cancellations which can occur in a difference between a cube and a square, in the simplest case. The result for polynomials is particularly clear since, unlike the case of integers, there is no extraneous undetermined constant floating around, and there is even $+1$ on the right-hand side.

The polynomial case as in Davenport and the Hall conjecture for integers are of course not independent. Examples in the polynomial case parametrize cases with integers when we substitute integers for the variable. Examples are given in [BCHS], one of them due to Birch being:

$$f(t) = t^6 + 4t^4 + 10t^2 + 6 \quad \text{and} \quad g(t) = t^9 + 6t^7 + 21t^5 + 35t^3 + \tfrac{63}{2}t,$$

whence

$$\text{degree}(f(t)^3 - g(t)^2) = \tfrac{1}{2}\deg f + 1.$$

Substituting large integral values of $t \equiv 2 \bmod 4$ gives examples of large values for $x^3 - y^2$. A fairly general construction is given by Danilov [Dan].

IV, §10. EXERCISES

1. Prove the statement that if a, b, c are relatively prime and $a + b = c$, then γ_2, γ_3 are relatively prime or their g.c.d. is 9.

2. Prove that the generalized Szpiro conjecture implies the *abc* conjecture.

3. **Conjecture.** There are infinitely many primes p such that $2^{p-1} \not\equiv 1 \bmod p^2$. (You know of course that $2^{p-1} \equiv 1 \bmod p$ if p is an odd prime.)
   (a) Let S be the set of primes such that $2^{p-1} \not\equiv 1 \bmod p^2$. If n is a positive integer, and p is a prime such that $2^n - 1 = pk$ with some integer k prime to p, then prove that p is in S.
   (b) Prove that the *abc* conjecture implies the above conjecture. (Silverman, *J. of Number Theory*, 1988.)

Remark. A conjecture of Lang–Trotter implies that the number of primes $p \leqq x$ such that $2^{p-1} \equiv 1 \bmod p^2$ is bounded by $C \log\log x$ for some constant $C > 0$. So most primes would satisfy the condition that $2^{p-1} \not\equiv 1 \bmod p^2$.

References

[Bak 98] A. Baker, Logarithmic forms and the *abc*-conjecture, *Number Theory*, Eds.: Győry/Pethö/Sós, Walter de Gruyter, Berlin-New York, 1998

[BCHS] B. Birch, S. Chowla, M. Hall, A. Schinzel, *On the difference* $x^3 - y^2$, Norske Vid. Selsk. Forrh. 38 (1965) pp. 65–69

[BLSTW] J. Brillhart, D. H. Lehmer, J. L. Selfridge, B. Tuckerman, and S. S. Wagstaff Jr., *Factorization of* $b^n \pm 1, b = 2, 3, 5, 6, 7, 10, 11$ *up to high powers*, Contemporary Mathematics Vol. 22, AMS, Providence, RI 1983

[Dan] L. V. Danilov, *The diophantine equation* $x^3 - y^2 = k$ *and Hall's conjecture*, Mat. Zametki Vol. 32 No. 3 (1982) pp. 273–275

[Dav] H. Davenport, *On* $f^3(t) - g^2(t)$, K. Norske Vod. Selskabs Farh. (Trondheim), 38 (1965) pp. 86–87

[Fr] G. Frey, *Links between stable elliptic curves and elliptic curves*, Number Theory, Lecture Notes Vol. 1380, Springer-Verlag, Ulm 1987 pp. 31–62

[Ha] M. Hall, *The diophantine equation* $x^3 - y^2 = k$, Computers in Number Theory, ed. by A. O. L. Atkin and B. Birch, Academic Press, London 1971 pp. 173–198

[La 85] S. Lang, *Math Encounters with High School Students*, Springer Verlag 1985

[La 90] S. Lang, *Old and new conjectured diophantine inequalities*, Bull. Amer. Math. Soc. (1990)

[La 99] S. Lang, *Math Talks for Undergraduates*, Springer Verlag, 1999

[Mas 83] R. C. Mason, *Diophantine Equations over Function Fields*, London Math Society Lecture Notes Series, Vol. 96, Cambridge University Press, 1984

[Sny 00] N. Snyder, An alternate proof for Mason's theorem, *Elemente Math.* **55** (2000) pp. 93–94

[StT 86] C. L. Stewart and R. Tijdeman, *On the Oesterle-Masser Conjecture*, Monatshefte fur Math. (1986) pp. 251–257

[Sto 81] W. Stothers, Polynomial identities and hauptmoduln, *Quart. Math. Oxford* (2) **32** (1981) pp. 349–370

[Ti 89] R. Tijdeman, *Diophantine Equations and Diophantine Approximations*, Banff lecture in Number Theory and its Applications, ed. R. A. Mollin, Kluwer Academic Press, 1989 see especially p. 234

[Za 95] U. Zannier, On Davenport's bound for the degree of $f^3 - g^2$ and Riemann's existence theorem, *Acta Arithm.* **LXXI.2** (1995) pp. 107–137

Readers acquainted with basic facts on vector spaces may jump immediately to the field theory of Chapter VII.

CHAPTER V

Vector Spaces and Modules

V, §1. VECTOR SPACES AND BASES

Let K be a field. A **vector space** V **over the field** K is an additive (abelian) group, together with a multiplication of elements of V by elements of K, i.e. an association

$$(x, v) \mapsto xv$$

of $K \times V$ into V, satisfying the following conditions:

VS 1. *If* 1 *is the unit element of* K, *then* $1v = v$ *for all* $v \in V$.

VS 2. *If* $c \in K$ *and* $v, w \in V$, *then* $c(v + w) = cv + cw$.

VS 3. *If* $x, y \in K$ *and* $v \in V$, *then* $(x + y)v = xv + yv$.

VS 4. *If* $x, y \in K$ *and* $v \in V$, *then* $(xy)v = x(yv)$.

Example 1. Let V be the set of continuous real-valued functions on the interval $[0, 1]$. Then V is a vector space over $\mathbf{R}$. The addition of functions is defined as usual: If f, g are functions, we define

$$(f + g)(t) = f(t) + g(t).$$

If $c \in \mathbf{R}$, we define $(cf)(t) = cf(t)$. It is then a simple routine matter to verify that all four conditions are satisfied.

Example 2. Let S be a non-empty set, and V the set of all maps of S into K. Then V is a vector space over K, the addition of maps and the multiplication of maps by elements of K being defined as for functions in the preceding example.

Example 3. Let K^n denote the product $K \times \cdots \times K$, i.e. the set of n-tuples of elements of K. (If $K = \mathbf{R}$, this is the usual Euclidean space.) We define addition of n-tuples componentwise, that is if

$$X = (x_1,\ldots,x_n) \qquad \text{and} \qquad Y = (y_1,\ldots,y_n)$$

are elements of K^n with x_i, $y_i \in K$, then we define

$$X + Y = (x_1 + y_1,\ldots,x_n + y_n).$$

If $c \in K$, we define

$$cX = (cx_1,\ldots,cx_n).$$

It is routine to verify that all four conditions of a vector space are satisfied by these operations.

Example 4. Taking $n = 1$ in Example 3, we see that K is a vector space over itself.

Let V be a vector space over the field K. Let $v \in V$. Then $0v = 0$.

Proof. $0v + v = 0v + 1v = (0 + 1)v = 1v = v$. Hence adding $-v$ to both sides shows that $0v = 0$.

If $c \in K$ and $cv = 0$, but $c \neq 0$, then $v = 0$.

To see this, multiply by c^{-1} to find $c^{-1}cv = 0$ whence $v = 0$.

We have $(-1)v = -v$.

Proof.

$$(-1)v + v = (-1)v + 1v = (-1 + 1)v = 0v = 0.$$

Hence $(-1)v = -v$.

Let V be a vector space, and W a subset of V. We shall say that W is a **subspace** if W is a subgroup (of the additive group of V), and if given $c \in K$ and $v \in W$ then cv is also an element of W. In other words, a subspace W of V is a subset satisfying the following conditions:

(i) If v, w are elements of W, their sum $v + w$ is also an element of W.

(ii) The element 0 of V is also an element of W.
(iii) If $v \in W$ and $c \in K$ then $cv \in W$.

Then W itself is a vector space. Indeed, properties **VS 1** through **VS 4**, being satisfied for all elements of V, are satisfied *a fortiori* for the elements of W.

Let V be a vector space, and $w_1, \dots, w_n$ elements of V. Let W be the set of all elements

$$x_1 w_1 + \cdots + x_n w_n$$

with $x_i \in K$. Then W is a subspace of V, as one verifies without difficulty. It is called the subspace **generated** by $w_1, \dots, w_n$, and we call $w_1, \dots, w_n$ **generators** for this subspace.

Let V be a vector space over the field K, and let $v_1, \dots, v_n$ be elements of V. We shall say that $v_1, \dots, v_n$ are **linearly dependent over** K if there exist elements $a_1, \dots, a_n$ in K not all equal to 0 such that

$$a_1 v_1 + \cdots + a_n v_n = 0.$$

If there do not exist such elements, then we say that $v_1, \dots, v_n$ are **linearly independent over** K. We often omit the words "over K".

Example 5. Let $V = K^n$ and consider the vectors

$$v_1 = (1, 0, \dots, 0)$$
$$\vdots$$
$$v_n = (0, 0, \dots, 1).$$

Then $v_1, \dots, v_n$ are linearly independent. Indeed, let $a_1, \dots, a_n$ be elements of K such that $a_1 v_1 + \cdots + a_n v_n = 0$. Since

$$a_1 v_1 + \cdots + a_n v_n = (a_1, \dots, a_n),$$

it follows that all $a_i = 0$.

Example 6. Let V be the vector space of all functions of a real variable t. Let $f_1(t), \dots, f_n(t)$ be n functions. To say that they are linearly dependent is to say that there exist n real numbers $a_1, \dots, a_n$ not all equal to 0 such that

$$a_1 f_1(t) + \cdots + a_n f_n(t) = 0$$

for *all* values of t.

The two functions e^t, e^{2t} are linearly independent. To prove this, suppose that there are numbers a, b such that

$$ae^t + be^{2t} = 0$$

(for all values of t). Differentiate this relation. We obtain

$$ae^t + 2be^{2t} = 0.$$

Subtract the first from the second relation. We obtain $be^{2t} = 0$, and hence $b = 0$. From the first relation, it follows that $ae^t = 0$, and hence $a = 0$. Hence e^t, e^{2t} are linearly independent.

Consider again an arbitrary vector space V over a field K. Let $v_1,\dots,v_n$ be linearly independent elements of V. Let $x_1,\dots,x_n$ and $y_1,\dots,y_n$ be numbers. Suppose that we have

$$x_1v_1 + \cdots + x_nv_n = y_1v_1 + \cdots + y_nv_n.$$

In other words, two linear combinations of $v_1,\dots,v_n$ are equal. Then we must have $x_i = y_i$ for each $i = 1,\dots,n$. Indeed, subtracting the right-hand side from the left-hand side, we get

$$x_1v_1 - y_1v_1 + \cdots + x_nv_n - y_nv_n = 0.$$

We can write this relation also in the form

$$(x_1 - y_1)v_1 + \cdots + (x_n - y_n)v_n = 0.$$

By definition, we must have $x_i - y_i = 0$ for all $i = 1,\dots,n$, thereby proving our assertion.

We define a **basis** of V over K to be a sequence of elements $\{v_1,\dots,v_n\}$ of V which generate V and are linearly independent.

The vectors $v_1,\dots,v_n$ of Example 5 form a basis of K^n over K.

Let W be the vector space of functions generated over **R** by the two functions e^t, e^{2t}. Then $\{e^t, e^{2t}\}$ is a basis of W over **R**.

Let V be a vector space, and let $\{v_1,\dots,v_n\}$ be a basis of V. The elements of V can be represented by n-tuples relative to this basis, as follows. If an element v of V is written as a linear combination

$$v = x_1v_1 + \cdots + x_nv_n$$

of the basis elements, then we call $(x_1,\dots,x_n)$ the **coordinates** of v with respect to our basis, and we call x_i the i-th coordinate. We say that the n-tuple $X = (x_1,\dots,x_n)$ is the **coordinate vector** of v with respect to the basis $\{v_1,\dots,v_n\}$.

For example, let V be the vector space of functions generated by the two functions e^t, e^{2t}. Then the coordinates of the function

$$3e^t + 5e^{2t}$$

with respect to the basis $\{e^t, e^{2t}\}$ are $(3, 5)$.

Example 7. Show that the vectors $(1, 1)$ and $(-3, 2)$ are linearly independent over $\mathbf{R}$.

Let a, b be two real numbers such that

$$a(1, 1) + b(-3, 2) = 0.$$

Writing this equation in terms of components, we find

$$a - 3b = 0,$$

$$a + 2b = 0.$$

This is a system of two equations which we solve for a and b. Subtracting the second from the first, we get $-5b = 0$, whence $b = 0$. Substituting in either equation, we find $a = 0$. Hence a, b are both 0, and our vectors are linearly independent.

Example 8. Find the coordinates of $(1, 0)$ with respect to the two vectors $(1, 1)$ and $(-1, 2)$.

We must find numbers a, b such that

$$a(1, 1) + b(-1, 2) = (1, 0).$$

Writing this equation in terms of coordinates, we find

$$a - b = 1,$$

$$a + 2b = 0.$$

Solving for a and b in the usual manner yields $b = -\frac{1}{3}$ and $a = \frac{2}{3}$. Hence the coordinates of $(1, 0)$ with respect to $(1, 1)$ and $(-1, 2)$ are $(\frac{2}{3}, -\frac{1}{3})$.

Let $\{v_1,\dots,v_n\}$ be a set of elements of a vector space V over a field K. Let r be a positive integer $\leqq n$. We shall say that $\{v_1,\dots,v_r\}$ is a **maximal** subset of linearly independent elements if $v_1,\dots,v_r$ are linearly independent, and if in addition, given any v_i with $i > r$, the elements $v_1,\dots,v_r$, v_i are linearly dependent.

The next theorem gives us a useful criterion to determine when a set of elements of a vector space is a basis.

Theorem 1.1. *Let $\{v_1,\dots,v_n\}$ be a set of generators of a vector space V. Let $\{v_1,\dots,v_r\}$ be a maximal subset of linearly independent elements. Then $\{v_1,\dots,v_r\}$ is a basis of V.*

Proof. We must prove that $v_1,\dots,v_r$ generate V. We shall first prove that each v_i (for $i > r$) is a linear combination of $v_1,\dots,v_r$. By hypothesis, given v_i, there exist $x_1,\dots,x_r$, $y \in K$, not all 0, such that

$$x_1v_1 + \cdots + x_rv_r + yv_i = 0.$$

Furthermore, $y \neq 0$, because otherwise, we would have a relation of linear dependence for $v_1,\dots,v_r$. Hence we can solve for v_i, namely

$$v_i = \frac{x_1}{-y} v_1 + \cdots + \frac{x_r}{-y} v_r,$$

thereby showing that v_i is a linear combination of $v_1,\dots,v_r$.

Next, let v be any element of V. There exist $c_1,\dots,c_n \in K$ such that

$$v = c_1 v_1 + \cdots + c_n v_n.$$

In this relation, we can replace each v_i $(i > r)$ by a linear combination of $v_1,\dots,v_r$. If we do this, and then collect terms, we find that we have expressed v as a linear combination of $v_1,\dots,v_r$. This proves that $v_1,\dots,v_r$ generate V, and hence form a basis of V.

Let V, W be vector spaces over K. A map

$$f: V \to W$$

is called a K-**linear map**, or a **homomorphism of vector spaces**, if f satisfies the following condition: For all $x \in K$ and v, $v' \in V$ we have

$$f(v + v') = f(v) + f(v'), \qquad f(xv) = xf(v).$$

Thus f is a homomorphism of V into W viewed as additive groups, satisfying the additional condition $f(xv) = xf(v)$. We usually say "linear map" instead of "K-linear map".

Let $f: V \to W$ and $g: W \to U$ be linear maps. Then the composite $g \circ f: V \to U$ is a linear map.

The verification is immediate, and will be left to the reader.

Theorem 1.2. *Let V, W be vector spaces, and $\{v_1,\dots,v_n\}$ a basis of V. Let $w_1,\dots,w_n$ be elements of W. Then there exists a unique linear map $f: V \to W$ such that $f(v_i) = w_i$ for all i.*

Proof. Such a K-linear map f is uniquely determined, because if

$$v = x_1 v_1 + \cdots + x_n v_n$$

is an element of V, with $x_i \in K$, then we must necessarily have

$$\begin{aligned} f(v) &= x_1 f(v_1) + \cdots + x_n f(v_n) \\ &= x_1 w_1 + \cdots + x_n w_n. \end{aligned}$$

The map f exists, for given an element v as above, we *define* $f(v)$ to be $x_1 w_1 + \cdots + x_n w_n$. We must then see that f is a linear map. Let

$$v' = y_1 v_1 + \cdots + y_n v_n$$

be an element of V with $y_i \in K$. Then

$$v + v' = (x_1 + y_1)v_1 + \cdots + (x_n + y_n)v_n.$$

Hence

$$\begin{aligned} f(v + v') &= (x_1 + y_1)w_1 + \cdots + (x_n + y_n)w_n \\ &= x_1 w_1 + y_1 w_1 + \cdots + x_n w_n + y_n w_n \\ &= f(v) + f(v'). \end{aligned}$$

If $c \in K$, then $cv = cx_1 v_1 + \cdots + cx_n v_n$, and hence

$$f(cv) = cx_1 w_1 + \cdots + cx_n w_n = cf(v).$$

This proves that f is linear, and concludes the proof of the theorem.

The **kernel** of a linear map is defined to be the kernel of the map viewed as an additive group-homomorphism. Thus $\operatorname{Ker} f$ is the set of $v \in V$ such that $f(v) = 0$. We leave to the reader to prove:

The kernel and image of a linear map are subspaces.

Let $f: V \to W$ be a linear map. Then f is injective if and only if $\operatorname{Ker} f = 0$.

Proof. Suppose f is injective. If $f(v) = 0$, then by definition and the fact that $f(0) = 0$ we must have $v = 0$. Hence $\operatorname{Ker} f = 0$. Conversely, suppose $\operatorname{Ker} f = 0$. Let $f(v_1) = f(v_2)$. Then $f(v_1 - v_2) = 0$ so $v_1 - v_2 = 0$ and $v_1 = v_2$. Hence f is injective. This proves our assertion.

Let $f: V \to W$ be a linear map. If f is bijective, that is injective and surjective, then f has an inverse mapping

$$g: W \to V.$$

If f is linear and bijective, then the inverse mapping $g: W \to V$ is also a linear map.

Proof. Let $w_1, w_2 \in W$. Since f is surjective, there exist $v_1, v_2 \in V$ such that $f(v_1) = w_1$ and $f(v_2) = w_2$. Then $f(v_1 + v_2) = w_1 + w_2$. By definition of the inverse map,

$$g(w_1 + w_2) = v_1 + v_2 = g(w_1) + g(w_2).$$

We leave to the reader the proof that $g(cw) = cg(w)$ for $c \in K$ and $w \in W$. This concludes the proof that g is linear.

As with groups, we say that a linear map $f: V \to W$ is an **isomorphism** (i.e. a vector space isomorphism) if it has a linear inverse, i.e. there exists a linear map $g: W \to V$ such that $g \circ f$ is the identity of V, and $f \circ g$ is the identity of W. The preceding remark shows that a linear map is an isomorphism if and only if it is bijective.

Let V, W be vector spaces over the field K. We let

$$\mathrm{Hom}_K(V, W) = \text{set of all linear maps of } V \text{ into } W.$$

Let f, $g: V \to W$ be linear maps. Then we can define the **sum** $f + g$, just as we define the sum of any mappings from a set into W. Thus by definition

$$(f + g)(v) = f(v) + g(v).$$

If $c \in K$ we define cf to be the map such that

$$(cf)(v) = cf(v).$$

With these definitions, it is then easily verified that

$$\mathrm{Hom}_K(V, W) \textit{ is a vector space over } K.$$

We leave the steps in the verification to the reader. In case $V = W$, we call the homomorphisms (or K-linear maps) of V into itself the **endomorphisms** of V, and we let

$$\mathrm{End}_K(V) = \mathrm{Hom}_K(V, V).$$

V, §1. EXERCISES

1. Show that the following vectors are linearly independent, over **R** and over **C**.
   (a) $(1, 1, 1)$ and $(0, 1, -1)$
   (b) $(1, 0)$ and $(1, 1)$
   (c) $(-1, 1, 0)$ and $(0, 1, 2)$
   (d) $(2, -1)$ and $(1, 0)$
   (e) $(\pi, 0)$ and $(0, 1)$
   (f) $(1, 2)$ and $(1, 3)$
   (g) $(1, 1, 0)$, $(1, 1, 1)$ and $(0, 1, -1)$
   (h) $(0, 1, 1)$, $(0, 2, 1)$ and $(1, 5, 3)$

2. Express the given vector X as a linear combination of the given vectors A, B and find the coordinates of X with respect to A, B.
   (a) $X = (1, 0)$, $A = (1, 1)$, $B = (0, 1)$
   (b) $X = (2, 1)$, $A = (1, -1)$, $B = (1, 1)$
   (c) $X = (1, 1)$, $A = (2, 1)$, $B = (-1, 0)$
   (d) $X = (4, 3)$, $A = (2, 1)$, $B = (-1, 0)$
   (You may view the above vectors as elements of $\mathbf{R}^2$ or $\mathbf{C}^2$. The coordinates will be the same.)

3. Find the coordinates of the vector X with respect to the vectors A, B, C.
   (a) $X = (1, 0, 0)$, $A = (1, 1, 1)$, $B = (-1, 1, 0)$, $C = (1, 0, -1)$
   (b) $X = (1, 1, 1)$, $A = (0, 1, -1)$, $B = (1, 1, 0)$, $C = (1, 0, 2)$
   (c) $X = (0, 0, 1)$, $A = (1, 1, 1)$, $B = (-1, 1, 0)$, $C = (1, 0, -1)$

4. Let (a, b) and (c, d) be two vectors in K^2. If $ad - bc = 0$, show that they are linearly dependent. If $ad - bc \neq 0$, show that they are linearly independent.

5. Prove that $1, \sqrt{2}$ are linearly independent over the rational numbers.

6. Prove that $1, \sqrt{3}$ are linearly independent over the rational numbers.

7. Let α be a complex number. Show that α is rational if and only if $1, \alpha$ are linearly dependent over the rational numbers.

V, §2. DIMENSION OF A VECTOR SPACE

The main result of this section is that any two bases of a vector space have the same number of elements. To prove this, we first have an intermediate result.

Theorem 2.1. *Let V be a vector space over the field K. Let $\{v_1,\dots,v_m\}$ be a basis of V over K. Let $w_1,\dots,w_n$ be elements of V, and assume that $n > m$. Then $w_1,\dots,w_n$ are linearly dependent.*

Proof. Assume that $w_1,\dots,w_n$ are linearly independent. Since $\{v_1,\dots,v_m\}$ is a basis, there exist elements $a_1,\dots,a_m \in K$ such that

$$w_1 = a_1 v_1 + \cdots + a_m v_m.$$

By assumption, we know that $w_1 \neq 0$, and hence some $a_i \neq 0$. After renumbering $v_1,\dots,v_m$ if necessary, we may assume without loss of generality that (say) $a_1 \neq 0$. We can then solve for v_1, and get

$$a_1 v_1 = w_1 - a_2 v_2 - \cdots - a_m v_m,$$

$$v_1 = a_1^{-1} w_1 - a_1^{-1} a_2 v_2 - \cdots - a_1^{-1} a_m v_m.$$

The subspace of V generated by $w_1, v_2,\dots,v_m$ contains v_1, and hence must be all of V since $v_1, v_2,\dots,v_m$ generate V. The idea is now to continue our procedure stepwise, and to replace successively $v_2, v_3,\dots$ by $w_2, w_3,\dots$ until all the elements $v_1,\dots,v_m$ are exhausted, and $w_1,\dots,w_m$ generate V. Let us now assume by induction that there is an integer r with $1 \leqq r < m$ such that, after a suitable renumbering of $v_1,\dots,v_m$, the elements $w_1,\dots,w_r, v_{r+1},\dots,v_m$ generate V. There exist elements $b_1,\dots,b_r, c_{r+1},\dots,c_m$ in K such that

$$w_{r+1} = b_1 w_1 + \cdots + b_r w_r + c_{r+1} v_{r+1} + \cdots + c_m v_m.$$

We cannot have $c_j = 0$ for $j = r+1,\dots,m$, for otherwise, we get a relation of linear dependence between $w_1,\dots,w_{r+1}$, contradicting our assumption. After renumbering $v_{r+1},\dots,v_m$ if necessary, we may assume without loss of generality that (say) $c_{r+1} \neq 0$. We then obtain

$$c_{r+1}v_{r+1} = w_{r+1} - b_1w_1 - \cdots - b_rw_r - c_{r+2}v_{r+2} - \cdots - c_mv_m.$$

Dividing by c_{r+1}, we conclude that v_{r+1} is in the subspace generated by $w_1,\dots,w_{r+1}, v_{r+2},\dots,v_m$. By our induction assumption, it follows that $w_1,\dots,w_{r+1}, v_{r+2},\dots,v_m$ generate V. Thus by induction, we have proved that $w_1,\dots,w_m$ generate V. If we write

$$w_{m+1} = x_1w_1 + \cdots + x_mw_m$$

with $x_i \in K$, we obtain a relation of linear dependence

$$w_{m+1} - x_1w_1 - \cdots - x_mw_m = 0,$$

as was to be shown.

Theorem 2.2. *Let V be a vector space over K, and let $\{v_1,\dots,v_n\}$ and $\{w_1,\dots,w_m\}$ be two bases of V. Then $m = n$.*

Proof. By Theorem 2.1, we must have $n \leqq m$ and $m \leqq n$, so $m = n$.

If a vector space has one basis, then every other basis has the same number of elements. This number is called the **dimension** of V (over K). If V is the zero vector space, we define V to have **dimension** 0.

Corollary 2.3. *Let V be a vector space of dimension n, and let W be a subspace containing n linearly independent elements. Then $W = V$.*

Proof. Let $v \in V$ and let $w_1,\dots,w_n$ be linearly independent elements of W. Then $w_1,\dots,w_n, v$ are linearly dependent, so there exist a, $b_1,\dots,b_n \in K$ not all zero such that

$$av + b_1w_1 + \cdots + b_nw_n = 0.$$

We cannot have $a = 0$, otherwise $w_1,\dots,w_n$ are linearly dependent. Then

$$v = -a^{-1}b_1w - \cdots - a^{-1}b_nw_n$$

is an element of W. This proves $V \subset W$, so $V = W$.

Theorem 2.4. *Let $f: V \to W$ be a homomorphism of vector spaces over K. Assume that V, W have finite dimension and that* $\dim V = \dim W$. *If* $\operatorname{Ker} f = 0$ *or if* $\operatorname{Im} f = W$ *then f is an isomorphism.*

Proof. Suppose $\operatorname{Ker} f = 0$. Let $\{v_1, \dots, v_n\}$ be a basis of V. Then $f(v_1), \dots, f(v_n)$ are linearly independent, for suppose $c_1, \dots, c_n \in K$ are such that

$$c_1 f(v_1) + \cdots + c_n f(v_n) = 0.$$

Then

$$f(c_1 v_1 + \cdots + c_n v_n) = 0,$$

and since f is injective, we have $c_1 v_1 + \cdots + c_n v_n = 0$. Hence $c_i = 0$ for $i = 1, \dots, n$ since $\{v_1, \dots, v_n\}$ is a basis of V. Hence $\operatorname{Im} f$ is a subspace of W of dimension n, whence $\operatorname{Im} f = W$ by Corollary 2.3. Hence f is also surjective, so f is an isomorphism.

We leave the other case to the reader, namely the proof that if f is surjective, then f is an isomorphism.

Let V be a vector space. We define an **automorphism** of V to be an invertible linear map

$$f: V \to V$$

of V with itself. We denote the set of automorphisms of V by

$$\operatorname{Aut}(V) \quad \text{or} \quad \operatorname{GL}(V).$$

The letters GL stand for "**General Linear**".

Theorem 2.5. *The set* $\operatorname{Aut}(V)$ *is a group.*

Proof. The multiplication is composition of mappings. Such composition is associative, we have seen that the inverse of a linear map is linear, and the identity is linear. Thus all the group axioms are satisfied.

The group $\operatorname{Aut}(V)$ is one of the most important groups in mathematics. In the next chapter we shall study it for finite dimensional vector spaces in terms of matrices.

V, §2. EXERCISES

1. Let V be a finite dimensional vector space over K. Let W be a subspace. Let $\{w_1, \dots, w_m\}$ be a basis of W. Show that there exist elements $w_{m+1}, \dots, w_n$ in V such that $\{w_1, \dots, w_n\}$ is a basis of V.

2. If f is a linear map, $f: V \to V'$, prove that

$$\dim V = \dim \operatorname{Im} f + \dim \operatorname{Ker} f.$$

3. Let U, W be subspaces of a vector space V.
   (a) Show that $U + W$ is a subspace.

(b) Define $U \times W$ to be the set of all pairs (u, w) with $u \in U$ and $w \in W$. Show how $U \times W$ is a vector space. If U, W are finite dimensional, show that

$$\dim(U \times W) = \dim U + \dim W.$$

(c) Prove that $\dim U + \dim W = \dim(U + W) + \dim(U \cap W)$. [*Hint*: Consider the linear map $f\colon U \times W \to U + W$ given by $f(u, w) = u - w$.]

V, §3. MATRICES AND LINEAR MAPS

Although we expect that the reader will have had some elementary linear algebra previously, we recall here some of these basic results to make the present book self contained. We go rapidly.

An $m \times n$ **matrix** $A = (a_{ij})$ is a doubly indexed family of elements a_{ij} in a field K, with $i = 1,\dots,m$ and $j = 1,\dots,n$. A matrix is usually written as a rectangular array

$$A = \begin{pmatrix} a_{11} & a_{12} & \cdots & a_{1n} \\ a_{21} & a_{22} & \cdots & a_{2n} \\ \vdots & \vdots & & \vdots \\ a_{m1} & a_{m2} & \cdots & a_{mn} \end{pmatrix}.$$

The elements a_{ij} are called the **components** of A. If $m = n$ then A is called a **square matrix**.

As a matter of notation, we let:

$$\mathrm{Mat}_{m \times n}(K) = \text{set of } m \times n \text{ matrices in } K,$$

$$\mathrm{Mat}_n(K) \text{ or } M_n(K) = \text{set of } n \times n \text{ matrices in } K.$$

Let $A = (a_{ij})$ and $B = (b_{ij})$ be $m \times n$ matrices in K. We define their **sum** to be the matrix whose ij-component is

$$a_{ij} + b_{ij}.$$

Thus we take the sum of matrices componentwise. Then $\mathrm{Mat}_{m \times n}(K)$ is an additive group under this addition. The verification is immediate.

Let $c \in K$. We define $cA = (ca_{ij})$, so we multiply each component of A by c. Then it is also immediately verified that $\mathrm{Mat}_{m \times n}(K)$ is a vector space over K. The zero element is the matrix

$$\begin{pmatrix} 0 & \cdots & 0 \\ \vdots & & \vdots \\ 0 & \cdots & 0 \end{pmatrix}$$

having components equal to 0.

Let $A = (a_{ij})$ be an $m \times n$ matrix and $B = (b_{jk})$ be an $n \times r$ matrix. Then we define the **product** AB to be the $m \times r$ matrix whose ik-component is

$$\sum_{j=1}^{n} a_{ij} b_{jk}.$$

It is easily verified that this multiplication satisfies the distributive law:

$$A(B + C) = AB + AC \quad \text{and} \quad (A + B)C = AC + BC$$

provided that the formulas make sense. For the first one, B and C must have the same size, and the products AB, AC must be defined. For the second one, A and B must have the same size, and AC, BC must be defined. Furthermore, if $(AB)C$ is defined, then $(AB)C = A(BC)$.

Given n we define the **unit** or **identity** $n \times n$ matrix to be

$$I_n = \begin{pmatrix} 1 & 0 & \cdots & 0 \\ 0 & 1 & \cdots & 0 \\ \vdots & \vdots & & \vdots \\ 0 & 0 & \cdots & 1 \end{pmatrix},$$

in other words, I_n is a square $n \times n$ matrix, having components 1 on the diagonal, and components 0 otherwise. From the definition of multiplication, if A is an $m \times n$ matrix, we get:

$$I_m A = A \quad \text{and} \quad A I_n = A.$$

Let $M_n(K)$ denote the set of all $n \times n$ matrices with components in K. Then $M_n(K)$ is a ring under the above addition and multiplication of matrices.

This statement is merely a summary of properties which we have already listed.

There is a natural map of K in $M_n(K)$, namely

$$c \mapsto cI_n = \begin{pmatrix} c & 0 & \cdots & 0 \\ 0 & c & \cdots & 0 \\ \vdots & \vdots & & \vdots \\ 0 & 0 & \cdots & c \end{pmatrix},$$

which sends an element $c \in K$ on the diagonal matrix having diagonal components equal to c and otherwise 0 components. We call cI_n a **scalar matrix**. The map

$$c \mapsto cI_n$$

is an isomorphism of K onto the K-vector space of scalar matrices.

We now describe the correspondence between matrices and linear maps.

Let $A = (a_{ij})$ be an $m \times n$ matrix in a field K. Then A gives rise to a linear map

$$L_A : K^n \to K^m \quad \text{by} \quad X \mapsto AX = L_A(X).$$

Theorem 3.1. *The association $A \mapsto L_A$ is an isomorphism between the vector space of $m \times n$ matrices and the space of linear maps $K^n \to K^m$.*

Proof. If A, B are $m \times n$ matrices and $L_A = L_B$ then $A = B$ because if E^j is the j-th unit vector

$$E^j = \begin{pmatrix} 0 \\ 0 \\ \vdots \\ 1 \\ 0 \\ \vdots \\ 0 \end{pmatrix}$$

with 1 in the j-th component and 0 otherwise, then $AE^j = A^j$ is the j-th column of A, so if $AE^j = BE^j$ for all $j = 1, \ldots, n$ we conclude that $A = B$.

Next we have to prove that $A \mapsto L_A$ is surjective. Let $L: K^n \to K^m$ be an arbitrary linear map. Let $\{U^1, \ldots, U^m\}$ be the unit vectors in K^m. Then there exist elements $a_{ij} \in K$ such that

$$L(E^j) = \sum_{i=1}^{m} a_{ij} U^i.$$

Let $A = (a_{ij})$. If

$$X = \sum_{j=1}^{n} x_j E^j,$$

then

$$L(X) = \sum_{j=1}^{n} x_j L(E^j) = \sum_{j=1}^{n} \sum_{i=1}^{m} x_j a_{ij} U^i$$

$$= \sum_{i=1}^{m} \left(\sum_{j=1}^{n} a_{ij} x_j \right) U^i.$$

This means that $L = L_A$ and concludes the proof of the theorem.

Next we give a slightly different formulation. Let V be a vector space of dimension n over K. Let $\{v_1, \ldots, v_n\}$ be a basis of V. Recall that we have an isomorphism

$$K^n \to V \quad \text{given by} \quad (x_1, \ldots, x_n) \mapsto x_1 v_1 + \cdots + x_n v_n.$$

Now let V, W be vector spaces over K of dimensions n, m respectively. Let

$$L: V \to W$$

be a linear map. Let $\{v_1, \ldots, v_n\}$ and $\{w_1, \ldots, w_m\}$ be bases of V and W respectively. Let $a_{ij} \in K$ be such that

$$L(v_j) = \sum_{i=1}^{m} a_{ij} w_i.$$

Then the matrix $A = (a_{ij})$ is said to be **associated** to L **with respect to the given bases.**

Theorem 3.2. *The association of the above matrix to L gives an isomorphism between the space of $m \times n$ matrices and the space of linear maps* $\mathrm{Hom}_K(V, W)$.

Proof. The proof is similar to the proof of Theorem 3.1 and is left to the reader.

Basically what is happening is that when we represent an element of V as a linear combination

$$v = x_1 v_1 + \cdots + x_n v_n,$$

and view

$$X = \begin{pmatrix} x_1 \\ x_2 \\ \vdots \\ x_n \end{pmatrix}$$

as its coordinate vector, then $L(v)$ is represented by AX in terms of the coordinates.

V, §3. EXERCISES

1. Exhibit a basis for the following vector spaces:
   (a) The space of all $m \times n$ matrices.
   (b) The space of symmetric $n \times n$ matrices. A matrix $A = (a_{ij})$ is said to be **symmetric** if $a_{ij} = a_{ji}$ for all i, j.

(c) The space of $n \times n$ triangular matrices. A matrix $A = (a_{ij})$ is said to be **upper triangular** if $a_{ij} = 0$ whenever $j < i$.

2. If $\dim V = n$ and $\dim W = m$, what is $\dim \mathrm{Hom}_K(V, W)$? Proof?

3. Let R be a ring. We define the **center** Z of R to be the subset of all elements $z \in R$ such that $zx = xz$ for all $x \in R$.
   (a) Show that the center of a ring R is a subring.
   (b) Let $R = \mathrm{Mat}_{n \times n}(K)$ be the ring of $n \times n$ matrices over the field K. Show that the center is the set of scalar matrices cI, with $c \in K$.

V, §4. MODULES

We may consider a generalization of the notion of vector space over a field, namely module over a ring. Let R be a ring. By a (left) **module** over R, or an R-**module**, one means an additive group M, together with a map $R \times M \to M$, which to each pair (x, v) with $x \in R$ and $v \in M$ associates an element xv of M, satisfying the four conditions:

MOD 1. *If e is the unit element of R, then $ev = v$ for all $v \in M$.*

MOD 2. *If $x \in R$ and $v, w \in M$, then $x(v + w) = xv + xw$.*

MOD 3. *If $x, y \in R$ and $v \in M$, then $(x + y)v = xv + yv$.*

MOD 4. *If $x, y \in R$ and $v \in M$, then $(xy)v = x(yv)$.*

Example 1. Every left ideal of R is a module. The additive group consisting of 0 alone is an R-module for every ring R.

As with vector spaces, we have $0v = 0$ for every $v \in M$. (Note that the 0 in $0v$ is the zero element of R, while the 0 on the other side of the equation is the zero element of the additive group M. However, there will be no confusion in using the same symbol 0 for all zero elements everywhere.) Also, we have $(-e)v = -v$, with the same proof as for vector spaces.

Let M be a module over R and let N be a subgroup of M. We say that N is a **submodule** of M if whenever $v \in N$ and $x \in R$ then $xv \in N$. It follows that N is then itself a module.

Example 2. Let M be a module and $v_1, \ldots, v_n$ elements of M. Let N be the subset of M consisting of all elements

$$x_1 v_1 + \cdots + x_n v_n$$

with $x_i \in R$. Then N is a submodule of M. Indeed,

$$0 = 0v_1 + \cdots + 0v_n$$

so $0 \in N$. If $y_1, \ldots, y_n \in R$, then

$$x_1 v_1 + \cdots + x_n v_n + y_1 v_1 + \cdots + y_n v_n = (x_1 + y_1)v_1 + \cdots + (x_n + y_n)v_n$$

is in N. Finally, if $c \in R$, then

$$c(x_1 v_1 + \cdots + x_n v_n) = cx_1 v_1 + \cdots + cx_n v_n$$

is in N, so we have proved that N is a submodule. It is called the submodule **generated** by $v_1, \ldots, v_n$, and we call $v_1, \ldots, v_n$ **generators** for N.

Example 3. Let M be an (abelian) additive group, and let R be a subring of $\mathrm{End}(M)$. (We defined $\mathrm{End}(M)$ in Chapter III, §1 as the ring of homomorphisms of M into itself.) Then M is an R-module, if to each $f \in R$ and $v \in M$ we associate the element $fv = f(v) \in M$. The verification of the four conditions for a module is trivially carried out.

Conversely, given a ring R and an R-module M, to each $x \in R$ we associate the mapping $\lambda_x \colon M \to M$ such that $\lambda_x(v) = xv$ for $v \in M$. Then the association

$$x \mapsto \lambda_x$$

is a ring-homomorphism of R into $\mathrm{End}(M)$, where $\mathrm{End}(M)$ is the ring of endomorphisms of M viewed as additive group. This is but another way of formulating the four conditions **MOD 1** through **MOD 4**. For instance, **MOD 4** in the present notation can be written

$$\lambda_{xy} = \lambda_x \lambda_y \qquad \text{or} \qquad \lambda_{xy} = \lambda_x \circ \lambda_y$$

since the multiplication in $\mathrm{End}(M)$ is composition of mappings.

Warning. It may be that the ring-homomorphism $x \mapsto \lambda_x$ is not injective, so that in general, when dealing with a module, we cannot view R as a subring of $\mathrm{End}(M)$.

Example 4. Let us denote by K^n the set of *column vectors*, that is column n-tuples

$$X = \begin{pmatrix} x_1 \\ x_2 \\ \vdots \\ x_n \end{pmatrix} \qquad \text{with components} \quad x_i \in K.$$

Then K^n is a module over the ring $M_n(K)$. Indeed, matrix multiplication defines a mapping

$$M_n(K) \times K^n \to K^n$$

by

$$(A, X) \mapsto AX.$$

This multiplication satisfies the four axioms **MOD 1** through **MOD 4** for a module. This is one of the most important examples of modules in mathematics.

Let R be a ring, and let M, M' be R-modules. By an R-**linear** map (or R-**homomorphism**) $f: M \to M'$ one means a map such that for all $x \in R$ and $v, w \in M$ we have

$$f(xv) = xf(v), \qquad f(v + w) = f(v) + f(w).$$

Thus an R-**linear** map is the generalization of a K-linear map when the module is a vector space over a field.

The set of all R-linear maps of M into M' will be denoted by $\mathrm{Hom}_R(M, M')$.

Example 5. Let M, M', M'' be R-modules. If

$$f: M \to M' \quad \text{and} \quad g: M' \to M''$$

are R-linear maps, then the composite map $g \circ f$ is R-linear.

In analogy with previous definitions, we say that an R-homomorphism $f: M \to M'$ is an **isomorphism** if there exists an R-homomorphism $g: M' \to M$ such that $g \circ f$ and $f \circ g$ are the identity mappings of M and M', respectively. We leave it to the reader to verify that:

An R-homomorphism is an isomorphism if and only if it is bijective.

As with vector spaces and additive groups, we have to consider very frequently the set of R-linear maps of a module M into itself, and it is convenient to have a name for these maps. They are called R-**endomorphisms** of M. The set of R-endomorphisms of M is denoted by

$$\mathrm{End}_R(M).$$

We often suppress the prefix R- when the reference to the ring R is clear.

Let $f: M \to M'$ be a homomorphism of modules over R. We define the **kernel** of f to be its kernel viewed as a homomorphism of additive groups.

In analogy with previous results, we have:

Let $f: M \to M'$ be a homomorphism of R-modules. Then the kernel of f and the image of f are submodules of M and M' respectively.

Proof. Let E be the kernel of f. Then we already know that E is an additive subgroup of M. Let $v \in E$ and $x \in R$. Then

$$f(xv) = xf(v) = x0 = 0,$$

so $xv \in E$, and this proves that the kernel of f is a submodule of M. We already know that the image of f is a subgroup of M'. Let v' be in the image of f, and $x \in R$. Let v be an element of M such that

$$f(v) = v'.$$

Then $f(xv) = xf(v) = xv'$ also lies in the image of M, which is therefore a submodule of M', thereby proving our assertion.

Example 6. Let R be a ring, and M a left ideal. Let $y \in M$. The map

$$r_y \colon M \to M$$

such that

$$r_y(x) = xy$$

is an R-linear map of M into itself. Indeed, if $x \in M$ then $xy \in M$ since $y \in M$ and M is a left ideal, and the conditions for R-linearity are reformulations of definitions. For instance,

$$\begin{aligned} r_y(x_1 + x_2) &= (x_1 + x_2)y = x_1y + x_2y \\ &= r_y(x_1) + r_y(x_2). \end{aligned}$$

Furthermore, for $z \in R$, $x \in M$,

$$r_y(zx) = zxy = zr_y(x).$$

We call r_y **right multiplication** by y. Thus r_y is an R-endomorphism of M.

Observe that any abelian group can be viewed as a module over the integers. Thus an R-module M is also a $\mathbf{Z}$-module, and any R-endomorphism of M is also an endomorphism of M viewed as abelian group. Thus $\mathrm{End}_R(M)$ is a subset of $\mathrm{End}(M) = \mathrm{End}_{\mathbf{Z}}(M)$.

In fact, $\mathrm{End}_R(M)$ *is a subring of* $\mathrm{End}(M)$, *so that* $\mathrm{End}_R(M)$ *is itself a ring.*

The proof is routine. For instance, if f, $g \in \mathrm{End}_R(M)$, and $x \in R$, $v \in M$, then

$$\begin{aligned}(f + g)(xv) &= f(xv) + g(xv)\\ &= xf(v) + xg(v)\\ &= x\big(f(v) + g(v)\big)\\ &= x(f + g)(v).\end{aligned}$$

So $f + g \in \mathrm{End}_R(M)$. Equally easily,

$$(f \circ g)(xv) = f\big(g(xv)\big) = f\big((xg(v)\big) = xf\big(g(v)\big).$$

The identity is in $\mathrm{End}_R(M)$. This proves that $\mathrm{End}_R(M)$ is a subring of $\mathrm{End}_{\mathbf{Z}}(M)$.

We now also see that M can be viewed as a module over $\mathrm{End}_R(M)$ since M is a module over $\mathrm{End}_{\mathbf{Z}}(M) = \mathrm{End}(M)$.

Let us denote $\mathrm{End}_R(M)$ by $R'(M)$ or simply R' for clarity of notation. Let $f \in R'$ and $x \in R$. Then by definition,

$$f(xv) = xf(v),$$

and consequently

$$f \circ \lambda_x(v) = \lambda_x \circ f(v).$$

Hence λ_x is an R'-linear map of M into itself, i.e. an element of $\mathrm{End}_{R'}(M)$. The association

$$\lambda\colon x \mapsto \lambda_x$$

is therefore a ring-homomorphism of R into $\mathrm{End}_{R'}(M)$, not only into $\mathrm{End}(M)$.

Theorem 4.1. *Let R be a ring, and M an R-module. Let J be the set of elements $x \in R$ such that $xv = 0$ for all $v \in M$. Then J is a two-sided ideal of R.*

Proof. If $x, y \in J$, then $(x + y)v = xv + yv = 0$ for all $v \in M$. If $a \in R$, then

$$(ax)v = a(xv) = 0 \qquad \text{and} \qquad (xa)v = x(av) = 0$$

for all $v \in M$. This proves the theorem.

We observe that the two-sided ideal of Theorem 4.1 is none other than the kernel of the ring-homomorphism

$$x \mapsto \lambda_x$$

described in the preceding discussion.

Theorem 4.2 (Wedderburn–Rieffel). *Let R be a ring, and L a non-zero left ideal, viewed as R-module. Let $R' = \mathrm{End}_R(L)$, and $R'' = \mathrm{End}_{R'}(L)$. Let*

$$\lambda: R \to R''$$

be the ring-homomorphism such that $\lambda_x(y) = xy$ for $x \in R$ and $y \in L$. Assume that R has no two-sided ideals other than 0 and R itself. Then λ is a ring-isomorphism.

Proof. (Rieffel) The fact that λ is injective follows from Theorem 4.1, and the hypothesis that L is non-zero. Therefore, the only thing to prove is that λ is surjective. By Example 6 of Chapter III, §2, we know that LR is a two-sided ideal, non-zero since R has a unit, and hence equal to R by hypothesis. Then

$$\lambda(R) = \lambda(LR) = \lambda(L)\lambda(R).$$

We now contend that $\lambda(L)$ is a left ideal of R''. To prove this, let $f \in R''$, and let $x \in L$. For all $y \in L$, we know from Example 6 that r_y is in R', and hence that

$$f \circ r_y = r_y \circ f.$$

This means that $f(xy) = f(x)y$. We may rewrite this relation in the form

$$f \circ \lambda_x(y) = \lambda_{f(x)}(y).$$

Hence $f \circ \lambda_x$ is an element of $\lambda(L)$, namely $\lambda_{f(x)}$. This proves that $\lambda(L)$ is a left ideal of R''. But then

$$R''\lambda(R) = R''\lambda(L)\lambda(R) = \lambda(L)\lambda(R) = \lambda(R).$$

Since $\lambda(R)$ contains the identity map, say e, it follows that for every $f \in R''$, the map $f \circ e = f$ is obtained in $\lambda(R)$, i.e. R'' is contained in $\lambda(R)$, and therefore $R'' = \lambda(R)$, as was to be proved.

The whole point of Theorem 4.2 is that it represents R as a ring of endomorphisms of some module, namely the left ideal L. This is important in the following case.

Let D be a ring. We shall say that D is a **division ring** if the set of non-zero elements of D is a multiplicative group (and so in particular, $1 \neq 0$ in the ring). Note that a commutative division ring is what we called a field.

Let R be a ring, and M a module over R. We shall say that M is a **simple module** if $M \neq \{0\}$, and if M has no submodules other than $\{0\}$ and M itself.

Theorem 4.3 (Schur's Lemma). *Let M be a simple module over the ring R. Then $\operatorname{End}_R(M)$ is a division ring.*

Proof. We know it is a ring, and we must prove that every non-zero element f has an inverse. Since $f \neq 0$, the image of f is a submodule of $M \neq 0$ and hence is equal to all of M, so that f is surjective. The kernel of f is a submodule of M and is not equal to M, so that the kernel of f is 0, and f is therefore injective. Hence f has an inverse as a group-homomorphism, and it is verified at once that this inverse is an R-homomorphism, thereby proving our theorem.

Example 7. Let R be a ring, and L a left ideal which is simple as an R-module (we say then that L is a **simple left ideal**). Then $\operatorname{End}_R(L) = D$ is a division ring. If it happens that D is commutative, then under the hypothesis of Theorem 4.2, we conclude that $R \approx \operatorname{End}_D(L)$ is the ring of all D-linear maps of L into itself, and L is a vector space over the field D. Thus we have a concrete picture concerning the ring R. See Exercises 23 and 24.

Example 8. In Exercise 21 you will show that the ring of endomorphisms of a finite dimensional vector space satisfies the hypothesis of Theorem 4.2. In other words, $\operatorname{End}_K(V)$ has no two-sided ideal other than $\{0\}$ and itself. Furthermore, V is simple as an $\operatorname{End}_K(V)$-module (Exercise 18). Theorem 4.2 gives some sort of converse to this, and shows that it is a typical example.

Just as with groups, one tries to decompose a module over a ring into simple parts. In Chapter II, §3, Exercises 15 and 16, you defined the direct sum of abelian groups. We have the same notion for modules as follows.

Let $M_1, \ldots, M_q$ be modules over R. We can form their **direct product**

$$\prod_{i=1}^{q} M_i = M_1 \times \cdots \times M_q$$

consisting of all q-tuples $(v_1, \dots, v_q)$ with $v_i \in M_i$. This direct product is the direct product of $M_1, \dots, M_q$ viewed as abelian groups, and we can define the multiplication by an element $c \in R$ componentwise, that is

$$c(v_1, \dots, v_q) = (cv_1, \dots, cv_q).$$

It is then immediately verified that the direct product is an R-module.

On the other hand, let M be a module, and let $M_1, \dots, M_q$ be submodules. We say that M is the **direct sum** of $M_1, \dots, M_q$ if every element $v \in M$ has a unique expression as a sum

$$v = v_1 + \cdots + v_q \qquad \text{with} \quad v_i \in M_i.$$

If M is such a direct sum, then we denote this sum by

$$M = M_1 \oplus \cdots \oplus M_q \qquad \text{or also} \qquad \bigoplus_{i=1}^{q} M_i.$$

For any module M with submodules $M_1, \dots, M_q$ there is a natural homomorphism from the direct product into M, namely

$$\prod_{i=1}^{q} M_i \to M \qquad \text{given by} \qquad (v_1, \dots, v_q) \mapsto v_1 + \cdots + v_q.$$

Proposition 4.4. *Let M be a module.*

(a) *Let M_1, M_2 be submodules. We have $M = M_1 \oplus M_2$ if and only if $M = M_1 + M_2$ and $M_1 \cap M_2 = \{0\}$.*
(b) *The module M is a direct sum of submodules $M_1, \dots, M_q$ if and only if the natural homomorphism from the product $\prod M_i$ into M is an isomorphism.*
(c) *The sum $\sum M_i$ is a direct sum of $M_1, \dots, M_q$ if and only if, given a relation*

$$v_1 + \cdots + v_q = 0 \qquad \textit{with} \quad v_i \in M_i$$

we have $v_i = 0$ for $i = 1, \dots, q$.

Proof. The proof will be left as a routine exercise to the reader. Note that condition (c) is similar to a condition of linear independence.

In §7 you will see an example of a direct sum decomposition for modules over principal rings, similar to the decomposition of an abelian group.

V, §4. EXERCISES

1. Let R be a ring. Show that R can be viewed as a module over itself, and has one generator.

2. Let R be a ring and M an R-module. Show that $\mathrm{Hom}_R(R, M)$ and M are isomorphic as additive groups, under the mapping $f \mapsto f(1)$.

3. Let E, F be R-modules. Show that $\mathrm{Hom}_R(E, F)$ is a module over $\mathrm{End}_R(F)$, the operation of the ring $\mathrm{End}_R(F)$ on the additive group $\mathrm{Hom}_R(E, F)$ being composition of mappings.

4. Let E be a module over the ring R, and let L be a left ideal of R. Let LE be the set of all elements $x_1v_1 + \cdots + x_nv_n$ with $x_i \in L$ and $v_i \in E$. Show that LE is a submodule of E.

5. Let R be a ring, E a module, and L a left ideal. Assume that L and E are simple.
   (a) Show that $LE = E$ or $LE = \{0\}$.
   (b) Assume that $LE = E$. Define the notion of isomorphism of modules. Prove that L is isomorphic to E as R-module. [*Hint*: Let $v_0 \in E$ be an element such that $Lv_0 \neq \{0\}$. Show that the map $x \mapsto xv_0$ establishes an R-isomorphism between L and E.]

6. Let R be a ring and let E, F be R-modules. Let $\sigma: E \to F$ be an isomorphism. Show that $\mathrm{End}_R(E)$ and $\mathrm{End}_R(F)$ are ring-isomorphic, under the map
$$f \mapsto \sigma \circ f \circ \sigma^{-1}$$
for $f \in \mathrm{End}_R(E)$.

7. Let E, F be simple modules over the ring R. Let $f: E \to F$ be a homomorphism. Show that f is 0 or f is an isomorphism.

8. Verify in detail the last assertion made in the proof of Theorem 4.3.

Let R be a ring, and E a module. We say that E is a **free** module if there exist elements $v_1, \ldots, v_n$ in E such that every element $v \in E$ has a unique expression of the form
$$v = x_1v_1 + \cdots + x_nv_n$$
with $x_i \in R$. If this is the case, then $\{v_1, \ldots, v_n\}$ is called a **basis** of E (over R).

9. Let E be a free module over the ring R, with basis $\{v_1, \ldots, v_n\}$. Let F be a module, and $w_1, \ldots, w_n$ elements of F. Show that there exists a unique homomorphism $f: E \to F$ such that $f(v_i) = w_i$ for $i = 1, \ldots, n$.

10. Let R be a ring, and S a set consisting of n elements, say $s_1, \ldots, s_n$. Let F be the set of mappings from S into R.
    (a) Show that F is a module.
    (b) If $x \in R$, denote by xs_i the function of S into R which associates x to s_i and 0 to s_j for $j \neq i$. Show that F is a free module, that $\{1s_1, \ldots, 1s_n\}$ is a basis for F over R, and that every element $v \in F$ has a unique expression of the form $x_1s_1 + \cdots + x_ns_n$ with $x_i \in R$.

11. Let K be a field, and $R = K[X]$ the polynomial ring over K. Let J be the ideal generated by X^2. Show that R/J is a K-space. What is its dimension?

12. Let K be a field and $R = K[X]$ the polynomial ring over K. Let $f(X)$ be a polynomial of degree $d > 0$ in $K[X]$. Let J be the ideal generated by $f(X)$. What is the dimension of R/J over K? Exhibit a basis of R/J over K. Show that R/J is an integral ring if and only if f is irreducible.

13. If R is a *commutative* ring, and E, F are modules, show that $\operatorname{Hom}_R(E, F)$ is an R-module in a natural way. Is this still true if R is not commutative?

14. Let K be a field, and R a vector space over K of dimension 2. Let $\{e, u\}$ be a basis of R over K. If a, b, c, d are elements of K, define the product

$$(ae + bu)(ce + du) = ace + (bc + ad)u.$$

Show that this product makes R into a ring. What is the unit element? Show that this ring is isomorphic to the ring $K[X]/(X^2)$ of Exercise 11.

15. Let the notation be as in the preceding exercise. Let $f(X)$ be a polynomial in $K[X]$. Show that

$$f(ae + u) = f(a)e + f'(a)u,$$

where f' is the formal derivative of f.

16. Let R be a ring, and let E', E, F be R-modules. If $f: E' \to E$ is an R-homomorphism, show that the map $\varphi \mapsto f \circ \varphi$ is a $\mathbf{Z}$-homomorphism

$$\operatorname{Hom}_R(F, E') \to \operatorname{Hom}_R(F, E),$$

and is an R-homomorphism if R is commutative.

17. A sequence of homomorphisms of abelian groups

$$A \xrightarrow{f} B \xrightarrow{g} C$$

is said to be **exact** if $\operatorname{Im} f = \operatorname{Ker} g$. Thus to say that $0 \to A \xrightarrow{f} B$ is exact means that f is injective. Let R be a ring. If

$$0 \to E' \xrightarrow{f} E \xrightarrow{g} E''$$

is an exact sequence of R-modules, show that for every R-module F

$$0 \to \operatorname{Hom}_R(F, E') \to \operatorname{Hom}_R(F, E) \to \operatorname{Hom}_R(F, E'')$$

is an exact sequence.

18. Let V be a finite dimensional vector space over a field K. Let $R = \operatorname{End}_K(V)$. Prove that V is a simple R-module. [*Hint*: Given $v, w \in V$ and $v \neq 0$, $w \neq 0$ use Theorem 1.2 to show that there exists $f \in R$ such that $f(v) = w$, so $Rv = V$.]

19. Let R be the ring of $n \times n$ matrices over a field K. Show that the set of matrices of type

$$\begin{pmatrix} a_1 & 0 & \cdots & 0 \\ \vdots & \vdots & & \vdots \\ a_n & 0 & \cdots & 0 \end{pmatrix}$$

having components equal to 0 except possibly on the first column, is a left ideal of R. Prove a similar statement for the set of matrices having components 0 except possibly on the j-th column.

20. Let A, B be $n \times n$ matrices over a field K, all of whose components are equal to 0 except possibly those of the first column. Assume $A \neq O$. Show that there exists an $n \times n$ matrix C over K such that $CA = B$. *Hint*: Consider first a special case where

$$A = \begin{pmatrix} 1 & 0 & \cdots & 0 \\ 0 & 0 & \cdots & 0 \\ \vdots & \vdots & & \vdots \\ 0 & 0 & \cdots & 0 \end{pmatrix}.$$

21. Let V be a finite dimensional vector space over the field K. Let R be the ring of K-linear maps of V into itself. Show that R has no two-sided ideals except $\{O\}$ and R itself. [*Hint*: Let $A \in R$, $A \neq O$. Let $v_1 \in V$, $v_1 \neq 0$, and $Av_1 \neq 0$. Complete v_1 to a basis $\{v_1, \ldots, v_n\}$ of V. Let $\{w_1, \ldots, w_n\}$ be arbitrary elements of V. For each $i = 1, \ldots, n$ there exists $B_i \in R$ such that

$$B_i v_i = v_1 \quad \text{and} \quad B_i v_j = 0 \quad \text{if} \quad j \neq i,$$

and there exists $C_i \in R$ such that $C_i A v_1 = w_i$ (justify these two existence statements in detail). Let $F = C_1 A B_1 + \cdots + C_n A B_n$. Show that $F(v_i) = w_i$ for all $i = 1, \ldots, n$. Conclude that the two-sided ideal generated by A is the whole ring R.]

22. Let V be a vector space over a field K and let R be a subring of $\mathrm{End}_K(V)$ containing all the scalar maps, i.e. all maps cI with $c \in K$. Let L be a left ideal of R. Let LV be the set of all elements $A_1 v_1 + \cdots + A_n v_n$ with $A_i \in L$ and $v_i \in V$, and all positive integers n. Show that LV is a subspace W of V such that $RW \subset W$. A subspace having this property is called R-**invariant**.

23. Let D be a division ring containing a field K as a subfield. We assume that K is contained in the center of D.
    (a) Verify that the addition and the multiplication in D allow us to view D as a vector space over K.
    (b) Assume that D is finite dimensional over K. Let $\alpha \in D$. Show that there exists a polynomial $f(t) \in K[t]$ of degree $\geqq 1$ such that $f(\alpha) = 0$. [*Hint*: For some n, the powers $1, \alpha, \alpha^2, \ldots, \alpha^n$ must be linearly dependent over K.]

For the rest of the exercises, remember Corollary 1.8 of Chapter IV.

(c) Assume that K is algebraically closed. Let D be a finite dimensional division ring over K as in parts (a) and (b). Prove that $D = K$, in other words, show that every element of D lies in K.

24. Let R be a ring containing a field K as a subfield with K in the center of R. Assume that K is algebraically closed. Assume that R has no two-sided ideal other than 0 and R. We also assume that R is of finite dimension > 0 over K. Let L be a left ideal of R, of smallest dimension > 0 over K.
 (a) Prove that $\mathrm{End}_R(L) = K$ (i.e. the only R-linear maps of L consist of multiplication by elements of K). [*Hint*: Cf. Schur's lemma and Exercise 23.]
 (b) Prove that R is ring-isomorphic to the ring of K-linear maps of L into itself. [*Hint*: Use Wedderburn-Rieffel.]

V, §5. FACTOR MODULES

We have already studied factor groups, and rings modulo a two-sided ideal. We shall now study the analogous notion for a module.

Let R be a ring, and M an R-module. By a **submodule** N we shall mean an additive subgroup of M which is such that for all $x \in R$ and $v \in N$ we have $xv \in N$. Thus N itself is a module (i.e. R-module).

We already know how to construct the factor group M/N. Since M is an abelian group, N is automatically normal in M, so this is an old story. The elements of the factor group are the cosets $v + N$ with $v \in M$. We shall now define a multiplication of these cosets by elements of R. This we do in the natural way. If $x \in R$, we define $x(v + N)$ to be the coset $xv + N$. If v_1 is another coset representative of $v + N$, then we can write $v_1 = v + w$ with $w \in N$. Hence

$$xv_1 = xv + xw,$$

and $xw \in N$. Consequently $xv_1 + N = xv + N$. Thus our definition is independent of the choice of representative v of the coset $v + N$. It is now trivial to verify that all the axioms of a module are satisfied by this multiplication. We call M/N the **factor module** of M by N, and also M **modulo** N.

We could also use the notation of congruences. If v, v' are elements of M, we write

$$v \equiv v' \pmod{N}$$

to mean that $v - v' \in N$. This amounts to saying that the cosets $v + N$ and $v' + N$ are equal. Thus a coset $v + N$ is nothing but the congruence class of elements of M which are congruent to $v \bmod N$. We can rephrase our statement that the multiplication of a coset by x is well defined as follows: If $v \equiv v' \pmod{N}$, then for all $x \in R$, we have $xv \equiv xv' \pmod{N}$.

Example 1. Let V be a vector space over the field K. Let W be a subspace. Then the factor module V/W is called the **factor space** in this case.

Let M be an R-module, and N a submodule. The map

$$f: M \to M/N$$

which to each $v \in M$ associates its congruences class $f(v) = v + N$ is obviously an R-homomorphism, because

$$f(xv) = xv + N = x(v + N) = xf(v)$$

by definition. It is called the **canonical** homomorphism. Its kernel is N.

Example 2. Let V be a vector space over the field K. Let W be a subspace. Then the canonical homomorphism $f: V \to V/W$ is a linear map, and is obviously surjective. Suppose that V is finite dimensional over K, and let W' be a subspace of V such that V is the direct sum, $V = W \oplus W'$. If $v \in V$, and we write $v = w + w'$ with $w \in W$ and $w' \in W'$, then $f(v) = f(w) + f(w') = f(w')$. Let us just consider the map f on W', and let us denote this map by f'. Thus for all $w' \in W'$ we have $f'(w') = f(w')$ by definition. Then f' maps W' onto V/W, and the kernel of f' is $\{0\}$, because $W \cap W' = \{0\}$. *Hence $f': W' \to V/W$ is an isomorphism between the complementary subspace W' of W and the factor space V/W.* We have such an isomorphism for any choice of complementary subspace W'.

V, §5. EXERCISES

1. Let V be a finite dimensional vector space over the field K, and let W be a subspace. Let $\{v_1, \ldots, v_r\}$ be a basis of W, and extend it to a basis $\{v_1, \ldots, v_n\}$ of V. Let $f: V \to V/W$ be the canonical map. Show that

$$\{f(v_{r+1}), \ldots, f(v_n)\}$$

is a basis of V/W.

2. Let V and W be as in Exercise 1. Let

$$A: V \to V$$

be a linear map such that $AW \subset W$, i.e. $Aw \in W$ for all $w \in W$. Show how to define a linear map

$$\bar{A}: V/W \to V/W,$$

by defining

$$\bar{A}(v + W) = Av + W.$$

(In the congruence terminology, if $v \equiv v' \pmod{W}$, then $Av \equiv Av' \pmod{W}$.) Write $\bar{v}$ instead of $v + W$. We call $\bar{A}$ the linear map induced by A on the factor space.

3. Let V be the vector space generated over $\mathbf{R}$ by the functions 1, t, t^2, e^t, te^t, t^2e^t. Let W be the subspace generated by 1, t, t^2, e^t, te^t. Let D be the derivative.
   (a) Show that D maps W into itself.
   (b) What is the linear map $\bar{D}$ induced by D on the factor space V/W?
4. Let V be the vector space over $\mathbf{R}$ consisting of all polynomials of degree $\leqq n$ (for some integer $n \geqq 1$). Let W be the subspace consisting of all polynomials of degree $\leqq n - 1$. What is the linear map $\bar{D}$ induced by the derivative D on the factor space V/W?
5. Let V, W be as in Exercise 1. Let $A: V \to V$ be a linear map, and assume that $AW \subset W$. Let $\{v_1, \dots, v_n\}$ be the basis of V as in Exercise 1.
   (a) Show that the matrix of A with respect to this basis is of type

$$\begin{pmatrix} M_1 & M_3 \\ O & M_2 \end{pmatrix},$$

   where M_1 is a square $r \times r$ matrix, and M_2 is a square $(n - r) \times (n - r)$ matrix.
   (b) In Exercise 2, show that the matrix of $\bar{A}$ with respect to the basis $\{\bar{v}_{r+1}, \dots, \bar{v}_n\}$ is precisely the matrix M_2.

V, §6. FREE ABELIAN GROUPS

We shall deal with commutative groups throughout this section. We wish to analyze under which conditions we can define the analogue of a basis for such groups.

Let A be an abelian group. By a **basis** for A we shall mean a set of elements $v_1, \dots, v_n$ $(n \geqq 1)$ of A such that every element of A has a unique expression as a sum

$$c_1v_1 + \cdots + c_nv_n$$

with integers $c_i \in \mathbf{Z}$. Thus a basis for an abelian group is defined in a manner entirely similar to a basis for vector spaces, except that the coefficients $c_1, \dots, c_n$ are now required to be integers.

Theorem 6.1. *Let A be an abelian group with a basis $\{v_1,\dots,v_n\}$. Let B be an abelian group, and let $w_1,\dots,w_n$ be elements of B. Then there exists a unique group-homomorphism $f: A \to B$ such that $f(v_i) = w_i$ for all $i = 1,\dots,n$.*

Proof. Copy the analogous proof for vector spaces, omitting irrelevant scalar multiplication, etc.

To avoid confusion when dealing with bases of abelian groups as above, and vector space bases, we shall call bases of abelian groups **Z-bases**.

Theorem 6.2. *Let A be a non-zero subgroup of $\mathbf{R}^n$. Assume that in any bounded region of space there exists only a finite number of elements of A. Let m be the maximal number of elements of A which are linearly independent over $\mathbf{R}$. Then we can select m elements of A which are linearly independent over $\mathbf{R}$, and form a $\mathbf{Z}$-basis of A.*

Proof. Let $\{w_1,\dots,w_m\}$ be a maximal set of elements of A linearly independent over $\mathbf{R}$. Let V be the vector space generated by these elements, and let V_{m-1} be the space generated by $w_1,\dots,w_{m-1}$. Let A_{m-1} be the intersection of A and V_{m-1}. Then certainly, in any bounded region of space, there exists only a finite number of elements of A_{m-1}. Therefore, if $m > 1$, we could have chosen inductively $\{w_1,\dots,w_m\}$ such that $\{w_1,\dots,w_{m-1}\}$ is a $\mathbf{Z}$-basis of A_{m-1}.

Now consider the set S of all elements of A which can be written in the form

$$t_1 w_1 + \cdots + t_m w_m$$

with $0 \leqq t_i < 1$ if $i = 1,\dots,m-1$ and $0 \leqq t_m \leqq 1$. This set S is certainly bounded, and hence contains only a finite number of elements (among which is w_m). We select an element v_m in this set whose last coordinate t_m is the smallest possible > 0. We shall prove that

$$\{w_1,\dots,w_{m-1}, v_m\}$$

is a $\mathbf{Z}$-basis for A. Write v_m as a linear combination of $w_1,\dots,w_m$ with real coefficients,

$$v_m = c_1 w_1 + \cdots + c_m w_m, \qquad 0 < c_m \leqq 1.$$

Let v be an element of A, and write

$$v = x_1 w_1 + \cdots + x_m w_m$$

with $x_i \in \mathbf{R}$. Let q_m be the integer such that

$$q_m c_m \leqq x_m < (q_m + 1)c_m.$$

Then the last coordinate of $v - q_m v_m$ with respect to $\{w_1, \dots, w_m\}$ is equal to $x_m - q_m c_m$, and

$$\begin{aligned} 0 &\leqq x_m - q_m c_m \\ &< (q_m + 1)c_m - q_m c_m = c_m \leqq 1. \end{aligned}$$

Let q_i $(i = 1, \dots, m-1)$ be integers such that

$$q_i \leqq x_i - q_m c_i < q_i + 1.$$

Then

$$v - q_m v_m - q_1 w_1 - \cdots - q_{m-1} w_{m-1} \tag{1}$$

is an element of S. If its last coordinate is not 0, then it would be an element with last coordinate smaller than c_m, contrary to the construction of v_m. Hence its last coordinate is 0, and hence the element of (1) lies in V_{m-1}. By induction, it can be written as a linear combination of $w_1, \dots, w_{m-1}$ with integer coefficients, and from this it follows at once that v can be written as a linear combination of $w_1, \dots, w_{m-1}, v_m$ with integral coefficients. Furthermore, it is clear that $w_1, \dots, w_{m-1}, v_m$ are linearly independent over $\mathbf{R}$, and hence satisfy the requirements of our theorem.

We can now apply our theorem to more general groups. Let A be an additive group, and let $f: A \to A'$ be an isomorphism of A with a group A'. If A' admits a basis, say $\{v'_1, \dots, v'_n\}$, and if v_i is the element of A such that $f(v_i) = v'_i$, then it is immediately verified that $\{v_1, \dots, v_n\}$ is a basis of A.

Theorem 6.3. *Let A be an additive group, having a basis with n elements. Let B be a subgroup $\neq \{0\}$. Then B has a basis with $\leqq n$ elements.*

Proof. Let $\{v_1, \dots, v_n\}$ be a basis for A. Let $\{e_1, \dots, e_n\}$ be the standard unit vectors of $\mathbf{R}^n$. By Theorem 6.1, there is a homomorphism

$$f: A \to \mathbf{R}^n$$

such that $f(v_i) = e_i$ for $i = 1, \dots, n$, and this homomorphism is obviously injective. Hence it gives an isomorphism of A with its image in $\mathbf{R}^n$. On the other hand, it is trivial to verify that in any bounded region of $\mathbf{R}^n$, there is only a finite number of elements of the image $f(A)$, because in any bounded region, the coefficients of a vector

$$(c_1,\dots,c_n)$$

are bounded. Hence by Theorem 6.2 we conclude that $f(B)$ has a **Z**-basis, whence B has a **Z**-basis, with $\leqq n$ elements.

Theorem 6.4. *Let A be an additive group having a basis with n elements. Then all bases of A have this same number of elements n.*

Proof. We look at our same homomorphism $f: A \to \mathbf{R}^n$ as in the proof of Theorem 6.3. Let $\{w_1,\dots,w_m\}$ be a basis of A. Each v_i is a linear combination with integer coefficients of $w_1,\dots,w_m$. Hence $f(v_i) = e_i$ is a linear combination with integer coefficients of $f(w_1),\dots,f(w_m)$. Hence $e_1,\dots,e_n$ are in the space generated by $f(w_1),\dots,f(w_m)$. By the theory of bases for *vector spaces*, we conclude that $m \geqq n$, whence $m = n$.

An abelian group is said to be **finitely generated** if it has a finite number of generators. It is said to be **free** if it has a basis.

Corollary 6.5. *Let $A \neq \{0\}$ be a finitely generated abelian group. Assume that A does not contain any element of finite period except the unit element. Then A has a basis.*

Proof. The proof will be left as an exercise, see Exercises 1 and 2 which also give you the main ideas for the proof.

Remark. We have carried out the above theory in Euclidean space to emphasize certain geometric aspects. Similar ideas can be carried out to prove Corollary 6.5 without recourse to Euclidean space. See for instance my *Algebra.*

Let A be an abelian group. In Chapter II, §7 we defined the **torsion subgroup** A_{tor} to be the subgroup consisting of all the elements of finite period in A. We refer to Chapter II, Theorem 7.1.

Theorem 6.6. *Suppose that A is a finitely generated abelian group. Then A/A_{tor} is a free abelian group, and A is the direct sum*

$$A = A_{\mathrm{tor}} \oplus F$$

where F is free.

Proof. Let $\{a_1,\dots,a_m\}$ be generators of A. If $a \in A$ let $\bar{a}$ be its image in the factor group A/A_{tor}. Then $\bar{a}_1,\dots,\bar{a}_m$ are generators of A/A_{tor}, which is therefore finitely generated. Let $\bar{a} \in A/A_{\text{tor}}$, and suppose $\bar{a}$ has finite period, say $d\bar{a} = 0$ for some positive integer d. This means that $da \in A_{\text{tor}}$, so there exists a positive integer d' such that $d'da = 0$, whence a itself lies in A_{tor} so $\bar{a} = 0$. Hence the torsion subgroup of A/A_{tor} is trivial. By Corollary 6.5 we conclude that A/A_{tor} is free.

Let $b_1,\ldots,b_r \in A$ be elements such that $\{\bar{b}_1,\ldots,\bar{b}_r\}$ is a basis of A/A_{tor}. Then $b_1,\ldots,b_r$ are linearly independent over $\mathbf{Z}$. Indeed, suppose $d_1,\ldots,d_r$ are integers such that

$$d_1b_1 + \cdots + d_rb_r = 0.$$

Then

$$d_1\bar{b}_1 + \cdots + d_r\bar{b}_r = 0,$$

whence $d_i = 0$ for all i by assumption that $\{\bar{b}_1,\ldots,\bar{b}_r\}$ is a basis of A/A_{tor}. Let F be the subgroup generated by $b_1,\ldots,b_r$. Then F is free. We now claim that

$$A = A_{\text{tor}} \oplus F.$$

Indeed, let $a \in A$. There exist integers $x_1,\ldots,x_r$ such that

$$\bar{a} = x_1\bar{b}_1 + \cdots + x_r\bar{b}_r.$$

Hence $\bar{a} - (x_1\bar{b}_1 + \cdots + x_r\bar{b}_r) = 0$, so $a - (x_1b_1 + \cdots + x_rb_r) \in A_{\text{tor}}$. This proves that

$$A = A_{\text{tor}} + F.$$

In addition, suppose $a \in A_{\text{tor}} \cap F$. The only element of finite order in a free group is the 0 element. Hence $a = 0$. This proves the theorem.

V, §6. EXERCISES

1. Let A be an abelian group with a finite number of generators, and assume that A does not contain any element of finite period except the unit element. We write A additively. Let d be a positive integer. Show that the map $x \mapsto dx$ is an injective homomorphism of A into itself, whose image is isomorphic to A.
2. Let the notation be as in Exercise 1. Let $\{a_1,\ldots,a_m\}$ be a set of generators of A. Let $\{a_1,\ldots,a_r\}$ be a maximal subset linearly independent over $\mathbf{Z}$. Let B be the subgroup generated by $a_1,\ldots,a_r$. Show that there exists a positive integer d such that dx lies in B for all x in A. Using Theorem 6.3 and Exercise 1, conclude that A has a basis.

V, §7. MODULES OVER PRINCIPAL RINGS

Throughout this section, we let R be a principal ring.

Let M be an R-module, $M \neq \{0\}$. If M is generated by one element v, then we say that M is **cyclic**. In this case, we have $M = Rv$. The map

$$x \mapsto xv$$

is a homomorphism of R into M, viewing both R and M as R-modules. Let J be the kernel of this homomorphism. Then J consists of all elements $x \in R$ such that $xv = 0$. Since R is assumed principal, either $J = \{0\}$ or there is some element $a \in R$ which generates J. Then we have an isomorphism of R-modules

$$R/J = R/aR \approx M.$$

The element a is uniquely determined up to multiple by a unit of R, and we shall say that a is a **period** of v. Any unit multiple of a will also be called a period of v.

Let M be an R-module. We say that M is a **torsion module** if given $v \in M$ there exists some element $a \in R$, $a \neq 0$ such that $av = 0$. Let p be a prime element of R. We denote by $M(p)$ the subset of elements $v \in M$ such that there exists some power p^r $(r \geqq 1)$ satisfying $p^r v = 0$. These definitions are analogous to those made for finite abelian groups in Chapter II, §7. A module is said to be **finitely generated** if it has a finite number of generators. As with abelian groups, we say that M has **exponent** a if every element of M has period dividing a, or equivalently $av = 0$ for all $v \in M$.

Observe that if M is finitely generated and is a torsion module, then there exists $a \in R$, $a \neq 0$, such that $aM = \{0\}$. (Proof?)

The following statements are used just as for abelian groups, and also illustrate some notions given more generally in §4 and §5.

Let p be a prime of R. Then R/pR is a simple module.

Let a, $b \in R$ be relatively prime. Let M be a module such that $aM = 0$. Then the map

$$v \mapsto bv$$

is an automorphism of M.

The proofs will be left as exercises.

The next theorem is entirely analogous to Theorem 7.1 of Chapter II.

Theorem 7.1. *Let M be a finitely generated torsion module over the principal ring R. Then M is the direct sum of its submodules $M(p)$ for all primes p such that $M(p) \neq \{0\}$. In fact, let $a \in R$ be such that*

$aM = 0$, and suppose we can write

$$a = bc \qquad \text{with } b, c \text{ relatively prime.}$$

Let M_b be the subset of M consisting of those elements v such that $bv = 0$, and similarly for M_c. Then

$$M = M_b \oplus M_c.$$

Proof. You can just copy the proof of Theorem 7.1 of Chapter II, since all the notions which we used there for abelian groups have been defined for modules over a principal ring. This kind of translation just continues what we did in Chapter IV, §6 for principal rings themselves.

Theorem 7.2. *Let M be a finitely generated torsion module over the principal ring R, and assume that there is a prime p such that $p^r M = \{0\}$ for some positive integer r. Then M is a direct sum of cyclic submodules*

$$M = \bigoplus_{i=1}^{q} Rv_i \qquad \text{where} \quad Rv_i \approx R/p^{r_i}R,$$

so v_i has period p^{r_i}. If we order these modules so that

$$r_1 \geqq r_2 \geqq \cdots \geqq r_s \geqq 1,$$

then the sequence of integers $r_1, \ldots, r_s$ is uniquely determined.

Proof. Again, the proof is similar to the proof of Theorem 7.2 in Chapter II. We repeat the proof here for convenience of the reader, so that you can see how to translate a slightly more involved proof from the integers to principal rings.

We start with a remark. Let $v \in M$, $v \neq 0$. Let k be an integer $\geqq 0$ such that $p^k v \neq 0$ and let p^m be a period of $p^k v$. Then v has period p^{k+m}. *Proof*: We certainly have $p^{k+m}v = 0$, and if $p^n v = 0$ then first $n \geqq k$, and second $n \geqq k + m$, otherwise the period of $p^k v$ would divide p^m and not be equal to p^m, contrary to the definition of p^m.

We shall now prove the theorem by induction on the number of generators. Suppose that M is generated by q elements. After reordering these elements if necessary, we may assume that one of these elements v_1 has maximal period. In other words, v_1 has period p^{r_1}, and if $v \in M$ then $p^r v = 0$ with $r \leqq r_1$. We let M_1 be the cyclic module generated by v_1.

Lemma 7.3. *Let $\bar{v}$ be an element of M/M_1, of period p^r. Then there exists a representative w of $\bar{v}$ in M which also has period p^r.*

Proof. Let v be any representative of $\bar{v}$ in M. Then $p^r v$ lies in M_1, say $p^r v = cv_1$ with $c \in R$. We note that the period of $\bar{v}$ divides the period of v. Write $c = p^k t$ where t is prime to p. Then tv_1 is also a generator of M_1 (proof?), and hence has period p^{r_1}. By our previous remark, the element v has period

$$p^{r+r_1-k},$$

whence by hypothesis, $r + r_1 - k \leqq r_1$ and $r \leqq k$. This proves that there exists an element $w_1 \in M_1$ such that $p^r v = p^r w_1$. Let $w = v - w_1$. Then w is a representative for $\bar{v}$ in M and $p^r w = 0$. Since the period of w is at least p^r we conclude that w has period equal to p^r. This proves the lemma.

We return to the main proof. The factor module M/M_1 is generated by $q - 1$ elements, and so by induction M/M_1 has a direct sum decomposition

$$M/M_1 = \bar{M}_2 \oplus \cdots \oplus \bar{M}_s$$

into cyclic modules with $\bar{M}_i \approx R/p^{r_i}R$. Let $\bar{v}_i$ be a generator for $\bar{M}_i$ ($i = 2,\ldots,s$), and let v_i be a representative in M of the same period as $\bar{v}_i$ according to the lemma. Let M_i be the cyclic module generated by v_i. We contend that M is the direct sum of $M_1,\ldots,M_s$, and we now prove this.

Given $v \in M$, let $\bar{v}$ denote its residue class in M/M_1. Then there exist elements $c_2,\ldots,c_s \in R$ such that

$$\bar{v} = c_2\bar{v}_2 + \cdots + c_s\bar{v}_s.$$

Hence $v - (c_2 v_2 + \cdots + c_s v_s)$ lies in M_1, and there exists an element $c_1 \in R$ such that

$$v = c_1 v_1 + c_2 v_2 + \cdots + c_s v_s.$$

Hence $M = M_1 + \cdots + M_s$.

Conversely, suppose $c_1,\ldots,c_s$ are elements of R such that

$$c_1 v_1 + \cdots + c_s v_s = 0.$$

Since v_i has period p^{r_i} ($i = 1,\ldots,s$), if we write $c_i = p^{m_i} t_i$ with $p \nmid t_i$, then we may suppose $m_i < r_i$. Putting a bar on this equation yields

$$c_2\bar{v}_2 + \cdots + c_s\bar{v}_s = 0.$$

Since $M/M_1 = \bar{M}$ is a direct sum of $\bar{M}_2, \ldots, \bar{M}_s$ we conclude that $c_j\bar{v}_j = 0$ for $j = 2, \ldots, s$. (See part (c) of Proposition 4.4.) Hence p^{r_j} divides c_j for $j = 2, \ldots, s$, whence also $c_j v_j = 0$ for $j = 2, \ldots, s$ since v_j and $\bar{v}_j$ have the same period. This proves that M is the direct sum of $M_1, \ldots, M_s$ and concludes the proof of the existence part of the theorem.

Now we prove uniqueness. If

$$M \approx R/p^{r_1}R \oplus \cdots \oplus R/p^{r_s}R$$

then we say that M has **type** $(p^{r_1}, \ldots, p^{r_s})$, just as we did for abelian groups. Suppose that M is written in two ways as a product of cyclic submodules, say of types

$$(p^{r_1}, \ldots, p^{r_s}) \quad \text{and} \quad (p^{m_1}, \ldots, p^{m_k})$$

with $r_1 \geqq r_2 \geqq \cdots \geqq r_s \geqq 1$, and $m_1 \geqq m_2 \geqq \cdots \geqq m_k \geqq 1$. Then pM is also a torsion module, of type

$$(p^{r_1 - 1}, \ldots, p^{r_s - 1}) \quad \text{and} \quad (p^{m_1 - 1}, \ldots, p^{m_k - 1}).$$

It is understood that if some exponent r_i or m_j is equal to 1, then the factor corresponding to

$$p^{r_i - 1} \quad \text{or} \quad p^{m_j - 1}$$

in pM is simply the trivial module 0. We now make an induction on the sum $r_1 + \cdots + r_s$, which may be called the **length** of the module. If this length is 1 in some representation of M as a direct sum, then this length is 1 in every representation, because R/pR is a simple module, whereas it is immediately verified that if the length is > 1, then there exists a submodule $\neq 0$ and $\neq R$, so the module cannot be simple. Thus the uniqueness is proved for modules of length 1.

By induction, the subsequence of $(r_1 - 1, \ldots, r_s - 1)$ consisting of those integers $\geqq 1$ is uniquely determined, and is the same as the corresponding subsequence of $(m_1 - 1, \ldots, m_k - 1)$. In other words, we have $r_i - 1 = m_i - 1$ for all those integers i such that $r_i - 1$ and $m_i - 1 \geqq 1$. Hence $r_i = m_i$ for all these integers i, and the two sequences

$$(p^{r_1}, \ldots, p^{r_s}) \quad \text{and} \quad (p^{m_1}, \ldots, p^{m_k})$$

can differ only in their last components which can be equal to p. These correspond to factors of type $(p, \ldots, p)$ occurring say ν times in the first sequence, and μ times in the second sequence. We have to show that $\nu = \mu$.

Let M_p be the submodule of M consisting of all elements $v \in M$ such that $pv = 0$. Since by hypothesis $pM_p = 0$, it follows that M_p is a module over R/pR, which is a field, so M_p is a vector space over R/pR. If N is a cyclic module, say $N = R/p^rR$, then $N_p = p^{r-1}R/p^rR \approx R/pR$, and so N_p has dimension 1 over R/pR. Hence

$$\dim_{R/pR} M_p = s \quad \text{and also} = k,$$

so $s = k$. But we have seen in the preceding paragraph that

$$(p^{r_1}, \dots, p^{r_s}) = (p^{r_1}, \dots, p^{r_{s-\nu}}, \underbrace{p, \dots, p}_{\nu \text{ times}}),$$

$$(p^{m_1}, \dots, p^{m_k}) = (p^{r_1}, \dots, p^{r_{k-\mu}}, \underbrace{p, \dots, p}_{\mu \text{ times}}).$$

and that $s - \nu = k - \mu$. Since $s = k$ it follows that $\nu = \mu$, and the theorem is proved.

Remark. In the analogous statement for abelian groups, we finished the proof by considering the order of groups. Here we use an argument having to do with the dimension of a vector space over R/pR. Otherwise, the argument is entirely similar.

V, §7. EXERCISES

1. Let a, b be relatively prime elements of R. Let M be an R-module, and denote by $a_M \colon M \to M$ multiplication by a. In other words, $a_M(v) = av$ for $v \in M$. Suppose that $a_M b_M = 0$. Prove that

$$\operatorname{Im} a_M = \operatorname{Ker} b_M$$

2. Let M be a finitely generated torsion module over R. Prove that there exists $a \in R$, $a \neq 0$ such that $a_M = 0$.

3. Let p be a prime of R. Prove that R/pR is a simple R-module.

4. Let $a \in R$, $a \neq 0$. If a is not prime, prove that R/aR is not a simple R-module.

5. Let $a, b \in R$ be relatively prime. Let M be a module such that $a_M = 0$. Prove that the map $v \mapsto bv$ is an automorphism of M.

V, §8. EIGENVECTORS AND EIGENVALUES

Let V be a vector space over a field K, and let

$$A \colon V \to V$$

be a linear map of V into itself. An element $v \in V$ is called an **eigenvector** of A if there exists $\lambda \in K$ such that $Av = \lambda v$. *If $v \neq 0$ then λ is uniquely determined*, because $\lambda_1 v = \lambda_2 v$ implies $\lambda_1 = \lambda_2$. In this case, we say that λ is an **eigenvalue** of A belonging to the eigenvector v. We also say that v is an eigenvector with the eigenvalue λ. Instead of eigenvector and eigenvalue, one also uses the terms **characteristic vector** and **characteristic value**.

If A is a square $n \times n$ matrix then an **eigenvector** of A is by definition an eigenvector of the linear map of K^n into itself represented by this matrix. Thus an eigenvector X of A is a (column) vector of K^n for which there exists $\lambda \in K$ such that $AX = \lambda X$.

Example 1. Let V be the vector space over $\mathbf{R}$ consisting of all infinitely differentiable functions. Let $\lambda \in \mathbf{R}$. Then the function f such that $f(t) = e^{\lambda t}$ is an eigenvector of the derivative d/dt because $df/dt = \lambda e^{\lambda t}$.

Example 2. Let

$$A = \begin{pmatrix} a_1 & \cdots & 0 \\ \vdots & \ddots & \vdots \\ 0 & \cdots & a_n \end{pmatrix}$$

be a diagonal matrix. Then every unit vector E^i $(i = 1, \dots, n)$ is an eigenvector of A. In fact, we have $AE^i = a_i E^i$:

$$\begin{pmatrix} a_1 & 0 & \cdots & 0 \\ 0 & a_2 & \cdots & 0 \\ \vdots & \vdots & & \vdots \\ 0 & 0 & \cdots & a_n \end{pmatrix} \begin{pmatrix} 0 \\ \vdots \\ 1 \\ \vdots \\ 0 \end{pmatrix} = \begin{pmatrix} 0 \\ \vdots \\ a_i \\ \vdots \\ 0 \end{pmatrix}.$$

Example 3. If $A: V \to V$ is a linear map, and v is an eigenvector of A, then for any non-zero scalar c, cv is also an eigenvector of A, with the same eigenvalue.

Theorem 8.1. *Let V be a vector space and let $A: V \to V$ be a linear map. Let $\lambda \in K$. Let V_λ be the subspace of V generated by all eigenvectors of A having λ as eigenvalue. Then every non-zero element of V_λ is an eigenvector of A having λ as eigenvalue.*

Proof. Let $v_1, v_2 \in V$ be such that $Av_1 = \lambda v_1$ and $Av_2 = \lambda v_2$. Then

$$A(v_1 + v_2) = Av_1 + Av_2 = \lambda v_1 + \lambda v_2 = \lambda(v_1 + v_2).$$

If $c \in K$ then $A(cv_1) = cAv_1 = c\lambda v_1 = \lambda c v_1$. This proves our theorem.

The subspace V_λ in Theorem 8.1 is called the **eigenspace** of A belonging to λ.

Note. If v_1, v_2 are eigenvectors of A with different eigenvalues $\lambda_1 \neq \lambda_2$ then of course $v_1 + v_2$ is *not* an eigenvector of A. In fact, we have the following theorem:

Theorem 8.2. *Let V be a vector space and let $A: V \to V$ be a linear map. Let $v_1, \ldots, v_m$ be eigenvectors of A, with eigenvalues $\lambda_1, \ldots, \lambda_m$ respectively. Assume that these eigenvalues are distinct, i.e.*

$$\lambda_i \neq \lambda_j \quad \text{if} \quad i \neq j.$$

Then $v_1, \ldots, v_m$ are linearly independent.

Proof. By induction on m. For $m = 1$, an element $v_1 \in V$, $v_1 \neq 0$ is linearly independent. Assume $m > 1$. Suppose that we have a relation

$$(*) \qquad c_1 v_1 + \cdots + c_m v_m = 0$$

with scalars c_i. We must prove all $c_i = 0$. We multiply our relation $(*)$ by λ_1 to obtain

$$c_1 \lambda_1 v_1 + \cdots + c_m \lambda_1 v_m = 0.$$

We also apply A to our relation $(*)$. By linearity, we obtain

$$c_1 \lambda_1 v_1 + \cdots + c_m \lambda_m v_m = 0.$$

We now subtract these last two expressions, and obtain

$$c_2(\lambda_2 - \lambda_1)v_2 + \cdots + c_m(\lambda_m - \lambda_1)v_m = 0.$$

Since $\lambda_j - \lambda_1 \neq 0$ for $j = 2, \ldots, m$ we conclude by induction that

$$c_2 = \cdots = c_m = 0.$$

Going back to our original relation, we see that $c_1 v_1 = 0$, whence $c_1 = 0$, and our theorem is proved.

Quite generally, let V be a finite dimensional vector space, and let

$$L: V \to V$$

be a linear map. Let $\{v_1,\dots,v_n\}$ be a basis of V. We say that this basis **diagonalizes** L if each v_i is an eigenvector of L, so $Lv_i = c_i v_i$ with some scalar c_i. Then the matrix representing L with respect to this basis is the diagonal matrix

$$A = \begin{pmatrix} c_1 & 0 & \cdots & 0 \\ 0 & c_2 & \cdots & 0 \\ \vdots & \vdots & \ddots & \vdots \\ 0 & 0 & \cdots & c_n \end{pmatrix}.$$

We say that the **linear map** L can be **diagonalized** if there exists a basis of V consisting of eigenvectors. We say that an $n \times n$ **matrix** A can be **diagonalized** if its associated linear map L_A can be diagonalized.

We shall now see how we can use determinants to find the eigenvalue of a matrix. We assume that readers are acquainted with determinants.

Theorem 8.3. *Let V be a finite dimensional vector space, and let λ be a number. Let $A: V \to V$ be a linear map. Then λ is an eigenvalue of A if and only if $A - \lambda I$ is not invertible.*

Proof. Assume that λ is an eigenvalue of A. Then there exists an element $v \in V$, $v \neq 0$ such that $Av = \lambda v$. Hence $Av - \lambda v = 0$, and $(A - \lambda I)v = 0$. Hence $A - \lambda I$ has a non-zero kernel, and $A - \lambda I$ cannot be invertible. Conversely, assume that $A - \lambda I$ is not invertible. By Theorem 2.4 we see that $A - \lambda I$ must have a non-zero kernel, meaning that there exists an element $v \in V$, $v \neq 0$ such that $(A - \lambda I)v = 0$. Hence $Av - \lambda v = 0$, and $Av = \lambda v$. Thus λ is an eigenvalue of A. This proves our theorem.

Let A be an $n \times n$ matrix, $A = (a_{ij})$. We define the **characteristic polynomial** P_A to be the determinant

$$P_A(t) = \operatorname{Det}(tI - A),$$

or written out in full,

$$P(t) = \begin{vmatrix} t - a_{11} & & -a_{ij} \\ & \ddots & \\ -a_{ij} & & t - a_{nn} \end{vmatrix}.$$

We can also view A as a linear map from K^n to K^n, and we also say that $P_A(t)$ is the **characteristic polynomial of this linear map.**

Example 4. The characteristic polynomial of the matrix

$$A = \begin{pmatrix} 1 & -1 & 3 \\ -2 & 1 & 1 \\ 0 & 1 & -1 \end{pmatrix}$$

is

$$\begin{vmatrix} t-1 & 1 & -3 \\ 2 & t-1 & -1 \\ 0 & -1 & t+1 \end{vmatrix},$$

which we expand according to the first column, to find

$$P_A(t) = t^3 - t^2 - 4t + 6.$$

For an arbitrary matrix $A = (a_{ij})$, the characteristic polynomial can be found by expanding according to the first column, and will always consist of a sum

$$(t - a_{11}) \cdots (t - a_{nn}) + \cdots.$$

Each term other than the one we have written down will have degree $< n$. Hence the characteristic polynomial is of type

$$P_A(t) = t^n + \text{terms of lower degree}.$$

For the next theorem, we assume that you know the following property of determinants:

A square matrix M over a field K is invertible if and only if its determinant is $\neq 0$.

Theorem 8.4. *Let A be an $n \times n$ matrix. An element $\lambda \in K$ is an eigenvalue of A if and only if λ is a root of the characteristic polynomial of A. If K is algebraically closed, then A has an eigenvalue in K.*

Proof. Assume that λ is an eigenvalue of A. Then $\lambda I - A$ is not invertible by Theorem 8.3 and hence $\operatorname{Det}(\lambda I - A) = 0$. Consequently λ is a root of the characteristic polynomial. Conversely, if λ is a root of the characteristic polynomial, then

$$\operatorname{Det}(\lambda I - A) = 0,$$

and hence we conclude that $\lambda I - A$ is not invertible. Hence λ is an eigenvalue of A by Theorem 8.3.

Theorem 8.5. *Let A, B be two $n \times n$ matrices, and assume that B is invertible. Then the characteristic polynomial of A is equal to the characteristic polynomial of $B^{-1}AB$.*

Proof. By definition, and properties of the determinant,

$$\begin{aligned}\operatorname{Det}(tI - A) &= \operatorname{Det}(B^{-1}(tI - A)B) = \operatorname{Det}(tB^{-1}B - B^{-1}AB) \\ &= \operatorname{Det}(tI - B^{-1}AB).\end{aligned}$$

This proves what we wanted.

Let

$$L: V \to V$$

be a linear map of a finite dimensional vector space into itself, so L is also called an endomorphism. Select a basis for V and let A be the matrix associated with L with respect to this basis. We then define the **characteristic polynomial of** L to be the characteristic polynomial of A. If we change basis, then A changes to $B^{-1}AB$ where B is invertible. By Theorem 8.5 this implies that the characteristic polynomial does not depend on the choice of basis.

Theorem 8.4 can be interpreted for L as stating:

Let K be algebraically closed.
Let V be a finite dimensional vector space over K of dimension > 0.
Let $L: V \to V$ be an endomorphism. Then L has a non-zero eigenvector and an eigenvalue in K.

V, §8. EXERCISES

1. Let V be an n-dimensional vector space and assume that the characteristic polynomial of a linear map $A: V \to V$ has n distinct roots. Show that V has a basis consisting of eigenvectors of A.

2. Let A be an invertible matrix. If λ is an eigenvalue of A show that $\lambda \neq 0$ and that λ^{-1} is an eigenvalue of A^{-1}.

3. Let V be the space generated over $\mathbf{R}$ by the two functions $\sin t$ and $\cos t$. Does the derivative (viewed as a linear map of V into itself) have any nonzero eigenvectors in V? If so, which?

4. Let D denote the derivative which we view as a linear map on the space of differentiable functions. Let k be an integer $\neq 0$. Show that the functions $\sin kx$ and $\cos kx$ are eigenvectors for D^2. What are the eigenvalues?

5. Let $A: V \to V$ be a linear map of V into itself, and let $\{v_1, \dots, v_n\}$ be a basis of V consisting of eigenvectors having distinct eigenvalues $c_1, \dots, c_n$. Show that any eigenvector v of A in V is a scalar multiple of some v_i.

6. Let A, B be square matrices of the same size. Show that the eigenvalues of AB are the same as the eigenvalues of BA.

7. **(Artin's theorem.)** Let G be a group, and let $f_1, \dots, f_n: G \to K^*$ be distinct homomorphisms of G into the multiplicative group of a field. In particular, $f_1, \dots, f_n$ are functions of G into K. Prove that these functions are linearly independent over K. [*Hint*: Use induction in a way similar to the proof of Theorem 8.2.]

V, §9. POLYNOMIALS OF MATRICES AND LINEAR MAPS

Let n be a positive integer. Let $\text{Mat}_n(K)$ denote the set of all $n \times n$ matrices with coefficients in a field K. Then $\text{Mat}_n(K)$ is a ring, which is a finite dimensional vector space over K, of dimension n^2. Let $A \in \text{Mat}_n(K)$. Then A generates a subring, which is commutative because powers of A commute with each other. Let $K[t]$ denote the polynomial ring over K. As a special case of the evaluation map, if $f(t) \in K[t]$ is a polynomial, we can evaluate f at A. Indeed:

$$\text{If } f(t) = a_n t^n + \cdots + a_0 \text{ then } f(A) = a_n A^n + \cdots + a_0 I.$$

We know that the evaluation map is a ring homomorphism, so we have the rules:

Let f, $g \in K[t]$, and $c \in K$. Then:

$$(f + g)(A) = f(A) + g(A),$$

$$(fg)(A) = f(A)g(A),$$

$$(cf)(A) = cf(A).$$

Example. Let $\alpha_1, \dots, \alpha_n$ be elements of K. Let

$$f(t) = (t - \alpha_1) \cdots (t - \alpha_n).$$

Then

$$f(A) = (A - \alpha_1 I) \cdots (A - \alpha_n I).$$

Let V be a vector space over K, and let $A: V \to V$ be an endomorphism (i.e. linear map of V into itself). Then we can form $A^2 = A \circ A = AA$, and in general $A^n =$ iteration of A taken n times for any positive integer n. We define $A^0 = I$ (where I now denotes the identity mapping). We have

$$A^{m+n} = A^m A^n$$

for all integers $m, n \geqq 0$. If f is a polynomial in $K[t]$, then we can form $f(A)$ the same way that we did for matrices. The same rules are satisfied, expressing the fact that $f \mapsto f(A)$ is a ring homomorphism. The image of $K[t]$ in $\mathrm{End}_K(V)$ under this homomorphism is the commutative subring denoted by $K[A]$.

Theorem 9.1. *Let A be an $n \times n$ matrix in a field K, or let $A: V \to V$ be an endomorphism of a vector space V of dimension n. Then there exists a non-zero polynomial $f \in K[t]$ such that $f(A) = O$.*

Proof. The vector space of $n \times n$ matrices over K is finite dimensional, of dimension n^2. Hence the powers

$$I, A, A^2, \ldots, A^N$$

are linearly dependent for $N > n^2$. This means that there exist numbers $a_0, \ldots, a_N \in K$ such that not all $a_i = 0$, and

$$a_N A^N + \cdots + a_0 I = O.$$

We let $f(t) = a_N t^N + \cdots + a_n$ to get what we want. The same proof applies when A is an endomorphism of V.

If we divide the polynomial f of Theorem 9.1 by its leading coefficient, then we obtain a polynomial g with leading coefficient 1 such that $g(A) = O$. It is usually convenient to deal with polynomials whose leading coefficient is 1, since it simplifies the notation.

The kernel of the evaluation map $f \mapsto f(A)$ is an ideal in $K[t]$, which is principal, and so generated by a unique monic polynomial which is called the **minimal polynomial** of A in $K[t]$. Since the ring generated by A over K may have divisors of zero, it is possible that this minimal polynomial is not irreducible. This is the first basic distinction which we encounter from the case when we evaluated polynomials in a field. We shall prove at the end of the section that if P_A is the characteristic polynomial, then $P_A(A) = O$. Therefore $P_A(t)$ is in the kernel of the map $f \mapsto f(A)$, and so the minimal polynomial divides the characteristic polynomial since the kernel is a principal ideal.

Let V be a vector space over K, and let $A: V \to V$ be an endomorphism. Then we may view V as a module over the polynomial ring $K[t]$ as follows. If $v \in V$ and $f(t) \in K[t]$, then $f(A): V \to V$ is also an endomorphism of V, and we **define**

$$f(t)v = f(A)v.$$

The properties needed to check that V is indeed a module over $K[t]$ are trivially verified. The big advantage of dealing with V as a module over $K[t]$ rather than over $K[A]$, for instance, is that $K[t]$ is a principal ring, and we can apply the results of §7.

Warning. The structure of $K[t]$-module depends of course on the choice of A. If we selected another endomorphism to define the operation of $K[t]$ on V, then this structure would also change. Thus we denote by V_A the module V over $K[t]$ determined by the endomorphism A as above. Theorem 9.1 can now be interpreted as stating:

The module V_A over $K[t]$ is a torsion module.

Therefore Theorems 7.1 and 7.2 apply to give us a description of V_A as a $K[t]$-module. We shall make this description more explicit.

Let W be a subspace of V. We shall say that W is an **invariant subspace under** A, or its A-**invariant**, if Aw lies in W for all $w \in W$, i.e. if AW is contained in W. It follows directly from the definitions that:

A subspace W is A-invariant if and only if W is a $K[t]$-submodule.

Example 1. Let v_1 be a non-zero eigenvector of A, and let V_1 be the 1-dimensional space generated by v_1. Then V_1 is an invariant subspace under A.

Example 2. Let λ be an eigenvalue of A, and let V_λ be the subspace of V consisting of all $v \in V$ such that $Av = \lambda v$. Then V_λ is an invariant subspace under A, called the **eigenspace** of λ.

Example 3. Let $f(t) \in K[t]$ be a polynomial, and let W be the kernel of $f(A)$. Then W is an invariant subspace under A.

Proof. Suppose that $f(A)w = 0$. Since $tf(t) = f(t)t$, we get

$$Af(A) = f(A)A,$$

whence

$$f(A)(Aw) = f(A)Aw = Af(A)w = 0.$$

Thus Aw is also in the kernel of $f(A)$, thereby proving our assertion.

Translating Theorem 7.1 into the present situation yields:

Theorem 9.2. *Let $f(t) \in K[t]$ be a polynomial, and suppose that $f = f_1 f_2$, where f_1, f_2 are polynomials of degree $\geqq 1$, and relatively* *prime. Let* $A: V \to V$ *be an endomorphism. Assume that* $f(A) = O$. *Let*

$$W_1 = \text{kernel of } f_1(A) \quad \textit{and} \quad W_2 = \text{kernel of } f_2(A).$$

Then V *is the direct sum of* W_1 *and* W_2. *In particular, suppose that* $f(t)$ *has a factorization*

$$f(t) = (t - \alpha_1)^{r_1} \cdots (t - \alpha_m)^{r_m}$$

with distinct roots $\alpha_1, \ldots, \alpha_m \in K$. *Let* W_i *be the kernel of* $(A - \alpha_i I)^{r_i}$. *Then* V *is the direct sum of the subspaces* $W_1, \ldots, W_m$.

Remark. If the field K is algebraically closed, then we can always factor the polynomial $f(t)$ into factors of degree 1 as above and the different powers $(t - \alpha_1)^{r_1}, \ldots, (t - \alpha_m)^{r_m}$ are relatively prime. This is of course the case over the complex numbers.

Example 4 (Differential equations). Let V be the space of (infinitely differentiable) solutions of the differential equation

$$D^n f + a_{n-1} D^{n-1} f + \cdots + a_0 f = 0,$$

with constant complex coefficients a_i. We shall determine a basis of V.

Theorem 9.3. *Let*

$$P(t) = t^n + a_{n-1} t^{n-1} + \cdots + a_0.$$

Factor $P(t)$ *as in Theorem* 9.2

$$P(t) = (t - \alpha_1)^{r_1} \cdots (t - \alpha_m)^{r_m}.$$

Then V *is the direct sum of the spaces of solutions of the differential equations*

$$(D - \alpha_i I)^{r_i} f = 0,$$

for $i = 1, \ldots, m$.

Proof. This is merely a direct application of Theorem 9.2.

Thus the study of the original differential equation is reduced to the study of the much simpler equation

$$(D - \alpha I)^r f = 0.$$

The solutions of this equation are easily found.

Theorem 9.4. *Let α be a complex number. Let W be the space of solutions of the differential equation*

$$(D - \alpha I)^r f = 0.$$

Then W is the space generated by the functions

$$e^{\alpha t}, te^{\alpha t}, \ldots, t^{r-1}e^{\alpha t}$$

and these functions form a basis for this space, which therefore has dimension r.

Proof. For any complex α we have

$$(D - \alpha I)^r f = e^{\alpha t} D^r(e^{-\alpha t} f).$$

(The proof is a simple induction.) Consequently, f lies in the kernel of $(D - \alpha I)^r$ if and only if

$$D^r(e^{-\alpha t} f) = 0.$$

The only functions whose r-th derivative is 0 are the polynomials of degree $\leqq r - 1$. Hence the space of solutions of $(D - \alpha I)^r f = 0$ is the space generated by the functions

$$e^{\alpha t}, te^{\alpha t}, \ldots, t^{r-1}e^{\alpha t}.$$

Finally these functions are linearly independent. Suppose we have a linear relation

$$c_0 e^{\alpha t} + c_1 t e^{\alpha t} + \cdots + c_{r-1} t^{r-1} e^{\alpha t} = 0$$

for all t, with constants $c_0, \ldots, c_{r-1}$. Let

$$Q(t) = c_0 + c_1 t + \cdots + c_{r-1} t^{r-1}.$$

Then $Q(t)$ is a non-zero polynomial, and we have

$$Q(t)e^{\alpha t} = 0 \qquad \text{for all } t.$$

But $e^{\alpha t} \neq 0$ for all t so $Q(t) = 0$ for all t. Since Q is a polynomial, we must have $c_i = 0$ for $i = 0, \ldots, r - 1$ thus concluding the proof.

We end this chapter by looking at the meaning of Theorem 7.2 for a finite dimensional vector space V over an algebraically closed field K. Let

$$A: V \to V$$

be an endomorphism as before. We first make explicit the cyclic case.

Lemma 9.5. *Let $v \in V$, $v \neq 0$. Suppose there exists $\alpha \in K$ such that $(A - \alpha I)^r v = 0$ for some positive integer r, and that r is the smallest such positive integer. Then the elements*

$$v, \quad (A - \alpha I)v, \ldots, (A - \alpha I)^{r-1}v$$

are linearly independent over K.

Proof. Let $B = A - \alpha I$ for simplicity. A relation of linear dependence between the above elements can be written

$$f(B)v = 0,$$

where f is a polynomial $\neq 0$ of degree $\leqq r - 1$, namely

$$c_0 v + c_1 Bv + \cdots + c_s B^s v = 0,$$

with $f(t) = c_0 + c_1 t + \cdots + c_s t^s$, and $s \leqq r - 1$. We also have $B^r v = 0$ by hypothesis. Let $g(t) = t^r$. If h is the greatest common divisor of f and g, then we can write

$$h = f_1 f + g_1 g,$$

where f_1, g_1 are polynomials, and thus $h(B) = f_1(B)f(B) + g_1(B)g(B)$. It follows that $h(B)v = 0$. But $h(t)$ divides t^r and is of degree $\leqq r - 1$, so that $h(t) = t^d$ with $d < r$. This contradicts the hypothesis that r is smallest, and proves the lemma.

The module V_A is cyclic over $K[t]$ if and only if there exists an element $v \in V$, $v \neq 0$ such that every element of V is of the form $f(A)v$ for some polynomial $f(t) \in K[t]$. Suppose that V_A is cyclic, and in addition that there is some element $\alpha \in K$ and a positive integer r such that $(A - \alpha I)^r v = 0$. Also let r be the smallest such positive integer. Then the minimal polynomial of A on V is precisely $(t - \alpha)^r$. Then Lemma 9.5 implies that

$$(*) \qquad \{(A - \alpha I)^{r-1}v, \ldots, (A - \alpha I)v, v\}$$

is a basis for V over K. With respect to this basis, the matrix of A is then particularly simple. Indeed, for each k we have

$$A(A - \alpha I)^k v = (A - \alpha I)^{k+1} v + \alpha(A - \alpha I)^k v.$$

By definition, it follows that the associated matrix for A with respect to this basis is equal to the triangular matrix

$$\begin{pmatrix} \alpha & 1 & 0 & \cdots & 0 & 0 \\ 0 & \alpha & 1 & \cdots & 0 & 0 \\ \vdots & \vdots & \ddots & \ddots & \vdots & \vdots \\ & & & & & 0 \\ 0 & 0 & 0 & \cdots & \alpha & 1 \\ 0 & 0 & 0 & \cdots & 0 & \alpha \end{pmatrix}.$$

This matrix has α on the diagonal, 1 above the diagonal, and 0 everywhere else. The reader will observe that $(A - \alpha I)^{r-1} v$ is an eigenvector for A, with eigenvalue α.

The basis (*) is called a **Jordan basis for** V **with respect to** A. Thus over an algebraically closed field, we have found a basis for a cyclic vector space as above such that the matrix of A with respect to this basis is particularly simple, and is almost diagonal. If $r = 1$, that is if $Av = \alpha v$, then the matrix is a 1×1 matrix, which is diagonal.

We now turn to the general case. We can reformulate Theorem 7.2 as follows.

Theorem 9.6. *Let V be a finite dimensional space over the algebraically closed field K, and $V \neq \{0\}$. Let $A: V \to V$ be an endomorphism. Then V is a direct sum of A-invariant subspaces*

$$V = V_1 \oplus \cdots \oplus V_q$$

such that each V_i is cyclic, generated over $K[t]$ by an element $v_i \neq 0$, and the kernel of the map

$$f(t) \mapsto f(A)v_i$$

is a power $(t - \alpha_i)^{r_i}$ for some positive integer r_i and $\alpha_i \in K$.

If we select a Jordan basis for each V_i, then the sequence of these bases forms a basis for V, again called a **Jordan basis for** V **with respect to** A. With respect to this basis, the matrix for A therefore splits into blocks (Fig. 1).

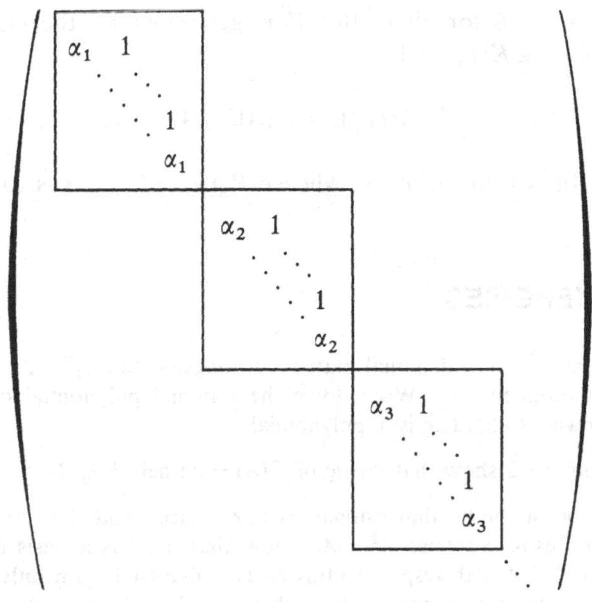

Figure 1

In each block we have an eigenvalue α_i on the diagonal. We have 1 above the diagonal, and 0 everywhere else. This matrix is called the **Jordan normal form for** A.

It is to be understood that if $r_i = 1$, then there is no 1 above the diagonal, and the eigenvalue α_i is simply repeated, the number of times being the dimension of the corresponding eigenspace.

The Jordan normal form also allows us to prove the **Cayley-Hamilton theorem** as a corollary, namely:

Theorem 9.7. *Let A be an $n \times n$ matrix over a field K, and let $P_A(t)$ be its characteristic polynomial. Then $P_A(A) = O$.*

Proof. We assume for this proof that K is contained in some algebraically closed field, and it will then suffice to prove the theorem under the assumption that K is algebraically closed. Then A represents an endomorphism of K^n, which we take as V. We denote the endomorphism by the same letter A. We decompose V into a direct sum as in Theorem 9.6. Then the characteristic polynomial of A is given by

$$P_A(t) = \prod_{i=1}^{q} (t - \alpha_i)^{r_i}$$

where $r_i = \dim_K(V_i)$. But $P_A(A) = \prod_{i=1}^{q} (A - \alpha_i I)^{r_i}$, and by Theorem 9.6,

$$(A - \alpha_i I)^{r_i} v_i = 0.$$

Hence $P_A(A)v_i = 0$ for all i. But V is generated by the elements $f(A)v_i$ for all i with $f \in K[t]$, and

$$P_A(A)f(A)v_i = f(A)P_A(A)v_i = 0.$$

Hence $P_A(A)v = 0$ for all $v \in V$, whence $P_A(A) = O$, as was to be proved.

V, §9. EXERCISES

1. Let M be an $n \times n$ diagonal matrix with eigenvalues $\lambda_1, \dots, \lambda_r$. Suppose that λ_i has multiplicity m_i. Write down the minimal polynomial of M, and also write down its characteristic polynomial.

2. In Theorem 9.2 show that image of $f_1(A) =$ kernel of $f_2(A)$.

3. Let V be a finite dimensional vector space, and let $A: V \to V$ be an endomorphism. Suppose $A^2 = A$. Show that there is a basis of V such that the matrix of A with respect to this basis is diagonal, with only 0 or 1 on the diagonal. Or, if you prefer, show that $V = V_0 \oplus V_1$ is a direct sum, where $V_0 = \text{Ker } A$ and V_1 is the $(+1)$-eigenspace of A.

4. Let $A: V \to V$ be an endomorphism, and V finite dimensional. Suppose that $A^3 = A$. Show that V is the direct sum

$$V = V_0 \oplus V_1 \oplus V_{-1},$$

where $V_0 = \text{Ker } A$, V_1 is the $(+1)$-eigenspace of A, and V_{-1} is the (-1)-eigenspace of A.

5. Let $A: V \to V$ be an endomorphism, and V finite dimensional. Suppose that the characteristic polynomial of A has the factorization

$$P_A(t) = (t - \alpha_1) \cdots (t - \alpha_n),$$

where $\alpha_1, \dots, \alpha_n$ are distinct elements of the field K. Show that V has a basis consisting of eigenvectors for A.

For the rest of the exercises, we suppose that $V \neq \{0\}$, and that V is finite dimensional over the algebraically closed field K. We let $A: V \to V$ be an endomorphism.

6. Prove that A is diagonalizable if and only if the minimal polynomial of A has all roots of multiplicity 1.

7. Suppose A is diagonalizable. Let W be a subspace of V such that $AW \subset W$. Prove that the restriction of A to W is diagonalizable as an endomorphism of W.

8. Let B be another endomorphism of V such that $BA = AB$. Prove that A, B have a common non-zero eigenvector.

9. Let B be another endomorphism of V. Assume that $AB = BA$ and both A, B are diagonalizable. Prove that A and B are simultaneously diagonalizable, that is V has a basis consisting of elements which are eigenvectors for both A and B.

10. Show that A can be written in the form $A = D + N$, where D is a diagonalizable endomorphism, N is nilpotent, and $DN = ND$.

11. Assume that V_A is cyclic, annihilated by $(A - \alpha I)^r$ for some $r > 0$ and $\alpha \in K$. Prove that the subspace of V generated by eigenvectors of A is one-dimensional.

12. Prove that V_A is cyclic if and only if the characteristic polynomial $P_A(t)$ is equal to the minimal polynomial of A in $K[t]$.

13. Assume that V_A is cyclic annihilated by $(A - \alpha I)^r$ for some $r > 0$. Let f be a polynomial. What are the eigenvalues of $f(A)$ in terms of those of A? Same question when V is not assumed cyclic.

14. Let P_A be the characteristic polynomial of A, and write it as a product

$$P_A(t) = \prod_{i=1}^{m} (t - \alpha_i)^{r_i},$$

where $\alpha_1, \ldots, \alpha_m$ are distinct. Let f be a polynomial. Express the characteristic polynomial $P_{f(A)}$ as a product of factors of degree 1.

15. If A is nilpotent and not O, show that A is not diagonalizable.

16. Suppose that A is nilpotent. Prove that V has a basis such that the matrix of A with respect to this basis has the form

$$\begin{pmatrix} N_1 & & & O \\ & N_2 & & \\ & & \ddots & \\ O & & & N_r \end{pmatrix} \quad \text{where } N_i = (0) \quad \text{or} \quad N_i = \begin{pmatrix} 0 & 1 & 0 & \cdots & 0 \\ 0 & 0 & 1 & \cdots & 0 \\ \vdots & \vdots & \vdots & \ddots & \vdots \\ \vdots & \vdots & \vdots & & 1 \\ 0 & 0 & 0 & \cdots & 0 \end{pmatrix}$$

The matrix on the right has components 0 except for 1's just above the diagonal.

Invariant subspaces

Let S be a set of endomorphisms of V. Let W be a subspace of V. We shall say that W is an **S-invariant subspace** if $BW \subset W$ for all $B \in S$. We shall say that V is a **simple** S-space if $V \neq \{0\}$ and if the only S-invariant subspaces are V itself and the zero subspace. Prove:

17. Let $A: V \to V$ be an endomorphism such that $AB = BA$ for all $B \in S$.
    (a) The image and kernel of A are S-invariant subspaces.
    (b) Let $f(t) \in K[t]$. Then $f(A)B = Bf(A)$ for all $B \in S$.
    (c) Let U, W be S-invariant subspaces of V. Show that $U + W$ and $U \cap W$ are S-invariant subspaces.

18. Assume that V is a simple S-space and that $AB = BA$ for all $B \in S$. Prove that either A is invertible or A is the zero map. Using the fact that V is finite dimensional and K algebraically closed, prove that there exists $\alpha \in K$ such that $A = \alpha I$.

19. Let V be a finite dimensional vector space over the field K, and let S be the set of all linear maps of V into itself. Show that V is a simple S-space.

20. Let $V = \mathbf{R}^2$, let S consist of the matrix $\begin{pmatrix} 1 & a \\ 0 & 1 \end{pmatrix}$ viewed as linear map of V into itself. Here, a is a fixed non-zero real number. Determine all S-invariant subspaces of V.

21. Let V be a vector space over the field K, and let $\{v_1, \dots, v_n\}$ be a basis of V. For each permutation σ of $\{1, \dots, n\}$ let $A_\sigma : V \to V$ be the linear map such that

$$A_\sigma(v_i) = v_{\sigma(i)}.$$

(a) Show that for any two permutations σ, τ we have

$$A_\sigma A_\tau = A_{\sigma\tau},$$

and $A_{\text{id}} = I$.

(b) Show that the subspace generated by $v = v_1 + \cdots + v_n$ is an invariant subspace for the set S_n consisting of all A_σ.

(c) Show that the element v of part (b) is an eigenvector of each A_σ. What is the eigenvalue of A_σ belonging to v?

(d) Let $n = 2$, and let σ be the permutation which is not the identity. Show that $v_1 - v_2$ generates a 1-dimensional subspace which is invariant under A_σ. Show that $v_1 - v_2$ is an eigenvector of A_σ. What is the eigenvalue?

22. Let V be a vector space over the field K, and let $A: V \to V$ be an endomorphism. Assume that $A^r = I$ for some integer $r \geqq 1$. Let

$$T = I + A + \cdots + A^{r-1}.$$

Let v_0 be an element of V. Show that the space generated by Tv_0 is an invariant subspace of A, and that Tv_0 is an eigenvector of A. If $Tv_0 \neq 0$, what is the eigenvalue?

23. Let (V, A) and (W, B) be pairs consisting of a vector space and endomorphism, over the same field K. We define a **morphism**

$$f: (V, A) \to (W, B)$$

to be a homomorphism $f: V \to W$ of K-vector spaces satisfying in addition the condition

$$B \circ f = f \circ A.$$

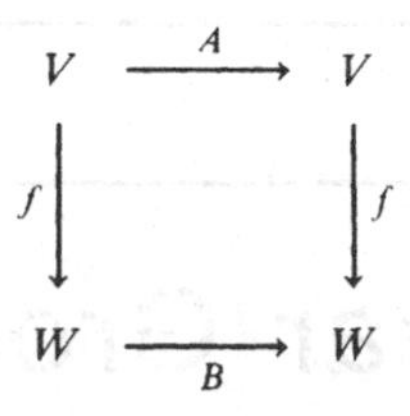

In other words, for all $v \in V$ we have $B(f(v)) = f(A(v))$. An **isomorphism** of pairs is a morphism which has an inverse.

Prove that (V, A) is isomorphic to (W, B) if and only if V_A and W_B are isomorphic as $K[t]$-modules. (The operation of $K[t]$ on V_A is the one determined by A, and the operation of $K[t]$ on W_B is the one determined by B.)

A direct sum decomposition of matrices

24. Let F be a field and $\text{Mat}_n(F) = M_n$ the ring of $n \times n$ matrices over F. Let E_{ij} for $i, j = 1, \ldots, n$ be the matrix with (ij)-component 1, and all other components 0. Then the set of elements E_{ij} is a basis for M_n. Let $D_n^* = D_n^*(F)$ be the multiplicative group of diagonal matrices with non-zero diagonal components. We write such matrices as $\text{diag}(a_1, \ldots, a_n) = a$. We define the **conjugation action** of D_n^* on M_n by

$$\mathbf{c}(a)X = aXa^{-1}.$$

(a) Show that $a \mapsto \mathbf{c}(a)$ is a homomorphism from D_n^* into the group of linear automorphisms of M_n.
(b) Show that each E_{ij} is an eigenvector for the action of $\mathbf{c}(a)$, the eigenvalue being given by $\chi_{ij}(a) = a_i/a_j$.

Thus M_n is a direct sum of eigenspaces. Each $\chi_{ij}\colon D^* \to F^*$ is a character, i.e. a homomorphism of D^* into the multiplicative group of F. A general context will be given in Chapter VI, §6.

25. For two matrices $X, Y \in M_n(F)$, define $[X, Y] = XY - YX$. Let $L_X\colon M_n \to M_n$ denote the map such that $L_X(Y) = [X, Y]$. One calls L_X the **bracket** (or **Lie**) **action** of X on M_n, and $[X, Y]$ the **Lie product** of X and Y.
(a) Show that for each X, the map $L_X\colon Y \mapsto [X, Y]$ is a linear map, satisfying the **Leibniz rule** for **derivations**, that is

$$[X, [Y, Z]] = [[X, Y], Z] + [Y, [X, Z]].$$

(b) Let D_n be the vector space of diagonal matrices. For each $H \in D$, show that E_{ij} is an eigenvector of L_H, with eigenvalue $\alpha_{ij}(H) = h_i - h_j$ (where $h_1, \ldots, h_n$ are the diagonal components of H). Show that $\alpha_{ij}\colon D \to F$ is linear. It is called an **eigencharacter** of the **bracket** or **Lie action.**
(c) For two linear maps A, B of a vector space, define $[A, B] = AB - BA$. Show that $L_{[X,Y]} = [L_X, L_Y]$, so L is also a homomorphism for the Lie product.

The next two chapters are logically independent. A reader interested first in field theory can omit the next chapter.

CHAPTER VI

Some Linear Groups

VI, §1. THE GENERAL LINEAR GROUP

The purpose of this first section is to make you think of multiplication of matrices in the context of group theory, and to work out basic examples accordingly in the exercises. Except for the logic involved, this section could have been placed as exercises in Chapter I.

Let R be any ring. We recall that the **units** of R are those elements $u \in R$ such that u has an inverse u^{-1} in R. By definition, the units in the ring $M_n(K)$ are the **invertible matrices**, that is the $n \times n$ matrices A which have an inverse A^{-1}. Such an inverse is an $n \times n$ matrix satisfying

$$AA^{-1} = A^{-1}A = I_n.$$

The set of units in any ring is a group, and therefore *the invertible* $n \times n$ *matrices form a group,* which we denote by

$$\mathrm{GL}_n(K).$$

This group is called the **general linear group** over K. If you know about determinants, you know that A is invertible if and only if $\det(A) \neq 0$. Computing the determinant gives you an effective way of determining whether a matrix is invertible or not.

VI, §1. EXERCISES

1. Let $A \in \mathrm{GL}_n(K)$ and $C \in K^n$. By the **affine map** determined by (A, C) we mean the map

$$f_{A,C}\colon K^n \to K^n \qquad \text{such that} \quad f_{A,C}(X) = AX + C.$$

(a) Show that the set of all affine maps is a group, called the **affine group**. We denote the affine group by G.
(b) Show that $\mathrm{GL}_n(K)$ is a subgroup. Is $\mathrm{GL}_n(K)$ a normal subgroup? Proof?
(c) Let $T_C: K^n \to K^n$ be the map $T_C(X) = X + C$. Such a map is called a **translation**. Show that the translations form a group, which is thus a subgroup of the affine group. Is the group of translations a normal subgroup of the affine group? Proof?
(d) Show that the map $f_{A,C} \mapsto A$ is a homomorphism of G onto $\mathrm{GL}_n(K)$. What is its kernel?

2. Determine the period of the following matrices:

$$\text{(a)}\ \begin{pmatrix} 0 & 1 & 0 \\ 0 & 0 & 1 \\ 1 & 0 & 0 \end{pmatrix} \qquad \text{(b)}\ \begin{pmatrix} 0 & 1 & 0 & 0 \\ 0 & 0 & 1 & 0 \\ 0 & 0 & 0 & 1 \\ 1 & 0 & 0 & 0 \end{pmatrix}$$

3. Let A, B be any $n \times n$ invertible matrices. Show that the periods of A and BAB^{-1} are the same.

4. Let A be an $n \times n$ matrix. By an **eigenvector** X for A we mean an element $X \in K^n$ such that there exists $c \in K$ satisfying $AX = cX$. If $X \neq O$ then c is called an **eigenvalue** of A.
(a) If X is an eigenvector for A with eigenvalue c, show that X is also an eigenvector for every power A^n, where n is a positive integer. What is the eigenvalue of A^n if $A^n X \neq O$?
(b) Suppose that A has finite (multiplicative) period. Show that an eigenvalue c is necessarily a root of unity, that is $c^n = 1$ for some positive integer n.

5. Show that the additive group of a field K is isomorphic to the multiplicative group of matrices of type

$$\begin{pmatrix} 1 & a \\ 0 & 1 \end{pmatrix} \quad \text{with } a \in K.$$

6. Let G be the group of matrices

$$\begin{pmatrix} a & b \\ 0 & d \end{pmatrix}$$

with $a, b, d \in K$ and $ad \neq 0$. Show that the map

$$\begin{pmatrix} a & b \\ 0 & d \end{pmatrix} \mapsto (a, d)$$

is a homomorphism of G onto the product $K^* \times K^*$ (where K^* is the multiplicative group of K). Describe the kernel. We could also view our homomorphism as being into the group of diagonal matrices

$$\begin{pmatrix} a & 0 \\ 0 & d \end{pmatrix}$$

which is isomorphic to $K^* \times K^*$.

7. A matrix $N \in M_n(K)$ is called **nilpotent** if there exists a positive integer n such that $N^n = O$. If N is nilpotent, show that the matrix $I + N$ is invertible. [*Hint*: Think of the geometric series.]

8. (a) Let $G = G_0$ be the group of 3×3 upper triangular matrices in a field K, consisting of all invertible triangular matrices

$$T = \begin{pmatrix} a_{11} & a_{12} & a_{13} \\ 0 & a_{22} & a_{23} \\ 0 & 0 & a_{33} \end{pmatrix}$$

so $a_{11}a_{22}a_{33} \neq 0$. Let G_1 be the set of matrices

$$\begin{pmatrix} 1 & a_{12} & a_{13} \\ 0 & 1 & a_{23} \\ 0 & 0 & 1 \end{pmatrix}.$$

Show that G_1 is a subgroup of G, and that it is the kernel of the homomorphism which to each triangular matrix T associates the diagonal matrix consisting of the diagonal elements of T.

(b) Let G_2 be the set of matrices

$$\begin{pmatrix} 1 & 0 & c \\ 0 & 1 & 0 \\ 0 & 0 & 1 \end{pmatrix}.$$

Show that G_2 is a subgroup of G_1.

(c) Generalize the above to the case of $n \times n$ matrices.

(d) Show that the map

$$\begin{pmatrix} 1 & a_{12} & a_{13} \\ 0 & 1 & a_{23} \\ 0 & 0 & 1 \end{pmatrix} \mapsto (a_{12}, a_{23})$$

is a homomorphism of the group G_1 onto the direct product

$$K \times K = K^2.$$

What is the kernel?

(e) Show that the group G_2 is isomorphic to K.

9. Let V be a vector space of dimension n over a field K. Let $\{V_1,\dots,V_n\}$ be a sequence of subspaces such that $\dim V_i = i$ and such that $V_i \subset V_{i+1}$. Let $A: V \to V$ be a linear map. We say that this sequence of subspaces is a **fan for** A if $AV_i \subset V_i$.
   (a) Let G be the set of all invertible linear maps of V for which $\{V_1,\dots,V_n\}$ is a fan. Show that G is a group.
   (b) Let G_i be the subset of G consisting of all linear maps A such that $Av = v$ for all $v \in V_i$. Show that G_i is a group.
   (c) By a **fan basis** we mean a basis $\{v_1,\dots,v_n\}$ of V such that $\{v_1,\dots,v_i\}$ is a basis for V_i. Describe the matrix associated with an element of G with respect to a fan basis. Also describe the matrix associated with an element of G_i.

10. Let F be a finite field with q elements. What is the order of the group of diagonal matrices:
   (a) $\begin{pmatrix} a & 0 \\ 0 & d \end{pmatrix}$ with $a, d \in F$, $ad \neq 0$
   (b) $\begin{pmatrix} a_1 & 0 & \dots & 0 \\ 0 & a_2 & \dots & 0 \\ \vdots & \vdots & & \vdots \\ 0 & 0 & \dots & a_n \end{pmatrix}$ with $a_1,\dots,a_n \in F$, $a_i \neq 0$ for all i

11. Let F be a finite field with q elements. Let G be the group of upper triangular matrices

$$\text{(a)} \begin{pmatrix} a_{11} & a_{12} \\ 0 & a_{22} \end{pmatrix} \quad \text{and} \quad \text{(b)} \begin{pmatrix} a_{11} & a_{12} & a_{13} \\ 0 & a_{22} & a_{23} \\ 0 & 0 & a_{33} \end{pmatrix}$$

with $a_{ij} \in F$ and $a_{11}a_{22}a_{33} \neq 0$. What is the order of G in each case (a) and (b)?

12. Let F be a finite field with q elements. Show that the order of $\mathrm{GL}_2(F)$ is $q(q^2-1)(q-1)$.

13. Let F be a finite field with q elements. Show that the order of $\mathrm{GL}_n(F)$ is

$$(q^n - 1)(q^n - q)\cdots(q^n - q^{n-1}) = q^{n(n-1)/2} \prod_{i=1}^{n} (q^i - 1).$$

[*Hint*: Let $\{v_1,\dots,v_n\}$ be a basis of F^n. Any element of $\mathrm{GL}_n(F)$ viewed as a linear map of F^n into itself is determined by its effect on this basis (Theorem 1.2 of Chapter V), and thus the order of $\mathrm{GL}_n(F)$ is equal to the number of all possible bases. If $A \in \mathrm{GL}_n(F)$, let $Av_i = w_i$. For w_1 we can select any of the $q^n - 1$ non-zero vectors in F^n. Suppose inductively that we have already chosen $w_1,\dots,w_r$ with $r < n$. These vectors generate a subspace of dimension r which has q^r elements. For w_{r+1} we can select any of the $q^n - q^r$ elements outside of this subspace. The formula drops out.]

VI, §2. STRUCTURE OF $GL_2(F)$

Let

$$\alpha = \begin{pmatrix} a & b \\ c & d \end{pmatrix}$$

be a 2×2 matrix with components in a field F. We define the **determinant**

$$\det(\alpha) = ad - bc.$$

You probably have already met determinants, but we won't use any properties which cannot be proved here directly by easy computations. In particular, by brute force, you can verify that if α, β are two 2×2 matrices, then

$$\det(\alpha\beta) = \det(\alpha)\det(\beta).$$

Also verify:

A 2×2 matrix α is invertible if and only if $\det(\alpha) \neq 0$.

To do this, simply solve for the inverse matrix:

$$\begin{pmatrix} a & b \\ c & d \end{pmatrix}\begin{pmatrix} x & y \\ z & w \end{pmatrix} = \begin{pmatrix} 1 & 0 \\ 0 & 1 \end{pmatrix}.$$

You will get two systems of two linear equations in two unknowns, which can be solved precisely when $ad - bc \neq 0$.

Let $G = GL_2(F)$ be the group of invertible 2×2 matrices in F.

From the above, we see that

$$\det\colon GL_2(F) \to F^*$$

is a homomorphism. We shall investigate its kernel in the next section. Here we note that this homomorphism is surjective, because the element $a \in F^*$ is the image of the matrix

$$\begin{pmatrix} a & 0 \\ 0 & 1 \end{pmatrix}.$$

Recall that for any group G, the **center** of G is the subgroup Z consisting of all elements $\gamma \in G$ such that $\gamma\alpha = \alpha\gamma$ for all $\alpha \in G$.

Lemma 2.1. *The center of* $GL_2(F)$ *is the group of scalar matrices*

$$\begin{pmatrix} a & 0 \\ 0 & a \end{pmatrix} \quad \text{with} \quad a \in F^*.$$

Proof. Each scalar matrix aI commutes with all matrices. Conversely, if α commutes with all matrices, show that α is a scalar matrix. For instance, use the commutation with matrices like

$$\begin{pmatrix} 1 & 1 \\ 0 & 1 \end{pmatrix}, \quad \begin{pmatrix} 1 & 0 \\ 1 & 1 \end{pmatrix}, \quad \text{and so forth.}$$

We leave the details as an exercise.

Let Z be the center of $GL_2(F)$. We define the **projective linear group**

$$PGL_2(F) = GL_2(F)/Z = G/Z.$$

Thus $PGL_2(F)$ is the factor group of $GL_2(F)$ by its center.

The Bruhat decomposition

We let the (standard) **Borel subgroup B** of $GL_2(F)$ be the group of all matrices

$$\begin{pmatrix} a & b \\ 0 & d \end{pmatrix} \quad \text{with} \quad ad \neq 0.$$

Lemma 2.2. *The Borel subgroup B is a maximal subgroup of G. That is, if H is a subgroup with $B \subset H \subset G$ then $H = B$ or $H = G$.*

The proof of this lemma will depend on an analysis of G as follows. Let

$$\tau = \begin{pmatrix} 0 & 1 \\ -1 & 0 \end{pmatrix}.$$

We let $B\tau B$ be the set of all elements $\alpha\tau\beta$ with $\alpha, \beta \in B$.

Lemma 2.3. *There is a decomposition*

$$G = B \cup B\tau B,$$

and B, $B\tau B$ have no elements in common.

Proof. Let $\alpha, \beta \in B$. Then a direct computation shows that

$$\alpha\tau\beta\begin{pmatrix} 1 \\ 0 \end{pmatrix} = \begin{pmatrix} * \\ x \end{pmatrix}$$

where $x \neq 0$. Since for all $\alpha \in B$ we have

$$\alpha\begin{pmatrix}1\\0\end{pmatrix} = \begin{pmatrix}*\\0\end{pmatrix},$$

it follows that $\alpha\tau\beta$ cannot be an element of B and conversely no element of B can be in $B\tau B$. Hence $B \cap B\tau B$ is empty. Furthermore, given $\gamma \in G$, write

$$\gamma = \begin{pmatrix}a & b\\c & d\end{pmatrix}.$$

If $c = 0$ then $\gamma \in B$. If $c \neq 0$ then for some $x \in K$ we get

$$\begin{pmatrix}1 & x\\0 & 1\end{pmatrix}\begin{pmatrix}a & b\\c & d\end{pmatrix} = \begin{pmatrix}0 & b'\\c & d\end{pmatrix}.$$

Let $\beta = \begin{pmatrix}1 & x\\0 & 1\end{pmatrix}$ so $\beta \in B$. Then

$$\tau\beta\gamma = \begin{pmatrix}c & d\\0 & -b'\end{pmatrix} \in B.$$

Since $\tau^2 = -I$ we get $\tau^{-1} = -\tau$, and $-I \in B$. Hence

$$\gamma \in \beta^{-1}\tau^{-1}B \subset B\tau B$$

thus proving that $G = B \cup B\tau B$. This concludes the proof of Lemma 2.3.

Now for the proof of Lemma 2.2 that B is a maximal subgroup of G. Let $B \subset H \subset G$ and $B \neq H$. By Lemma 2.3 there exists an element $\gamma \in H$ such that

$$\gamma = \alpha\tau\beta, \qquad \text{with} \quad \alpha, \beta \in B.$$

Since H contains B, it follows that H also contains

$$B\gamma B = B\tau B.$$

Hence $H = G$ by Lemma 2.3. This concludes the proof of Lemma 2.2.

VI, §2. EXERCISES

1. Let F be a finite field. What is the order of $\mathrm{PGL}_2(F)$?
2. Show that the center of $\mathrm{GL}_n(F)$ is the group of scalar non-zero matrices.
3. Let Z be the center of $\mathrm{GL}_n(F)$. What is the order of $\mathrm{PGL}_n(F)$, which is defined to be $\mathrm{GL}_n(F)/Z$?

4. Let $F = \mathbf{F}_2 = \mathbf{Z}/2\mathbf{Z}$ be the field with 2 elements. The group $GL_2(F)$ is isomorphic to some group which you already have encountered. Which one? What about $PGL_2(F)$?

5. Let $F = \mathbf{F}_3 = \mathbf{Z}/3\mathbf{Z}$. The group $PGL_2(F)$ is isomorphic to some group which you already have encountered. Which one?

6. Let ζ be a primitive n-th root of unity, for instance $\zeta = e^{2\pi i/n}$ in the complex numbers. Let G be the subgroup of all 2×2 matrices generated by the matrices

$$\alpha = \begin{pmatrix} 0 & 1 \\ 1 & 0 \end{pmatrix} \quad \text{and} \quad z = \begin{pmatrix} \zeta & 0 \\ 0 & \zeta^{-1} \end{pmatrix}.$$

Show that G has order $2n$. What is $\alpha z \alpha^{-1}$?

7. Let ζ be a primitive n-th root of unity where n is an odd integer. Let G be the subgroup of all 2×2 matrices generated by the matrices

$$w = \begin{pmatrix} 0 & -1 \\ 1 & 0 \end{pmatrix} \quad \text{and} \quad z = \begin{pmatrix} \zeta & 0 \\ 0 & \zeta^{-1} \end{pmatrix}.$$

Show that G has order $4n$ What is wzw^{-1}?

VI, §3. $SL_2(F)$

We define $SL_2(F)$ to be the kernel of the determinant map

$$\det: GL_2(F) \to F^*.$$

Thus $SL_2(F)$ consists of the matrices with determinant 1. It is a normal subgroup of $GL_2(F)$. The S stands for "special" and $SL_2(F)$ is called the **special linear group**.

Lemma 3.1. *The center of* $SL_2(F)$ *is* $\pm I$.

This is proved just as for the center of $GL_2(F)$, using the commutation rule with special matrices like

$$\begin{pmatrix} 1 & x \\ 0 & 1 \end{pmatrix} \quad \text{or} \quad \begin{pmatrix} 1 & 0 \\ y & 1 \end{pmatrix}.$$

We leave the computation to the reader.

We let $PSL_2(F) = SL_2(F)/Z$ where Z is the center of $SL_2(F)$. We call $PSL_2(F)$ the **projective special linear group**. The main result of this section will be that if F has at least four elements, then $PSL_2(F)$ is a simple group.

Lemma 3.2. *For $x, y \in F$ we let*

$$u(x) = \begin{pmatrix} 1 & x \\ 0 & 1 \end{pmatrix} \quad \textit{and} \quad v(y) = \begin{pmatrix} 1 & 0 \\ y & 1 \end{pmatrix}.$$

Then the set of matrices $u(x)$ and $v(y)$ for $x, y \in F$ generate $\mathrm{SL}_2(F)$.

Proof. Multiplication on the left by $u(x)$ adds x times the second row to the first row. Multiplication of $u(x)$ on the right adds x times the first column to the second column. And similarly for multiplication with $v(y)$. Thus multiplication with elements $u(x)$ and $v(y)$ carries out row and column operations. By such multiplications, a given matrix in $\mathrm{SL}_2(F)$ can be brought to diagonal form, that is

$$\begin{pmatrix} a & 0 \\ 0 & d \end{pmatrix},$$

and $ad = 1$, so $d = a^{-1}$. Let $w(a) = u(a)v(-a^{-1})$. Then

$$v(-a^{-1})w(a)w(-1)u(-1) = \begin{pmatrix} a & 0 \\ 0 & a^{-1} \end{pmatrix},$$

thus concluding the proof that the elements $u(x)$, $v(y)$ generate $\mathrm{SL}_2(F)$.

As for $\mathrm{GL}_2(F)$, we let B_S be the **Borel subgroup** of $\mathrm{SL}_2(F)$, that is B_S is the group of matrices

$$\begin{pmatrix} a & b \\ 0 & a^{-1} \end{pmatrix}$$

with $a \in F^*$ and $b \in F$. Or in other words,

$$B_S = B \cap \mathrm{SL}_2(F).$$

Lemma 3.3. $\mathrm{SL}_2(F) = B_S \cup B_S \tau B_S$, *and* B_S, $B_S \tau B_S$ *are disjoint.*

Proof. The proof is similar to that of Lemma 2.3 and will be left to the reader.

Lemma 3.4. *The Borel subgroup B_S is a maximal subgroup of* $\mathrm{SL}_2(F)$.

Proof. Same as Lemma 2.2.

Lemma 3.5. *The intersection of all subgroups conjugate to B_S, that is all subgroups $\alpha B_S \alpha^{-1}$ with $\alpha \in \mathrm{SL}_2(F)$, is the center of* $\mathrm{SL}_2(F)$.

Proof. Note that

$$\tau = \begin{pmatrix} 0 & 1 \\ -1 & 0 \end{pmatrix}$$

is an element of $\mathrm{SL}_2(F)$. By a direct computation, you can see that

$$\tau B \tau^{-1} = \bar{B}$$

is the group of lower triangular matrices

$$\begin{pmatrix} x & 0 \\ y & z \end{pmatrix},$$

and therefore that

$$B_S \cap \tau B_S \tau^{-1} \subset \text{subgroup of matrices in } \mathrm{SL}_2(F) \text{ which are both upper and lower triangular.}$$

A matrix in $\mathrm{SL}_2(F)$ which is both upper and lower triangular has the form

$$\begin{pmatrix} a & 0 \\ 0 & a^{-1} \end{pmatrix}.$$

If we conjugate such a matrix by $\begin{pmatrix} 1 & 1 \\ 0 & 1 \end{pmatrix}$ we get

$$\begin{pmatrix} 1 & -1 \\ 0 & 1 \end{pmatrix}\begin{pmatrix} a & 0 \\ 0 & a^{-1} \end{pmatrix}\begin{pmatrix} 1 & 1 \\ 0 & 1 \end{pmatrix} = \begin{pmatrix} a & a - a^{-1} \\ 0 & a^{-1} \end{pmatrix}.$$

If such a matrix lies in the intersection of all conjugates of B_S then we must have $a - a^{-1} = 0$ so $a = a^{-1}$ and $a^2 = 1$. This implies that the matrix is $\pm I$, thus proving the lemma.

Let G be a group. By the **commutator group** G^c **we mean the group** generated by all elements of the form

$$\alpha\beta\alpha^{-1}\beta^{-1} \quad \text{with} \quad \alpha, \beta \in G.$$

Lemma 3.6. *If F has at least four elements, then* $\mathrm{SL}_2(F)$ *is equal to its own commutator group.*

Proof. Let

$$s(a) = \begin{pmatrix} a & 0 \\ 0 & a^{-1} \end{pmatrix} \quad \textbf{for} \quad a \in F^*.$$

We have the commutator relation

$$s(a)u(b)s(a)^{-1}u(b)^{-1} = u(ba^2 - b) = u(b(a^2 - 1))$$

for all $a \in F^*$ and $b \in F$. Let $G = \mathrm{SL}_2(F)$. Let G^c be its commutator group, and let B_S^c be the commutator group of B_S. From the hypothesis that F has at least four elements, we can find an element $a \neq 0$ in F such that $a^2 \neq 1$, whence the commutator relation shows that B_S^c is the group of all matrices

$$\begin{pmatrix} 1 & b \\ 0 & 1 \end{pmatrix} \quad \text{with} \quad b \in F.$$

Denote this group by U. It follows that $G^c \supset U$, and since G^c is normal (prove this as an exercise), we get

$$G^c \supset \tau U \tau^{-1} = \bar{U},$$

where $\bar{U}$ is the group of all matrices

$$\begin{pmatrix} 1 & 0 \\ c & 1 \end{pmatrix} \quad \text{with} \quad c \in F.$$

From Lemma 3.2 we conclude that $G^c = G$, thus proving Lemma 3.6.

Lemma 3.7. *Let $G = \mathrm{SL}_2(F)$. Let H be a normal subgroup of G. Then either $H \subset Z$ (where Z is the center) or $H \supset G^c$.*

Proof. Write B instead of B_S for simplicity. By the maximality of B we must have

$$HB = B \quad \text{or} \quad HB = G.$$

If $HB = B$ then $H \subset B$. Since H is normal, we conclude that H is contained in every conjugate of B, whence in the center by Lemma 3.5. On the other hand, suppose that $HB = G$. Write

$$\tau = h\beta \quad \text{with} \quad h \in H \quad \text{and} \quad \beta \in B.$$

Then

$$\tau U \tau^{-1} = \bar{U} = h\beta U \beta^{-1} h^{-1} = hUh^{-1} \subset HU$$

because H is normal, so $HU = UH$. Since $\bar{U} \subset HU$ and U, $\bar{U}$ generate G by Lemma 3.2, it follows that $HU = G$. Let

$$f\colon G = HU \to G/H = HU/H$$

be the canonical homomorphism. The $f(h) = 1$ for all $h \in H$. Since U is commutative, it follows that $f(G) = f(U)$ and that G/H is a homomorphic image of the commutative group U, whence G/H is abelian. This implies that H contains G^c, thus proving the lemma.

Theorem 3.8. *Let F be a field with at least four elements. Let Z be the center of $\mathrm{SL}_2(F)$. Then $\mathrm{SL}_2(F)/Z$ is simple.*

Proof. Let $G = \mathrm{SL}_2(F)$. Let

$$g: G \to G/Z$$

be the canonical homomorphism. Let $\bar{H}$ be a normal subgroup of G/Z and let

$$H = g^{-1}(\bar{H}).$$

Then H is a subgroup of G which contains the center Z. If $H = Z$ then $\bar{H}$ is just the unit element of G/Z. If $H \neq Z$ then $H = G$ by Lemma 3.7 and Lemma 3.6 which says that $G^c = G$. Hence $\bar{H} = G/Z$. This concludes the proof.

VI, §3. EXERCISES

Throughout Exercises 1, 2, 3, *we let* $G = \mathrm{SL}_2(\mathbf{R})$ *where* $\mathbf{R}$ *is the field of real numbers.*

1. Let $\mathbf{H}$ be the upper half plane, that is the set of all complex numbers

$$z = x + iy$$

with $y > 0$. Let

$$\alpha = \begin{pmatrix} a & b \\ c & d \end{pmatrix} \in G.$$

Define

$$\alpha(z) = \frac{az + b}{cz + d}.$$

Prove by explicit computation that $\alpha(z) \in \mathbf{H}$ and that:
(a) If $\alpha, \beta \in G$ then $\alpha(\beta(z)) = (\alpha\beta)(z)$.
(b) If $\alpha = \pm I$ then $\alpha(z) = z$.
In other words, we have defined an operation of $\mathrm{SL}_2(\mathbf{R})$ on $\mathbf{H}$, according to the definition of Chapter II, §8.

2. Given a real number θ, let

$$r(\theta) = \begin{pmatrix} \cos\theta & \sin\theta \\ -\sin\theta & \cos\theta \end{pmatrix}.$$

(a) Show that $\theta \mapsto r(\theta)$ is a homomorphism of $\mathbf{R}$ into G. We denote by K the set of all such matrices $r(\theta)$. So K is a subgroup of G.
(b) Show that if $\alpha = r(\theta)$ then $\alpha(i) = i$, where $i = \sqrt{-1}$.
(c) Show that if $\alpha \in G$, and $\alpha(i) = i$ then there is some θ such that $\alpha = r(\theta)$.
In the terminology of the operation of a group, we note that K is the isotropy group of i in G.

3. Let A be the subgroup of G consisting of all matrices

$$s(a) = \begin{pmatrix} a & 0 \\ 0 & a^{-1} \end{pmatrix} \quad \text{with} \quad a > 0.$$

(a) Show that the map $a \mapsto s(a)$ is a homomorphism of $\mathbf{R}^+$ into G. Since this homomorphism is obviously injective, this homomorphism gives an imbedding of $\mathbf{R}^+$ into G.
(b) Let U be the subgroup of G consisting of all elements

$$u(x) = \begin{pmatrix} 1 & x \\ 0 & 1 \end{pmatrix} \quad \text{with} \quad x \in \mathbf{R}.$$

Thus $u \mapsto u(x)$ gives an imbedding of $\mathbf{R}$ into G. Then UA is a subset of G. Show that UA is a subgroup. How does it differ from the Borel subgroup of G? Show that U is normal in UA.
(c) Show that the map

$$UA \to \mathbf{H}$$

given by

$$\beta \mapsto \beta(i)$$

gives a bijection of UA onto $\mathbf{H}$.
(d) Show that every element of $\mathrm{SL}_2(\mathbf{R})$ admits a unique expression as a product

$$u(x)s(a)r(\theta),$$

so in particular, $G = UAK$.

4. Let $G = \mathrm{GL}_2(F)$ where $F = \mathbf{Z}/3\mathbf{Z}$, and let $V = F \times F$ be the vector space of pairs of elements of F, having dimension 2 over F.
(a) Show that G operates as a permutation group of the subspaces of V of dimension 1. How many such subspaces are there?
(b) From (a), establish an isomorphism $G/\pm 1 \approx S_4$ (where S_4 is the symmetric group on 4 elements).
(c) Establish an isomorphism $\mathrm{SL}_2(\mathbf{Z}/3\mathbf{Z})/\pm 1 \approx A_4$, where A_4 is the alternating subgroup of S_4.

5. Let F be a finite field of characteristic p. Let U be the subgroup of $GL_n(F)$ consisting of upper triangular matrices whose diagonal elements are all equal to 1. Prove that U is a p-Sylow subgroup of $GL_n(F)$.

6. Again let F be a finite field of characteristic p. Prove that the p-Sylow subgroups of $SL_n(F)$ and $GL_n(F)$ are the same.

7. Let R be a principal ring. By $GL_n(R)$ we mean the set of matrices with components in R such that the determinant is a unit in R. We assume that you know determinants from a course in linear algebra. Show that $GL_n(R)$ is a group.

8. Let $(x_{11}, \ldots, x_{1n})$ be an n-tuple of elements in a principal ring R, and assume that they are relatively prime, that is, the ideal generated by them in R is R itself (the unit ideal). Show that there exist elements x_{ij} $(i, j = 1, \ldots, n)$ such that the matrix $X = (X_{ij})$ is in $GL_n(R)$. Do this by induction, starting with $n = 2$.

9. Let $SL_n(R)$ be the subset of $GL_n(R)$ consisting of those matrices with determinant 1. Show that $SL_n(R)$ is a subgroup.

10. Let F be the quotient field of the principal ring R. Let $B_n(F)$ be the subset of $GL_n(F)$ consisting of the upper triangular matrices (arbitrary on the diagonal, but non-zero determinant). Show by induction that

$$GL_n(F) = SL_n(R)B_n(F).$$

[*Hint*: First do the case $n = 2$ using Exercise 8. Next let $n > 2$. Let $X = (x_{ij})$ be an unknown matrix in $SL_n(R)$, and let X_n be its bottom row. Let $A^1, \ldots, A^n$ be the columns of a given matrix $A \in GL_n(F)$. We want to solve for $X_n \cdot A^i = 0$ for $i = 1, \ldots, n-1$, so that XA has its last row equal to 0 except for the lower right corner. Consider the R-module consisting of R-vectors X_n satisfying these orthogonality relations. It has a non-zero element, and by unique factorization it has an element X_n whose components are relatively prime. Use Exercise 8. For some $A' \in GL_{n-1}(F)$, XA is a matrix with 0 in the bottom row except for the lower right hand corner, which is a unit in R. Use induction again to get a matrix $Y \in SL_{n-1}(R)$ such that YA' is upper triangular. Conclude.]

VI, §4. SL$_n$(R) AND SL$_n$(C) IWASAWA DECOMPOSITIONS

Let F be a field. By $SL_n(F)$ we mean the group of $n \times n$ matrices with components in F, having determinant 1. We shall give decompositions valid over the real and complex numbers in terms of special subgroups.

Let $G = SL_n(\mathbf{R})$. Note that the subset consisting of the two elements $I, -I$ is a subgroup. Also note that $SL_n(\mathbf{R})$ is a subgroup of the group $GL_n(\mathbf{R})$ (all real matrices with non-zero determinant).

Let:

$U =$ subgroup of upper triangular matrices with 1's on the diagonal,

$$u(x) = \begin{pmatrix} 1 & x_{12} & \cdots & x_{1n} \\ 0 & 1 & \cdots & x_{2n} \\ \vdots & \vdots & \ddots & \vdots \\ 0 & 0 & \cdots & 1 \end{pmatrix} \quad \text{called } \textbf{unipotent}.$$

$A =$ subgroup of positive diagonal elements:

$$a = \begin{pmatrix} a_1 & & & \\ & a_2 & & \\ & & \ddots & \\ & & & a_n \end{pmatrix} \quad \text{with} \quad a_i > 0 \text{ for all } i.$$

$K =$ subgroup of real unitary matrices k, satisfying ${}^t k = k^{-1}$.

Theorem 4.1 (Iwasawa decomposition). *The product map $U \times A \times K \to G$ given by*

$$(u, a, k) \mapsto uak$$

is a bijection.

Proof. Let $e_1, \ldots, e_n$ be the standard unit vectors of $\mathbf{R}^n$ (vertical). Let $g = (g_{ij}) \in G$. Then we have

$$ge_i = \begin{pmatrix} g_{11} & \cdots & g_{1n} \\ \vdots & & \vdots \\ g_{n1} & \cdots & g_{nn} \end{pmatrix} \begin{pmatrix} 0 \\ \vdots \\ 1_i \\ \vdots \\ 0 \end{pmatrix} = \begin{pmatrix} g_{1i} \\ \vdots \\ g_{ni} \end{pmatrix} = g^{(i)} = \sum_{q=1}^{n} g_{qi} e_q.$$

There exists an upper triangular matrix $B = (b_{ij})$, so with $b_{ij} = 0$ if $i > j$, such that

$$\begin{aligned}
&b_{11} g^{(1)} && = e'_1 \\
&b_{12} g^{(1)} + b_{22} g^{(2)} && = e'_2 \\
&\vdots && \vdots \\
&b_{1j} g^{(1)} + b_{2j} g^{(2)} + \cdots + b_{jj} g^{(j)} && = e'_j \\
&\vdots && \vdots \\
&b_{1n} g^{(1)} + b_{2n} g^{(2)} + \quad \cdots \quad + b_{nn} g^{(n)} && = e'_n,
\end{aligned}$$

such that the diagonal elements are positive, that is $b_{11}, \ldots, b_{nn} > 0$, and such that the vectors $e'_1, \ldots, e'_n$ are mutually perpendicular unit vectors. Getting such a matrix B is merely applying the usual Gram Schmidt orthogonalization process, subtracting a linear combination of previous vectors to get orthogonality, and then dividing by the norms to get unit vectors. Thus

$$e'_j = \sum_{i=1}^{j} b_{ij} g^{(i)} = \sum_{i=1}^{n} \sum_{q=1}^{n} g_{qi} b_{ij} e_q = \sum_{q=1}^{n} \sum_{i=1}^{n} g_{qi} b_{ij} e_q.$$

Let $gB = k \in K$. Then $ke_i = e'_i$, so k maps the orthogonal unit vectors $e_1, \ldots, e_n$ to the orthogonal unit vectors $e'_1, \ldots, e'_n$. Therefore k is unitary, and $g = kB^{-1}$. Then

$$g^{-1} = Bk^{-1} \quad \text{and} \quad B = au$$

where a is the diagonal matrix with $a_i = b_{ii}$ and u is unipotent, $u = a^{-1}B$. This proves the surjection $G = UAK$. For uniqueness of the decomposition, if $g = uak = u'a'k'$, let $u_1 = u^{-1}u'$, so using $g^t g$ you get $a^{2t}u_1^{-1} = u_1 a'^2$. These matrices are lower and upper triangular respectively, with diagonals a^2, a'^2, so $a = a'$, and finally $u_1 = I$, proving uniqueness.

The elements of U are called **unipotent** because they are of the form

$$u(X) = I + X,$$

where X is strictly upper triangular, and $X^{n+1} = 0$. Thus $X = u - I$ is called **nilpotent**. Let

$$\exp Y = \sum_{j=0}^{\infty} \frac{Y^j}{j!} \quad \text{and} \quad \log(I + X) = \sum_{i=1}^{\infty} (-1)^{i+1} \frac{X^i}{i}.$$

Let $\mathfrak{n}$ denote the space of all strictly upper triangular matrices. Then

$$\exp\colon \mathfrak{n} \to U, \qquad Y \mapsto \exp Y$$

is a bijection, whose inverse is given by the log series, $Y = \log(I + X)$. Note that, because of the nilpotency, the exp and log series are actually polynomials, defining inverse polynomial mappings between U and $\mathfrak{n}$. The bijection actually holds over any field of characteristic 0. The relations

$$\exp \log(I + X) = I + X \quad \text{and} \quad \log \exp Y = \log(I + X) = Y$$

hold as identities of formal power series. Cf. my *Complex Analysis*, Chapter II, §3, Exercise 2.

We now recall Chapter II, §8.

Let X be a set. A bijective map $\sigma: X \to X$ of X with itself is called a **permutation.** You can verify at once that the set of permutations of X is a group, denoted by $\text{Perm}(X)$. By an **action** of a group G on X we mean a map

$$G \times X \to X \qquad \text{denoted by} \quad (g, x) \mapsto gx,$$

satisfying the two properties:

If e is the unit element of G, then $ex = x$ for all $x \in X$.
For all $g_1, g_2 \in G$ and $x \in X$ we have $g_1(g_2x) = (g_1g_2)x$.

This is just a general formulation of action, of which we have seen an example above. Given $g \in G$, the map $x \mapsto gx$ of X into itself is a permutation of X. You can verify this directly from the definition, namely the inverse permutation is given by $x \mapsto g^{-1}x$. Let $\sigma(g)$ denote the permutation associated with g. Then you can also verify directly from the definition that

$$g \mapsto \sigma(g)$$

is a homomorphism of G into the group of permutations of X. Conversely, such a homomorphism gives rise to an action of G on X.

Let $x \in X$. By the **isotropy group** of x (in G) we mean the subset of elements $g \in G$ such that $gx = x$. This subset is immediately verified to be a subgroup, denoted by G_x.

Geometric interpretation in dimension 2

Let $\mathbf{H}^2$ be the upper half plane of complex numbers $z = x + \mathbf{i}y$ with $x, y \in \mathbf{R}$ and $y > 0$, $y = y(z)$. For

$$g = \begin{pmatrix} a & b \\ c & d \end{pmatrix} \in G = \text{SL}_2(\mathbf{R})$$

define

$$g(z) = (az + b)(cz + d)^{-1}.$$

It is routinely verified by computation that this defines an action of G on the upper half plane. Also note the property

If $g(z) = z$ *for all* $z \in \mathbf{H}^2$ *then* $g = \pm I$.

In other words, the kernel of the action homomorphism $G \to \text{Aut}(\mathbf{H}^2)$ is the subgroup $\{\pm I\}$.

To see that if $z \in \mathbf{H}^2$ then $g(z) \in \mathbf{H}^2$ also, you will need to check the transformation formula

$$y(g(z)) = \frac{y(z)}{|cz+d|^2}.$$

These statements are proved by easy brute force.

Furthermore, by direct computation, you can verify:

Theorem 4.2. *The isotropy group of* $\mathbf{i}$ *is* K, *i.e.,* K *is the subgroup of elements* $k \in G$ *such that* $k(\mathbf{i}) = \mathbf{i}$. *This is the group of matrices of the form*

$$\begin{pmatrix} \cos\theta & \sin\theta \\ -\sin\theta & \cos\theta \end{pmatrix}.$$

Or equivalently, $a = d$, $c = -b$, $a^2 + b^2 = 1$.

For $x \in \mathbf{R}$ and $a_1 > 0$, let

$$u(x) = \begin{pmatrix} 1 & x \\ 0 & 1 \end{pmatrix} \quad \text{and} \quad a = \begin{pmatrix} a_1 & 0 \\ 0 & a_2 \end{pmatrix} \quad \text{with} \quad a_2 = a_1^{-1}.$$

If $g = uak$, then $u(x)(z) = z + x$, so putting $y = a_1^2$, we get $a(\mathbf{i}) = y\mathbf{i}$,

$$g(\mathbf{i}) = uak(\mathbf{i}) = ua(\mathbf{i}) = y\mathbf{i} + x = x + \mathbf{i}y.$$

Thus G acts transitively, and we have a description of the action in terms of the Iwasawa decomposition and the coordinates of the upper half plane.

Geometric interpretation in dimension 3

We shall use the quaternions, whose elements are linear combinations

$$z = x_1 + x_2\mathbf{i} + x_3\mathbf{j} + x_4\mathbf{k} \quad \text{with} \quad x_1, x_2, x_3, x_4 \in \mathbf{R}$$

and $\mathbf{i}^2 = \mathbf{j}^2 = \mathbf{k}^2 = -1$, $\mathbf{ij} = \mathbf{k}$, $\mathbf{jk} = \mathbf{i}$, $\mathbf{ki} = \mathbf{j}$; also $x \in \mathbf{R}$ commutes with $\mathbf{i}, \mathbf{j}, \mathbf{k}$. Define

$$\bar{z} = x_1 - x_2\mathbf{i} - x_3\mathbf{j} - x_4\mathbf{k}.$$

Then

$$z\bar{z} = x_1^2 + x_2^2 + x_3^2 + x_4^2,$$

and we define $|z| = (z\bar{z})^{1/2}$.

Let $\mathbf{H}^3$ be the upper half space consisting of elements z whose $\mathbf{k}$-component is 0, and $x_3 > 0$, so we write

$$z = x_1 + x_2\mathbf{i} + y\mathbf{j} \quad \text{with} \quad y > 0.$$

Let $G = \mathrm{SL}_2(\mathbf{C})$, so elements of G are matrices

$$g = \begin{pmatrix} a & b \\ c & d \end{pmatrix} \quad \text{with} \quad a, b, c, d \in \mathbf{C} \text{ and } ad - bc = 1.$$

As in the case of $\mathbf{H}^2$, define

$$g(z) = (az + b)(cz + d)^{-1}.$$

Verify by brute force that if $z \in \mathbf{H}^3$ then $g(z) \in \mathbf{H}^3$, and that G acts on $\mathbf{H}^3$, namely the two properties listed in the previous example are also satisfied here. Since the quaternions are not commutative, we have to use the quotient as written $(az + b)(cz + d)^{-1}$. Also note that the y-coordinate transformation formula for $z \in \mathbf{H}^3$ reads the same as for $\mathbf{H}^2$, namely

$$y(g(z)) = \frac{y(z)}{|cz + d|^2}.$$

For the Iwasawa decomposition, we use the groups:

U = group of elements $u(x) = \begin{pmatrix} 1 & x \\ 0 & 1 \end{pmatrix}$ with $x \in \mathbf{C}$;
A = *same group as before* in the case of $\mathrm{SL}_2(\mathbf{R})$;
K = complex unitary group of elements k such that ${}^t\bar{k} = k^{-1}$.

The group $G = \mathrm{SL}_2(\mathbf{C})$ has the Iwasawa decomposition $G = UAK$. Each element of G has a unique decomposition $g = uak$ with $u \in U$, $a \in A$, $k \in K$.

The previous proof works the same way, BUT you can verify directly:

Theorem 4.3. *The isotropy group $G_{\mathbf{j}}$ is K.*

If $g = uak$ with $u \in U$, $a \in A$, $k \in K$, $u = u(x)$ and $y = y(a)$, then

$$g(\mathbf{j}) = x + y\mathbf{j}.$$

Thus G acts transitively, and the Iwasawa decomposition also follows trivially from this group action (see below). Thus the orthogonalization-type proof can be completely avoided. Of course, it can be similarly avoided for $\mathbf{H}^2$ and $\mathrm{SL}_2(\mathbf{R})$, using $x \in \mathbf{R}$.

Proof of the Iwasawa decomposition from the above two properties. Let $g \in G$ and $g(\mathbf{j}) = x + y\mathbf{j}$. Let $u = u(x)$ and a be such that $y = a_1/a_2 = a_1^2$.

Let $g' = ua$. Then by the second property, we get $g(\mathbf{j}) = g'(\mathbf{j})$, so $\mathbf{j} = g^{-1}g'(\mathbf{j})$. By the first property, we get $g^{-1}g' = k$ for some $k \in K$, so

$$g'k^{-1} = uak^{-1} = g,$$

concluding the proof.

We now come to the general case of $\mathrm{SL}_n(\mathbf{C})$.

Let $G = \mathrm{SL}_n(\mathbf{C})$. Let:

$U = U(\mathbf{C})$ be the subgroup of elements $I + Z$ such that Z is strictly upper triangular with components in $\mathbf{C}$.

D = subgroup of diagonal matrices.

A = same A as for $\mathrm{SL}_n(\mathbf{R})$, namely the subgroup of diagonal matrices with positive diagonal elements.

K = subgroup of elements k such that ${}^t\bar{k} = k^{-1}$. Then K is also called the **(complex) unitary subgroup**. Its elements are called **unitary**.

For $g \in G$, define

$$g^* = {}^t\bar{g}.$$

You can verify directly and easily that

$$(g_1g_2)^* = g_2^*g_1^*.$$

Note that K is the subgroup of G consisting of those elements k such that $k^* = k^{-1}$.

Let $\langle \cdot , \cdot \rangle$ be the standard hermitian scalar product on $\mathbf{C}^n$, that is: if $z \in \mathbf{C}^n$,

$$z = {}^t(z_1, \ldots, z_n) \quad \text{and} \quad w = {}^t(w_1, \ldots, w_n) \quad \text{with} \quad z_i, w_i \in \mathbf{C},$$

then

$$\langle z, w \rangle = \sum_{i=1}^{n} z_i \bar{w}_i.$$

The element g^* satisfies

$$\langle gz, w \rangle = \langle z, g^*w \rangle \qquad \text{for all} \quad z, w \in \mathbf{C}^n,$$

so it is called the **adjoint** (which you should know from a course in linear algebra).

We also have the **Iwasawa decomposition**, *with the same A as in the real case*, but complexified U and K, as follows.

Theorem 4.4. *Let $G = \mathrm{SL}_n(\mathbf{C})$. Then with U, A, K defined as above, the product map*

$$U \times A \times K \to UAK = G$$

gives a bijection with G.

The proof is the same as in the real case, using orthogonalization with respect to the hermitian product.

VI, §5. OTHER DECOMPOSITIONS

The star operation is of course defined for all matrices. A complex $n \times n$ matrix Z is called **hermitian** if and only if $Z = Z^*$.

Let $G = \mathrm{SL}_n(\mathbf{C})$. We define the **quadratic map** on G by

$$g \mapsto gg^*.$$

Then gg^* is hermitian positive definite, i.e., for every $v \in \mathbf{C}^n$, we have $\langle gg^*v, v\rangle \geqq 0$, and $= 0$ only if $v = 0$.

We denote by $\mathbf{P} = \mathrm{SPos}_n(\mathbf{C})$ the set of all hermitian positive definite $n \times n$ matrices with determinant 1.

Theorem 5.1. *The set* $\mathbf{P} = \mathrm{SPos}_n(\mathbf{C})$ *is the set of elements* kak^{-1} *with* $a \in A$ *and* $k \in K$, *in other words, it is* $\mathbf{c}(K)A$ *(image of A under K-conjugation).*

Proof. We use a fact which you should have learned in a course on linear algebra, that a hermitian matrix can be diagonalized. In fact, if g is hermitian, then there exists a basis consisting of orthogonal unit vectors such that each basis vector is an eigenvector of g. An equivalent way of putting this is that there exists a unitary matrix k such that kgk^{-1} is diagonal. Since p is assumed positive definite, putting $g = p$, it follows that the diagonal elements of kpk^{-1} are real positive. Conversely, let $a \in A$, $k \in K$. Write $a = b^2$ with $b \in A$. Then for any vector v,

$$\langle kb^2k^{-1}v, v\rangle = \langle bk^{-1}v, bk^{-1}v\rangle > 0,$$

which proves the positive definiteness, and the theorem.

Theorem 5.2. *Let* p, q *be hermitian and commute. Then there is a basis of* $\mathbf{C}^n$ *consisting of vectors which are also eigenvectors for both* p *and* q, *i.e.,* p, q *can be simultaneously diagonalized.*

Proof. Let $E = \mathbf{C}^n$. There exists a basis of E consisting of p-eigenvectors. For each eigenvalue c of p, let $E_c(p)$ be the c-eigenspace, that is, the subspace of eigenvectors with eigenvalue c. Then E is the direct sum of the subspaces $E_c(p)$, taken over the eigenvalues c ($\neq 0$ since p is invertible). Directly from the commutativity $pq = qp$, we see that q maps an eigenspace $E_c(p)$ into itself, namely

$$\text{if } pv = cv \text{ then } pqv = qpv = qcv = cqv.$$

Since q is hermitian, there is a basis $\{w_1, \ldots, w_r\}$ of $E_c(p)$ consisting of eigenvectors of q. Thus p, q can be simultaneously diagonalized, as desired. By orthogonalization and multiplication by positive scalars, one can of course achieve that the basis is orthonormal.

Corollary 5.3. *Let p be hermitian positive definite. Then p has a unique square root in $\mathbf{P}$.*

Proof. We write $E = \bigoplus E_c(p)$ as a sum of eigenspaces as in the theorem. Let $q^2 = p$, with $q \in \mathbf{P}$. Since q maps $E_c(p)$ into itself, it suffices to prove that its restriction to $E_c(p)$ is unique. Let $\{v_1, \ldots, v_m\}$ be a basis of $E_c(p)$ consisting of eigenvectors for q, so $pv_i = cv_i$ and $qv_i = b_i v_i$ for all i. By the positive definiteness, we have c, b_i real 0. From $q^2 = p$ we get $b_i^2 = c$, whence b_i is the unique positive square root of c, thus proving the corollary.

Theorem 5.4 (Cartan Decomposition). *The quadratic map $g \mapsto gg^*$ induces a bijection*

$$G/K \to \mathrm{SPos}_n(\mathbf{C}) = \mathbf{P}.$$

In fact, we have a product decomposition $G = \mathbf{P}K$, each $g \in G$ having a unique product decomposition $g = pk$ with $p \in \mathbf{P}$ and $k \in K$.

Proof. Since by Corollary 5.3, every $p \in \mathbf{P}$ can be written as q^2 with $q \in \mathbf{P}$, the quadratic map is surjective. For injectivity on G/K, suppose that

$$g_1 g_1^* = g_2 g_2^*.$$

Then $g_2^{-1} g_1 = g_2^*(g_1^{-1})^* = (g_1^{-1} g_2)^*$. Hence $g_1^{-1} g_2 \in K$, so $g_2 \in g_1 K$, which proves the injectivity on G/K. By Theorem 5.1, given $g \in G$ there exists $k \in K$ and $b^2 \in A$ such that

$$gg^* = kb^2k^{-1} = (kbk^{-1})^2.$$

Let $p = kbk^{-1}$. Then $p \in \mathbf{P}$ and $pp^* = gg^*$, whence by the first part of the theorem, $pK = gK$, which proves $G = \mathbf{P}K$. For the uniqueness, suppose $p_1 k_1 = p_2 k_2$. Taking the quadratic map yields $p_1^2 = p_2^2$, and Corollary 5.3 shows $p_1 = p_2$, whence $k_1 = k_2$, thus proving the theorem.

Theorem 5.5. *The group G has the decomposition called* **polar**

$$G = KAK.$$

If $g \in G$ is written as a product $g = k_1 b k_2$ with $k_1, k_2 \in K$ and $b \in A$, then b is uniquely determined up to a permutation of the diagonal elements.

Proof. From $G = \mathbf{P}K$ (Theorem 5.4) and $\mathbf{P} = \mathbf{c}(K)A$ (Corollary 5.3), we get $G = KAK$. Then $gg^* = k_1 b^2 k_1^{-1}$. Hence the roots of the characteristic polynomial of gg^* are the diagonal elements $b_1^2, \ldots, b_n^2$, uniquely determined up to a permutation. Then $b_1, \ldots, b_n$ are also uniquely determined up to a permutation, as was to be shown.

The linear structure of the permutations of the diagonal elements will be further analyzed in the next section (Theorems 6.1 and 6.2).

VI, §5. EXERCISES

1. Verify in detail that Theorems 5.1, 5.2, 5.3, 5.4 are valid for $\mathrm{SL}_n(\mathbf{R})$, with exactly the same proofs, replacing "hermitian" by "symmetric" and using $\mathrm{SPos}_n(\mathbf{R})$. For further results, cf. for instance J. Jorgenson and S. Lang, *Spherical Inversion* on $\mathrm{SL}_n(\mathbf{R})$, Chapter I and Chapter V, §2 (the Bruhat decomposition). That book gives an exposition on SL_n of what is usually regarded as more advanced topics. One sees how some algebraic structures are used as backdrop for analysis on certain types of groups.

2. Let $\mathrm{Her} = \mathrm{Her}_n(\mathbf{C})$ be the real vector space of hermitian $n \times n$ matrices, and $\mathrm{Sk} = \mathrm{Sk}_n(\mathbf{C})$ the real vector space of skew hermitian matrices, that is, matrices Z satisfying ${}^t\bar{Z} = -Z$. Show that $\mathrm{Mat}_n(\mathbf{C})$ is the direct sum $\mathrm{Her} + \mathrm{Sk}$.

3. Let L be hermitian, and let v, v' be eigenvectors with distinct eigenvalues for L. Show that v, v' are orthogonal.

VI, §6. THE CONJUGATION ACTION

Let G be a group. The **conjugation action** of G on itself is defined for $g, g' \in G$ by

$$\mathbf{c}(g)g' = gg'g^{-1}.$$

It is immediately verified that the map $g \mapsto \mathbf{c}(g)$ is a homomorphism of G into $\mathrm{Aut}(G)$ (the group of automorphisms of G). Then G also acts on spaces naturally associated to G. We describe such spaces.

Consider the special case when $G = \mathrm{SL}_n(F)$ with a field F. Let:

$\mathfrak{d}$ = vector space of diagonal matrices $\mathrm{diag}(h_1, \ldots, h_n)$ with trace 0, $\sum h_i = 0$.

$\mathfrak{n}$ = vector space of strictly upper triangular matrices (h_{ij}) with $h_{ij} = 0$ if $i \geqq j$.

${}^t\mathfrak{n}$ = vector space of strictly lower triangular matrices.

$\mathfrak{g}$ = vector space of $n \times n$ matrices in $\mathrm{Mat}_n(F)$ with trace 0.

To denote the dependence on F, we may write $\mathfrak{d}(F), \mathfrak{n}(F), {}^t\mathfrak{n}(F)$, and $\mathfrak{g}(F) = \mathfrak{sl}_n(F)$.

We shall now use the diagonal group:

$D =$ group of diagonal matrices in F with determinant 1.

Then $\mathfrak{g}$ is the direct sum,

$$\mathfrak{g} = \mathfrak{d} + \mathfrak{n} + {}^t\mathfrak{n}. \tag{1}$$

Furthermore, D acts by conjugation on $\mathfrak{g}$. One can then decompose $\mathfrak{g}$ into a direct sum of eigenspaces for this action as follows. Let $\alpha_{ij}: D \to F^*$ be the homomorphism (called a **character**) defined on a diagonal element $a = \text{diag}(a_1, \dots, a_n)$ by

$$a^{\alpha_{ij}} = a_i/a_j.$$

Let E_{ij} $(i < j)$ be the matrix with ij-component 1 and all other components 0. Then for $a \in D$, we have

$$\mathbf{c}(a)E_{ij} = aE_{ij}a^{-1} = (a_i/a_j)E_{ij} = a^{\alpha_{ij}}E_{ij}$$

by direct computation. Thus α_{ij} (written exponentially) is a homomorphism of D into F^*. The set of such homomorphisms will be called the set of **regular characters**, denoted by $\mathscr{R}(\mathfrak{n})$. Note that $\mathfrak{n}$ is the direct sum of the 1-dimensional eigenspaces having basis E_{ij} $(i < j)$. We write

$$\mathfrak{n} = \bigoplus_{\alpha \in \mathscr{R}(\mathfrak{n})} \mathfrak{n}_\alpha = \bigoplus_{\alpha \in \mathscr{R}(\mathfrak{n})} FE_\alpha, \tag{2}$$

where $\mathfrak{n}_\alpha$ is the subspace of elements $X \in \mathfrak{n}$ such that $aXa^{-1} = a^\alpha X$. We have similarly

$${}^t\mathfrak{n} = \bigoplus_\alpha ({}^t\mathfrak{n})_{-\alpha}. \tag{3}$$

Note that $\mathfrak{d}$ is the 1-eigenspace for the conjugation action of D.

If $F = \mathbf{R}$ or $\mathbf{C}$, then instead of D and $\mathfrak{d}$, we may consider only the group A of diagonal matrices with positive diagonal elements, and $\mathfrak{a}$ the real vector space of real matrices with trace 0. The regular characters are given by the same formula as above.

We have expressed $\mathfrak{g}$ as a direct sum of eigenspaces. Over the arbitrary field F, these eigenspaces can be taken to have dimension 1. This is clear from (2) and (3) for $\mathfrak{n}$ and ${}^t\mathfrak{n}$. For $\mathfrak{d}$ itself, basis elements of any basis will do. The usual, and most natural basis, consists of the elements $(H_1, \dots, H_{n-1})$ where

$$H_i = \text{diag}(0, \dots, 0, 1_i, -1_{i+1}, 0, \dots, 0).$$

By an **algebra** we mean a vector space with a bilinear map into itself, called a product. We make $\mathfrak{g}$ into an algebra by defining the **Lie product** of $X, Y \in \mathfrak{g}$ to be

$$[X, Y] = XY - YX.$$

It is immediately verified that this product is bilinear but not associative. We call $\mathfrak{g}$ the **Lie algebra** of G. Let the space of linear maps of $\mathfrak{g}$ into itself be denoted by $\operatorname{End}(\mathfrak{g})$, whose elements are called **endomorphisms** of $\mathfrak{g}$. By definition the **regular representation** of $\mathfrak{g}$ on itself is the map

$$\mathfrak{g} \to \operatorname{End}(\mathfrak{g})$$

which to each $X \in \mathfrak{g}$ associates the endomorphism $L(X)$ of $\mathfrak{g}$ such that

$$L(X)(Y) = [X, Y].$$

Note that $X \mapsto L(X)$ is a linear map. We also write L_X for $L(X)$.

Verify the derivation property for all $Y, Z \in \mathfrak{g}$,

$$L_X[Y, Z] = [L_X Y, Z] + [Y, L_X Z].$$

Using only the bracket notation, this looks like

$$[X, [Y, Z]] = [[X, Y], Z] + [Y, [X, Z]].$$

We use α also to denote the additive linear character on $\mathfrak{d}$ given on a diagonal matrix $H = \operatorname{diag}(h_1, \ldots, h_n)$ by

$$\alpha_{ij}(H) = h_i - h_j \qquad \text{so} \qquad \alpha_{ij}\colon \mathfrak{d} \to F.$$

This is the additive version of the multiplicative character previously considered multiplicatively on D. Then each $\mathfrak{n}_\alpha$ is also the α-eigenspace for the additive character α, namely for $H \in \mathfrak{d}$, we have

$$[H, E_\alpha] = \alpha(H)E_\alpha,$$

which you can verify at once from the definition of multiplication of matrices.

We let $\operatorname{Nor}(D)$ be the normalizer of D in G, that is, the subgroup of elements $g \in G$ such that $gDg^{-1} = D$. We let $\operatorname{Cen}(D)$ be the centralizer of D, i.e. $g \in G$ such that $gzg^{-1} = z$ for all $z \in D$. Verify that $\operatorname{Cen}(D)$ is normal in $\operatorname{Nor}(D)$.

Theorem 6.1. *For $g \in \mathrm{Nor}(D)$, conjugation $\mathbf{c}(g)$ restricted to D permutes the diagonal elements. The map $g \mapsto \mathbf{c}(g)$ induces an isomorphism of* $\mathrm{Nor}(D)/\mathrm{Cen}(D)$ *with the permutation group of the diagonal.*

Proof. The characteristic polynomial is $P(t) = \prod (t - z_i)$, where $z_1, \ldots, z_n$ are the diagonal elements. It is unchanged under conjugation, so we get an injection of $\mathrm{Nor}(D)/\mathrm{Cen}(D)$ into the permutation group. To show the surjectivity, you have to prove that given a permutation, there is $g \in \mathrm{Nor}(D)$ such that $\mathbf{c}(g)$ induces the permutation. Do this as an exercise.

Actually, you may want to work out Theorem 6.1 first in the case $F = \mathbf{R}$, $G = \mathrm{SL}_n(\mathbf{R})$, in which case we can use A instead of D. Thus we now let M' be the subgroup of K consisting of the matrices which normalize A, that is k such that $kAk^{-1} = A$. Then we let M be the centralizer of A in K. The group M'/M is called the **Weyl group**.

Theorem 6.2. *Elements of the Weyl group act as a group of permutations of the diagonal elements. For $k \in M'$, let w_k be the corresponding permutation. Then $k \mapsto w_k$ gives an isomorphism*

$$W = M'/M \xrightarrow{\approx} \text{Permutation group of the diagonal.}$$

Proof. Exercise.

You will find that the "permutation matrices" have components which are 0 or ± 1, and can then be used over any field F of any characteristic. Check the case $n = 2$ first. You can see the exercise worked out in Jorgenson-Lang, *Spherical Inversion on* $\mathrm{SL}_n(\mathbf{R})$, Springer Verlag 2001, Chapter III, Theorem 3.4.

VI, §6. EXERCISES

1. A diagonal $n \times n$ matrix H is called **regular** if its diagonal elements are distinct. Let $X \in \mathrm{Mat}_n(F)$ (with a field F). If X commutes with one regular H, show that X is diagonal. Note that H is regular if and only if $\alpha(H) \neq 0$ for all regular characters α.

2. Let $F = \mathbf{R}$, and K the real unitary subgroup of $\mathrm{SL}_n(\mathbf{R})$. Let M be the subgroup of K consisting of elements commuting with all elements of $\mathfrak{a}$, that is, the centralizer of $\mathfrak{a}$ in K. Then M consists of the diagonal matrices with ± 1 on the diagonal.

3. Prove Theorems 6.1 and 6.2. Do it first when $n = 2$. Then for arbitrary n, determine the matrix inducing the transposition of h_i and h_{i+1} in a diagonal matrix $\mathrm{diag}(h_1, \ldots, h_n)$.

CHAPTER VII

Field Theory

VII, §1. ALGEBRAIC EXTENSIONS

Let F be a subfield of a field E. We also say that E is an **extension** of F and we denote the extension by E/F. Let F be a field. An element α in some extension of F is said to be **algebraic** over F if there exists a non-zero polynomial $f \in F[t]$ such that $f(\alpha) = 0$, i.e. if α satisfies a polynomial equation

$$a_n\alpha^n + \cdots + a_0 = 0$$

with coefficients in F, not all 0. If F is a subfield of E, and every element of E is algebraic over F, we say that E is **algebraic** over F.

Example 1. If $\alpha^2 = 2$, i.e. if α is one of the two possible square roots of 2, then α is algebraic over the rational numbers $\mathbf{Q}$. Similarly, a cube root of 2 is algebraic. Any one of the numbers $e^{2\pi i/n}$ (with n integer $\geqq 1$) is algebraic over $\mathbf{Q}$, since it is a root of $t^n - 1$. It is known (but hard to prove) that neither e nor π is algebraic over $\mathbf{Q}$.

Let E be an extension of F. We may view E as a vector space over F. We shall say that E is a **finite extension** if E is a finite dimensional vector space over F.

Example 2. $\mathbf{C}$ is a finite extension of $\mathbf{R}$, and $\{1, i\}$ is a basis of $\mathbf{C}$ over $\mathbf{R}$. The real numbers are not a finite extension of $\mathbf{Q}$.

Theorem 1.1. *If E is a finite extension of F, then every element of E is algebraic over F.*

Proof. The powers of an element α of E, namely $1, \alpha, \alpha^2, \dots, \alpha^n$ cannot be linearly independent over F, if $n > \dim E$. Hence there exist elements $a_0, \dots, a_n \in F$ not all 0 such that $a_n\alpha^n + \cdots + a_0 = 0$. This means that α is algebraic over F.

Proposition 1.2. *Let α be algebraic over F. Let J be the ideal of polynomials in $F[t]$ of which α is a root, i.e. polynomials f such that $f(\alpha) = 0$. Let $p(t)$ be a generator of J, with leading coefficient 1. Then p is irreducible.*

Proof. Suppose that $p = gh$ with $\deg g < \deg p$ and $\deg h < \deg p$. Since $p(\alpha) = 0$, we have $g(\alpha) = 0$ or $h(\alpha) = 0$. Say $g(\alpha) = 0$. Since $\deg g < \deg p$, this is impossible, by the assumption on p.

The irreducible polynomial p (with leading coefficient 1) is uniquely determined by α in $F[t]$, and will be called **the irreducible polynomial of α over F**. Its degree will be called the **degree** of α over F. We shall immediately give another interpretation for this degree.

Theorem 1.3. *Let α be algebraic over F. Let n be the degree of its irreducible polynomial over F. Then the vector space generated over F by $1, \alpha, \dots, \alpha^{n-1}$ is a field, and the dimension of that vector space is n.*

Proof. Let f be any polynomial in $F[t]$. We can find $q, r \in F[t]$ such that $f = qp + r$, and $\deg r < \deg p$. Then

$$f(\alpha) = q(\alpha)p(\alpha) + r(\alpha) = r(\alpha).$$

Hence if we denote the vector space generated by $1, \alpha, \dots, \alpha^{n-1}$ by E, we find that the product of two elements of E is again in E. Suppose that $f(\alpha) \neq 0$. Then f is not divisible by p. Hence there exist polynomials $g, h \in F[t]$ such that

$$gf + hp = 1.$$

We obtain $g(\alpha)f(\alpha) + h(\alpha)p(\alpha) = 1$, whence $g(\alpha)f(\alpha) = 1$. Thus every nonzero element of E is invertible, and hence E is a field.

The field generated by the powers of α over F as in Theorem 1.3 will be denoted by $F(\alpha)$.

If E is a finite extension of F, we denote by

$$[E : F]$$

the dimension of E viewed as vector space over F, and call it the **degree** of E over F.

Remark. If $[E:F]=1$ then $E=F$. Proof?

Theorem 1.4. *Let E_1 be a finite extension of F, and let E_2 be a finite extension of E_1. Then E_2 is a finite extension of F, and*

$$[E_2:F]=[E_2:E_1][E_1:F].$$

Proof. Let $\{\alpha_1,\dots,\alpha_n\}$ be a basis of E_1 over F, and $\{\beta_1,\dots,\beta_m\}$ a basis of E_2 over E_1. We prove that the elements $\{\alpha_i\beta_j\}$ form a basis of E_2 over F. Let v be an element of E_2. We can write

$$v=\sum_j w_j\beta_j=w_1\beta_1+\cdots+w_m\beta_m$$

with some elements $w_j\in E_1$. We write each w_j as a linear combination of $\alpha_1,\dots,\alpha_n$ with coefficients in F, say

$$w_j=\sum_i c_{ij}\alpha_i.$$

Substituting, we find

$$v=\sum_j\sum_i c_{ij}\alpha_i\beta_j.$$

Hence the elements $\alpha_i\beta_j$ generate E_2 over F. Assume that we have a relation

$$0=\sum_j\sum_i x_{ij}\alpha_i\beta_j$$

with $x_{ij}\in F$. Thus

$$\sum_j\left(\sum_i x_{ij}\alpha_i\right)\beta_j=0.$$

From the linear independence of $\beta_1,\dots,\beta_m$ over E_1, we conclude that

$$\sum_i x_{ij}\alpha_i=0$$

for each j, and from the linear independence of $\alpha_1,\dots,\alpha_n$ over F we conclude that $x_{ij}=0$ for all i, j as was to be shown.

Let α, β be algebraic over F. We suppose α, β are contained in some extension E of F. Then *a fortiori*, β is algebraic over $F(\alpha)$. We can form the field $F(\alpha)(\beta)$. Any field which contains F and α, β, will contain $F(\alpha)(\beta)$.

Hence $F(\alpha)(\beta)$ is the smallest field containing F and both α, β. Furthermore, by Theorem 1.4, $F(\alpha)(\beta)$ is finite over F, being decomposed in the inclusions

$$F \subset F(\alpha) \subset F(\alpha)(\beta).$$

Hence by Theorem 1.1, the field $F(\alpha)(\beta)$ is algebraic over F and in particular $\alpha\beta$ and $\alpha + \beta$ are algebraic over F. Furthermore, it does not matter whether we write $F(\alpha)(\beta)$ or $F(\beta)(\alpha)$. Thus we shall denote this field by $F(\alpha, \beta)$.

Inductively, if $\alpha_1, \ldots, \alpha_r$ are algebraic over F, and contained in some extension of F, we let $F(\alpha_1, \ldots, \alpha_r)$ be the smallest field containing F and $\alpha_1, \ldots, \alpha_r$. It can be expressed as $F(\alpha_1)(\alpha_2) \cdots (\alpha_r)$. It is algebraic over F by repeated applications of Theorem 1.4. We call it the field obtained by **adjoining** $\alpha_1, \ldots, \alpha_r$ to F, and we say that $F(\alpha_1, \ldots, \alpha_r)$ is **generated by** $\alpha_1, \ldots, \alpha_r$ over F, or that $\alpha_1, \ldots, \alpha_r$ are **generators** over F.

If S is a set of elements in some field containing F, then we denote by $F(S)$ the smallest field containing S and F. If, for instance, S consists of elements $\alpha_1, \alpha_2, \alpha_3, \ldots$, then $F(S)$ is the union of all the fields

$$\bigcup_{r=1}^{\infty} F(\alpha_1, \ldots, \alpha_r).$$

Let $\boldsymbol{\mu}_n$ be the group of n-th roots of unity in some extension of F. Then we shall often consider the field

$$F(\boldsymbol{\mu}_n).$$

Since $\boldsymbol{\mu}_n$ is cyclic by Theorem 1.10 of Chapter IV, it follows that if ζ is a generator of $\boldsymbol{\mu}_n$, then

$$F(\boldsymbol{\mu}_n) = F(\zeta).$$

Remark. An extension E of F can be algebraic without its being finite. This can happen of course only if E is generated by an infinite number of elements. For instance, let

$$E = \mathbf{Q}(2^{1/2}, 2^{1/3}, \ldots, 2^{1/n}, \ldots),$$

so E is obtained by adjoining all the n-th roots of 2 for all positive integers n. Then E is not finite over $\mathbf{Q}$.

Warning. We have just used the notation $2^{1/n}$ to denote an n-th root of 2. Of course, we can mean the real n-th root of 2, and one usually means this real root. You should know that in the complex numbers, there are other elements α such that $\alpha^n = 2$, and these also have a right to be called n-th roots of 2. Therefore in general, I strongly advise against using the notation $a^{1/n}$ to denote an n-th root of an element a in a field, because such an element is not uniquely determined. One should use a letter, for instance α, to denote an element whose n-th power is a, that is $\alpha^n = a$. The totality of such elements can be denoted by $\alpha_1, \ldots, \alpha_n$.

Whenever we have a sequence of extension fields

$$F \subset E_1 \subset E_2 \subset \cdots \subset E_r$$

we call such a sequence a **tower of fields**. One could express Theorem 1.4 by saying that:

The degree is multiplicative in towers.

We could call the field $E = \mathbf{Q}(2^{1/2}, 2^{1/3}, \ldots, 2^{1/n}, \ldots)$ an **infinite tower**, defined by the sequence of fields

$$\mathbf{Q} \subset \mathbf{Q}(2^{1/2}) \subset \mathbf{Q}(2^{1/2}, 2^{1/3}) \subset \mathbf{Q}(2^{1/2}, 2^{1/3}, 2^{1/4}) \subset \cdots.$$

We now return to possibly infinite algebraic extensions in general.

Let A be the set of all complex numbers which are algebraic over $\mathbf{Q}$. Then A is a field which is algebraic over $\mathbf{Q}$ but is not finite over $\mathbf{Q}$. (Exercise 16).

We define a field A to be **algebraically closed** if every polynomial of degree $\geqq 1$ with coefficients in A has a root in A. It follows that if f is such a polynomial, then f has a factorization

$$f(t) = c(t - \alpha_1) \cdots (t - \alpha_n)$$

with $c \neq 0$ and $\alpha_1, \ldots, \alpha_n \in A$.

Let F be a field. By an **algebraic closure** of F we mean a field A which is algebraically closed and algebraic over F. We shall prove later that an algebraic closure exists, and is uniquely determined up to isomorphism.

Example. It will be proved later in this book that the complex numbers $\mathbf{C}$ are algebraically closed. The set of numbers in $\mathbf{C}$ which are algebraic over $\mathbf{Q}$ form a subfield, which is an algebraic closure of $\mathbf{Q}$.

Let E_1, E_2 be extensions of a field F, *and suppose that* E_1, E_2 *are contained in some larger field* K. By the **composite** E_1E_2 we mean the smallest subfield of K containing E_1 and E_2. This composite exists, and is the intersection of all subfields of K containing E_1 and E_2. Since there is at least one such subfield, namely K itself, the intersection is not empty. If $E_1 = F(\alpha)$ and $E_2 = F(\beta)$ then $E_1E_2 = F(\alpha, \beta)$.

Remark. If E_1, E_2 are not contained in some larger field, then the notion of "composite" field does not make sense, and we don't use it. Given a finite extension E of a field F, there may be several ways this extension can be embedded in a larger field containing F. For instance, let $E = \mathbf{Q}(\alpha)$ where α is a root of the polynomial $t^3 - 2$. Then there are three ways E can be embedded in $\mathbf{C}$, as we shall see later. The element α may get mapped on the three roots of this polynomial. Once we deal with subfields of $\mathbf{C}$, then of course we can form their composite in $\mathbf{C}$. We shall see in Chapter IX that there are other natural fields in which we can embed finite extensions of $\mathbf{Q}$, namely the "p-adic completions" of $\mathbf{Q}$. A priori, there is no natural intersection between such a p-adic completion and the real numbers, except $\mathbf{Q}$ itself.

Let E_1, E_2 be finite extensions of F, contained in some larger field. We are interested in the degree $[E_1E_2 : F]$. Suppose that $E_1 \cap E_2 = F$. It does not necessarily follow that $[E_1E_2 : E_2] = [E_1 : F]$.

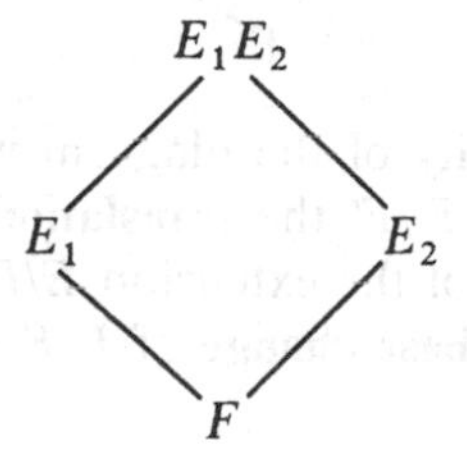

Example. Let α be the real cube root of 2, and let β be a complex cube root of 2, that is $\alpha = \zeta\beta$, where $\zeta = e^{2\pi i/3}$ for example. Let $E_1 = \mathbf{Q}(\alpha)$ and $E_2 = \mathbf{Q}(\beta)$. Then

$$E_1 \cap E_2 = \mathbf{Q}$$

because $[E_1 : E_1 \cap E_2]$ divides 3 by Theorem 1.4, and cannot equal 1 since $E_1 \neq E_2$. Hence by Theorem 1.4, $[E_1 \cap E_2 : \mathbf{Q}] = 1$, so $E_1 \cap E_2 = \mathbf{Q}$. Note that

$$E_1E_2 = \mathbf{Q}(\alpha, \beta) = \mathbf{Q}(\alpha, \sqrt{-3})$$

as you can verify easily since

$$\zeta = \frac{-1 + \sqrt{-3}}{2}.$$

We have $[E_1 : \mathbf{Q}] = [E_2 : \mathbf{Q}] = 3$, and $[E_1E_2 : E_1] = 2$. However, as you will see in Exercise 12, this dropping of degrees cannot occur if $[E_1 : \mathbf{Q}]$

and $[E_2:\mathbf{Q}]$ are relatively prime. But the dropping of degrees is compatible with a general fact:

Proposition 1.5. *Let E/F be a finite extension, and let F' be any extension of F. Suppose that E, F' are contained in some field, and let EF' be the composite. Then*

$$[EF':F'] \leqq [E:F].$$

Proof. Exercise 11. Write $E = F(\alpha_1, \ldots, \alpha_r)$ and use induction. The field diagram illustrating the proposition is as follows.

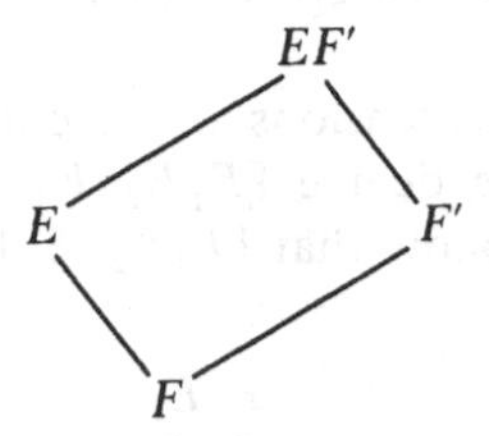

Because of the similarity of the diagram with those in plane geometry, one calls the extension EF'/F' the **translation of** E/F **over** F'. Sometimes, F is called the **base field** of the extension E/F, in which case the extension EF'/F' is also called the **base change** of E/F **over** F'.

We have now met three basic constructions of field extensions:

Translation of an extension.
Tower of extensions.
Composite of extensions.

We shall find these constructions occurring systematically through the rest of field theory, in various contexts.

So far, we have been dealing with arbitrary fields. We now come to a result for which it is essential to make some restrictions. Indeed, let K be a field and let $f(t) \in K[t]$ be a polynomial. We have seen how to define the derivative $f'(t)$. If K has characteristic p, then this derivative may be identically zero. For instance, let

$$f(t) = t^p.$$

Then f has degree p, and $f'(t) = pt^{p-1} = 0$ because $p = 0$ in K. Such a phenomenon cannot happen in characteristic 0.

Theorem 1.6. *Let F be a subfield of an algebraically closed field A of characteristic* 0. *Let f be an irreducible polynomial in $F[t]$. Let $n = \deg f \geqq 1$. Then f has n distinct roots in A.*

Proof. We can write

$$f(t) = a_n(t - \alpha_1) \cdots (t - \alpha_n)$$

with $\alpha_i \in A$. Let α be a root of f. It will suffice to prove that α has multiplicity 1. We note that f is the irreducible polynomial of α over F. We also note that the formal derivative f' has degree $< n$. (Cf. Chapter IV, §3.) Hence we cannot have $f'(\alpha) = 0$, because f' is not the zero polynomial (immediate from the definition of the formal derivative—the leading coefficient of f' is $na_n \neq 0$). Hence α has multiplicity 1 by Theorem 3.6 of Chapter IV. This concludes the proof of the theorem.

Remark. The same proof applies to the following statement, under a weaker hypothesis than characteristic zero.

Let F be any field and f an irreducible polynomial in $F[t]$ of degree $n \geqq 1$. If the characteristic is p and $p \nmid n$, then every root of f has multiplicity 1.

VII, §1. EXERCISES

1. Let $\alpha^2 = 2$. Show that the field $\mathbf{Q}(\alpha)$ is of degree 2 over $\mathbf{Q}$.

2. Show that a polynomial $(t - a)^2 + b^2$ with a, b rational, $b \neq 0$, is irreducible over the rational numbers.

3. Show that the polynomial $t^3 - p$ is irreducible over the rational numbers for each prime number p.

4. What is the degree of the following fields over $\mathbf{Q}$? Prove your assertion.
   (a) $\mathbf{Q}(\alpha)$ where $\alpha^3 = 2$
   (b) $\mathbf{Q}(\alpha)$ where $\alpha^3 = p$ (prime)
   (c) $\mathbf{Q}(\alpha)$ where α is a root of $t^3 - t - 1$
   (d) $\mathbf{Q}(\alpha, \beta)$ where α is a root of $t^2 - 2$ and β is a root of $t^2 - 3$
   (e) $\mathbf{Q}(\sqrt{-1}, \sqrt{3})$.

5. Show that the cube root of unity $\zeta = e^{2\pi i/3}$ is the root of a polynomial of degree 2 over $\mathbf{Q}$. Show that $\mathbf{Q}(\zeta) = \mathbf{Q}(\sqrt{-3})$.

6. What is the degree over $\mathbf{Q}$ of the number $\cos(2\pi/6)$? Prove your assertion.

7. Let F be a field and let $a, b \in F$. Let $\alpha^2 = a$ and $\beta^2 = b$. Assume that α, β have degree 2 over F. Prove that $F(\alpha) = F(\beta)$ if and only if there exists $c \in F$ such that $a = c^2 b$.

8. Let E_1, E_2 be two extensions of a field F. Assume that $[E_2 : F] = 2$, and that $E_1 \cap E_2 = F$. Let $E_2 = F(\alpha)$. Show that $E_1(\alpha)$ has degree 2 over E_1.

9. Let $\alpha^3 = 2$, let ζ be a complex cube root of unity, and let $\beta = \zeta\alpha$. What is the degree of $\mathbf{Q}(\alpha, \beta)$ over $\mathbf{Q}$? Prove your assertion.

10. Let E_1 have degree p over F and E_2 have degree p' over F, where p, p' are prime numbers. Show that either $E_1 = E_2$ or $E_1 \cap E_2 = F$.

11. Prove Proposition 1.5.

12. Let E_1, E_2 be finite extensions of a field F, and assume E_1, E_2 contained in some field. If $[E_1 : F]$ and $[E_2 : F]$ are relatively prime, show that

$$[E_1E_2 : F] = [E_1 : F][E_2 : F] \quad \text{and} \quad [E_1E_2 : E_2] = [E_1 : F].$$

In Exercises 13 and 14, assume F has characteristic 0.

13. Let E be an extension of degree 2 of a field F. Show that E can be written $F(\alpha)$ for some root α of a polynomial $t^2 - a$, with $a \in F$. [*Hint*: Use the high school formula for the solution of a quadratic equation.]

14. Let $t^2 + bt + c$ be a polynomial of degree 2 with b, c in F. Let α be a root. Show that $F(\alpha)$ has degree 2 over F if $b^2 - 4c$ is not a square in F, and otherwise, that $F(\alpha)$ has degree 1 over F, i.e. $\alpha \in F$.

15. Let $a \in \mathbf{C}$, and $a \neq 0$. Let α be a root of $t^n - a$. Show that all roots of $t^n - a$ are of type $\zeta\alpha$, where ζ is an n-th root of unity, i.e.

$$\zeta = e^{2\pi ik/n}, \qquad k = 0, \ldots, n - 1.$$

16. Let A be the set of algebraic numbers over $\mathbf{Q}$ in the complex numbers. Prove that A is a field.

17. Let $E = F(\alpha)$ where α is algebraic over F, of odd degree. Show that $E = F(\alpha^2)$.

18. Let α, β be algebraic over the field F. Let f be the irreducible polynomial of α over F, and let g be the irreducible polynomial of β over F. Suppose $\deg f$ and $\deg g$ are relatively prime. Show that g is irreducible over $F(\alpha)$.

19. Let α, β be complex numbers such that $\alpha^3 = 2$ and $\beta^4 = 3$. What is the degree of $\mathbf{Q}(\alpha, \beta)$ over $\mathbf{Q}$? Prove your assertion.

20. Let $F = \mathbf{Q}[\sqrt{3}]$ be the set of all elements $a + b\sqrt{3}$, where a, b are rational numbers.
    (a) Show that F is a field and that $1, \sqrt{3}$ is a basis of F over $\mathbf{Q}$.
    (b) Let
    $$M = \begin{pmatrix} 0 & 3 \\ 1 & 0 \end{pmatrix}.$$
    Show that $M^2 = \begin{pmatrix} 3 & 0 \\ 0 & 3 \end{pmatrix}$.
    (c) Let $I = I_2$ be the unit 2×2 matrix. Show that the map
    $$a + b\sqrt{3} \mapsto aI + bM$$
    is an isomorphism of F onto the subring of $\mathrm{Mat}_2(\mathbf{Q})$ generated by M over $\mathbf{Q}$.

21. Let α be a root of the polynomial $t^3 + t^2 + 1$ over the field $F = \mathbf{Z}/2\mathbf{Z}$. What is the degree of α over F? Prove your assertion.

22. Let t be a variable and let $K = \mathbf{C}(t)$ be the field of rational functions in t. Let $f(X) = X^n - t$ for some positive integer n. Let α be a root of f in some field containing $\mathbf{C}(t)$. What is the degree of $K(\alpha)$ over K? Prove your assertion.

23. Let F be a field and let t be transcendental over F, that is, not algebraic over F. Let $x \in F(t)$ and suppose $x \notin F$. Prove that $F(t)$ is algebraic over $F(x)$. If you express x as a quotient of relatively prime polynomials, $x = f(t)/g(t)$, how would you describe the degree $[F(t) : F(x)]$ in terms of the degrees of f and g? Prove all assertions.

24. Let F be a field and $p(t)$ an irreducible polynomial in $F[t]$. Let $g(t)$ be an arbitrary polynomial in $F[t]$. Suppose that there exists an extension field E of F and an element $\alpha \in E$ which is a root of both $p(t)$ and $g(t)$. Prove that $p(t)$ divides $g(t)$ in $F[t]$. Is this conclusion still true if we do not assume $p(t)$ irreducible?

VII, §2. EMBEDDINGS

Let F be a field, and L another field. By an **embedding** of F in L, we shall mean a mapping

$$\sigma: F \to L$$

which is a ring-homomorphism. Since $\sigma(1) = 1$ it follows that σ is not the zero mapping. If $x \in F$ and $x \neq 0$ then

$$xx^{-1} = 1 \quad \Rightarrow \quad \sigma(x)\sigma(x)^{-1} = 1,$$

so σ is a homomorphism both for the additive group of F and the multiplicative group of non-zero elements of F. Furthermore, the kernel of σ viewed as homomorphism of the additive group is 0. It follows that σ is injective, i.e. $\sigma(x) \neq \sigma(y)$ if $x \neq y$. This is the reason for calling σ an embedding. We shall often write σx instead of $\sigma(x)$, and σF instead of $\sigma(F)$.

An embedding $\sigma: F \to F'$ is said to be an **isomorphism** if the image of σ is F'. (One should specify an isomorphism of **fields**, or a **field-isomorphism** but the context will always make our meaning clear.) If $\sigma: F \to L$ is an embedding, then the image σF of F under σ is obviously a subfield of L, and thus σ gives rise to an isomorphism of F with σF. If $\sigma: F \to F'$ is an isomorphism, then one can define an inverse isomorphism $\sigma^{-1}: F' \to F$ in the usual way.

Let $f(t)$ be a polynomial in $F[t]$. Let $\sigma: F \to L$ be an embedding. Write

$$f(t) = a_n t^n + \cdots + a_0.$$

We define σf to be the polynomial

$$\sigma f(t) = \sigma(a_n)t^n + \cdots + \sigma(a_0).$$

Then it is trivially verified that for two polynomials f, g in $F[t]$, we have

$$\sigma(f + g) = \sigma f + \sigma g \quad \text{and} \quad \sigma(fg) = (\sigma f)(\sigma g).$$

If $p(t)$ is an irreducible polynomial in $F[t]$, then σp is irreducible over σF.

This is an important fact. Its proof is easy, for if we have a factorization

$$\sigma p = gh$$

over σF, then

$$p = \sigma^{-1}\sigma p = (\sigma^{-1}g)(\sigma^{-1}h)$$

has a factorization over F.

Let $f(t) \in F[t]$, and let α be algebraic over F. Let $\sigma: F(\alpha) \to L$ be an embedding into some field L. Then

$$(\sigma f)(\sigma\alpha) = \sigma(f(\alpha)).$$

This is immediate from the definition of an embedding, for if $f(t)$ is as above, then

$$f(\alpha) = a_n\alpha^n + \cdots + a_0,$$

whence

$$(*) \qquad \sigma(f(\alpha)) = \sigma(a_n)\sigma(\alpha)^n + \cdots + \sigma(a_0).$$

In particular, we obtain two important properties of embeddings:

Property 1. *If α is a root of f, then $\sigma(\alpha)$ is a root of σf.*

Property 2. *If σ is an embedding of $F(\alpha)$ whose effect is known on F and on α, then the effect of σ is uniquely determined on $F(\alpha)$ by $(*)$.*

Let $\sigma: F \to L$ be an embedding. Let E be an extension of F. An embedding $\tau: E \to L$ is said to be an **extension** of σ if $\tau(x) = \sigma(x)$ for all $x \in F$. We also say that σ is a **restriction** of τ to F, or that τ is **over** σ.

Let E be an extension of F and let σ be an isomorphism, or embedding of E which restricts to the identity on F. Then σ is said to be an isomorphism or embedding of E **over** F.

Theorem 2.1. *Let $\sigma: F \to L$ be an embedding. Let $p(t)$ be an irreducible polynomial in $F[t]$. Let α be a root of p in some extension of F, and let β be a root of σp in L. Then there exists an embedding $\tau: F(\alpha) \to L$ which is an extension of σ, and such that $\tau\alpha = \beta$. Conversely, every extension τ of σ to $F(\alpha)$ is such that $\tau\alpha$ is a root of σp.*

Proof. The second assertion follows from Property 1. To prove the existence of τ, let f be any polynomial in $F[t]$, and define τ on the element $f(\alpha)$ to be $(\sigma f)(\beta)$. The same element $f(\alpha)$ has many representations as values $g(\alpha)$, for many polynomials g in $F[t]$. Thus we must show that our definition of τ does not depend on the choice of f. Suppose that $f, g \in F[t]$ are such that $f(\alpha) = g(\alpha)$. Then $(f - g)(\alpha) = 0$. Hence there exists a polynomial h in $F[t]$ such that $f - g = ph$. Then

$$\sigma f = \sigma g + (\sigma p)(\sigma h).$$

Hence

$$\begin{aligned}(\sigma f)(\beta) &= (\sigma g)(\beta) + (\sigma p)(\beta) \cdot (\sigma h)(\beta)\\ &= (\sigma g)(\beta).\end{aligned}$$

This proves that our map is well defined. We used that fact that p is irreducible in an essential way! It is now a triviality to verify that τ is an embedding, and we leave it to the reader.

Special case. Suppose that σ is the identity on F, and let α, β be two roots of an irreducible polynomial in $F[t]$. Then there exists an isomorphism

$$\tau: F(\alpha) \to F(\beta)$$

which is the identity of F and which maps α on β.

There is another way of describing the isomorphism in Theorem 2.1, or the special case.

Let $p(t)$ be irreducible in $F[t]$. Let α be a root of p in some field. Then (p) is the kernel of the homomorphism

$$F[t] \to F(\alpha)$$

which is the identity on F and maps t on α. Indeed, $p(\alpha) = 0$ so $p(t)$ is in the kernel. Conversely, if $f(t) \in F[t]$ and $f(\alpha) = 0$ then $p \mid f$, otherwise the greatest common divisor of p, f is 1 since p is irreducible, and 1 does not vanish on α, so this is impossible. Hence we get an isomorphism

$$\sigma: F[t]/(p) \to F(\alpha).$$

Similarly, we have an isomorphism $\tau\colon F[t]/(p) \to F(\beta)$. The isomorphism

$$\tau\sigma^{-1}\colon F(\alpha) \to F(\beta)$$

maps α on β and is the identity on F. We can represent $\tau\sigma^{-1}$ as in the following diagram.

$$F(\alpha) \xleftarrow{\sigma} F[t]/(p) \xrightarrow{\tau} F(\beta).$$

So far, we have been given a root of the irreducible polynomial in some field. The above presentation suggests how to prove the existence of such a root.

Theorem 2.2. *Let F be a field and $p(t)$ an irreducible polynomial of degree $\geqq 1$ in $F[t]$. Then there exists an extension field E of F in which p has a root.*

Proof. The ideal (p) in $F[t]$ is maximal, and the residue class ring

$$F[t]/(p)$$

is a field. Indeed, if $f(t) \in F[t]$ and $p \nmid f$, then $(f, p) = 1$ so there exist polynomials $g(t)$, $h(t) \in F[t]$ such that

$$gf + hp = 1.$$

This means that $gf \equiv 1 \bmod p$, whence $f \bmod p$ is invertible in $F[t]/(p)$. Hence every non-zero element of $F[t]/(p)$ is invertible, so $F[t]/(p)$ is a field. This field contains the image of F under the homomorphism which to each polynomial assigns its residue class $\bmod p(t)$. Since a field has no ideals other than (0) and itself, and since $1 \not\equiv 0 \bmod p(t)$ because $\deg p \geqq 1$, we conclude that the natural homomorphism

$$\sigma\colon F \to F[t]/(p(t))$$

is an injection. Thus $F[t]/(p(t))$ is a finite extension of σF. If we then identify F with σF, we may view $F[t]/(p(t))$ as a finite extension of F itself, and the polynomial p has a root α in this extension, which is equal to the residue class of $t \bmod p(t)$. Thus we have constructed an extension E of F in which p has a root.

In the next section, we shall show how to get a field in which "all" the roots of a polynomial are contained, in a suitable sense.

Theorem 2.3 (Extension Theorem). *Let E be a finite extension of F. Let $\sigma\colon F \to A$ be an embedding of F into an algebraically closed field A. Then there exists an extension of σ to an embedding $\bar{\sigma}\colon E \to A$.*

Proof. Let $E = F(\alpha_1,\dots,\alpha_n)$. By Theorem 2.1 there exists an extension σ_1 of σ to an embedding $\sigma_1\colon F(\alpha_1) \to A$. By induction on the number of generators, there exists an extension of σ_1 to an embedding $\bar{\sigma}\colon E \to A$, as was to be shown.

Next, we look more closely at the embeddings of a field $F(\alpha)$ over F.

Let α be algebraic over F. Let $p(t)$ be the irreducible polynomial of α over F. Let $\alpha_1,\dots,\alpha_n$ be the roots of p. Then we call these roots the **conjugates** of α over F. By Theorem 2.1, for each α_i, there exists an embedding σ_i of $F(\alpha)$ which maps α on α_i, and which is the identity on F. This embedding is uniquely determined.

Example 1. Consider a root α of the polynomial $t^3 - 2$. We take α to be the real cube root of 2, written $\alpha = \sqrt[3]{2}$. Let $1, \zeta, \zeta^2$ be the three cube roots of unity. The polynomial $t^3 - 2$ is irreducible over $\mathbf{Q}$, because it has no root in $\mathbf{Q}$ (cf. Exercise 2 of Chapter IV, §3). Hence there exist three embeddings of $\mathbf{Q}(\alpha)$ into $\mathbf{C}$, namely the three embeddings $\sigma_1, \sigma_2, \sigma_3$ such that

$$\sigma_1\alpha = \alpha, \qquad \sigma_2\alpha = \zeta\alpha, \qquad \sigma_3\alpha = \zeta^2\alpha.$$

Example 2. If $\alpha = 1 + \sqrt{2}$, there exist two embeddings of $\mathbf{Q}(\alpha)$ into $\mathbf{C}$, namely those sending α to $1 + \sqrt{2}$ and $1 - \sqrt{2}$ respectively.

Example 3. Let F be a field of characteristic p, and let $a \in F$. Consider the polynomial $t^p - a$. Suppose α is a root of this polynomial in some extension field E. Then α is the only root; there is no other root, because

$$(t - \alpha)^p = t^p - \alpha^p = t^p - a,$$

and we apply the unique factorization in the polynomial ring $E[t]$. Hence the only isomorphism of E over F in some field containing E must map α on α, and hence is the identity on $F(\alpha)$.

Because of the above phenomenon, the rest of this section will be under assumptions which preclude the phenomenon.

For the rest of this section, we let A denote an algebraically closed field of characteristic 0.

The reason for this assumption lies in the following corollary, which makes use of the key Theorem 1.6.

Corollary 2.4. *Let E be a finite extension of F. Let n be the degree of E over F. Let $\sigma\colon F \to A$ be an embedding of F into A. Then the number of extensions of σ to an embedding of E into A is equal to n.*

Proof. Suppose first $E = F(\alpha)$ and $f(t) \in F[t]$ is the irreducible polynomial of α over F, of degree n. Then σf has n distinct roots in A by Theorem 1.6, so our assertion follows from Theorem 2.1. In general, we can write E in the form $E = F(\alpha_1,\dots,\alpha_r)$. Consider the tower

$$F \subset F(\alpha_1) \subset F(\alpha_1, \alpha_2) \subset \cdots \subset F(\alpha_1,\dots,\alpha_r).$$

Let $E_{r-1} = F(\alpha_1,\dots,\alpha_{r-1})$. Suppose that we have proved by induction that the number of extensions of σ to E_{r-1} is equal to the degree $[E_{r-1}:F]$. Let $\sigma_1,\dots,\sigma_m$ be the extensions of σ to E_{r-1}. Let d be the degree of α_r over E_{r-1}. For each $i = 1,\dots,m$ we can find precisely d extensions of σ_i to E, say $\sigma_{i1},\dots,\sigma_{id}$. Then it is clear that the set $\{\sigma_{ij}\}$ $(i = 1,\dots,m$ and $j = 1,\dots,d)$ is the set of distinct extensions of σ to E. This proves our corollary.

Theorem 2.5 (Primitive element theorem). *Let F be a field of characteristic 0. Let E be a finite extension of F. Then there exists an element γ of E such that $E = F(\gamma)$.*

Proof. It will suffice to prove that if $E = F(\alpha, \beta)$ with two elements α, β algebraic over F, then we can find γ in E such that $E = F(\gamma)$, for we can then proceed inductively. Let $[E:F] = n$. Let $\sigma_1,\dots,\sigma_n$ be the n distinct embeddings of E into A extending the identity map on F. We shall first prove that we can find an element $c \in F$ such that the elements

$$\sigma_i\alpha + c\sigma_i\beta = \sigma_i(\alpha + c\beta)$$

are distinct, for $i = 1,\dots,n$. We consider the polynomial

$$\prod_{i=1}^{n} \prod_{j \neq i} [\sigma_j\alpha - \sigma_i\alpha + t(\sigma_j\beta - \sigma_i\beta)].$$

It is not the zero polynomial, since each factor is different from 0. This polynomial has a finite number of roots. Hence we can certainly find an element c of F such that when we substitute c for t we don't get the value 0. This element c does what we want.

Now we assert that $E = F(\gamma)$, where $\gamma = \alpha + c\beta$. In fact, by construction, we have n distinct embeddings of $F(\gamma)$ into A extending the identity on F, namely $\sigma_1,\dots,\sigma_n$ restricted to $F(\gamma)$. Hence $[F(\gamma):F] \geqq n$ by Corollary 2.4. Since $F(\gamma)$ is a subspace of E over F, and has the same dimension as E, it follows that $F(\gamma) = E$, and our theorem is proved.

Example 4. Prove as an exercise that if $\alpha^3 = 2$ and β is a square root of 2, then $\mathbf{Q}(\alpha, \beta) = \mathbf{Q}(\gamma)$, where $\gamma = \alpha + \beta$.

Remark. In arbitrary characteristic, Corollary 2.4, and Theorem 2.5 are not necessarily true. In order to obtain an analogous theory resulting from the properties of extensions expressed in those corollaries, it is necessary to place a restriction on the types of finite extensions that are considered. Namely, we define a polynomial $f(t) \in F[t]$ to be **separable** if the number of distinct roots of f is equal to the degree of f. Thus if f has degree d, then f has d distinct roots. An element α of E is defined to be **separable** over F if its irreducible polynomial over F is separable, or equivalently if α is the root of a separable polynomial in $F[t]$. We define a finite extension E of F to be **separable** if it satisfies the property forming the conclusion of Corollary 2.4, that is: the number of possible extensions of an embedding $\sigma: F \to A$ to an embedding of E into A is equal to the degree $[E:F]$. You can now prove:

Let the characteristic be arbitrary. Let E be a finite extension of F.

(a) *If E is separable over F and $\alpha \in E$, then the irreducible polynomial of α over F is separable, and so every element of E is separable over F.*

(b) *If $E = F(\alpha_1, \ldots, \alpha_r)$ and if each α_i is separable over F, then E is separable over F.*

(c) *If E is separable over F and F' is any extension of F such that E, F' are subfields of some larger field, then EF' is separable over F'.*

(d) *If $F \subset E_1 \subset E_2$ is a tower of extensions such that E_1 is separable over F and E_2 is separable over E_1, then E_2 is separable over F.*

(e) *If E is separable over F and E_1 is a subfield of E containing F, then E_1 is separable over F.*

The proofs are easy, and consist merely of further applications of the techniques we have already seen. However, in first reading, it may be psychologically preferable for you just to assume characteristic 0, to get into the main theorems of Galois theory right away. The technical complication arising from lack of separability can then be absorbed afterwards. Therefore we shall omit the proof of the above statements, which you can look up in a more advanced text, e.g. my *Algebra* if you are so inclined.

Note that the substitute for Theorem 2.5 in general can be formulated as follows.

Let E be a finite separable extension of F. Then there exists an element γ of E such that $E = F(\gamma)$.

The proof is the same as that of Theorem 2.5. Only separability was needed in the proof, except for finite fields, in which case we use Theorem 1.10 of Chapter IV.

The same principle will apply later in the Galois theory. We shall define the splitting field of a polynomial. In arbitrary characteristic, we define a Galois extension to be the splitting field of a separable polynomial. Then the statements and proofs of the Galois theory go through unchanged. Thus the hypothesis of separability replaces the too restrictive hypothesis of characteristic 0 throughout.

VII, §2. EXERCISES

All fields in these exercises are assumed to be of characteristic 0.

1. In each case, find an element γ such that $\mathbf{Q}(\alpha, \beta) = \mathbf{Q}(\gamma)$. Always prove all assertions which you make.
   (a) $\alpha = \sqrt{-5}$, $\beta = \sqrt{2}$ (b) $\alpha = \sqrt[3]{2}$, $\beta = \sqrt{2}$
   (c) $\alpha =$ root of $t^3 - t + 1$, $\beta =$ root of $t^2 - t - 1$
   (d) $\alpha =$ root of $t^3 - 2t + 3$, $\beta =$ root of $t^2 + t + 2$
   Determine the degrees of the fields $\mathbf{Q}(\alpha, \beta)$ over $\mathbf{Q}$ in each one of the cases.

2. Suppose β is algebraic over F but not in F, and lies in some algebraic closure A of F. Show that there exists an embedding of $F(\beta)$ over F which is not the identity on β.

3. Let E be a finite extension of F, of degree n. Let $\sigma_1, \ldots, \sigma_n$ be all the distinct embeddings of E over F into A. For $\alpha \in E$, define the **trace** and **norm** of α respectively (from E to F), by

$$\mathrm{Tr}_F^E(\alpha) = \sum_{i=1}^{n} \sigma_i \alpha = \sigma_1 \alpha + \cdots + \sigma_n \alpha,$$

$$N_F^E(\alpha) = \prod_{i=1}^{n} \sigma_i \alpha = \sigma_1 \alpha \cdots \sigma_n \alpha.$$

   (a) Show that the norm and trace of α lie in F.
   (b) Show that the trace is an additive homomorphism, and that the norm is a multiplicative homomorphism.

4. Let α be algebraic over the field F, and let

$$p(t) = t^n + a_{n-1}t^{n-1} + \cdots + a_0$$

   be the irreducible polynomial of α over F. Show that

$$N(\alpha) = (-1)^n a_0 \quad \text{and} \quad \mathrm{Tr}(\alpha) = -a_{n-1}.$$

   (The norm and trace are taken from $F(\alpha)$ to F.)

5. Let E be a finite extension of F, and let a be an element of F. Let

$$[E : F] = n.$$

   What are the norm and trace of a from E to F?

6. Let α be algebraic over the field F. Let $m_\alpha: F(\alpha) \to F(\alpha)$ be multiplication by α, which is an F-linear map. We assume you know about determinants of linear maps. Let $D(\alpha)$ be the determinant of m_α, and let $T(\alpha)$ be the trace of m_α. (The trace of a linear map is the sum of the diagonal elements in a matrix representing the linear map with respect to a basis.) Show that

$$D(\alpha) = N(\alpha) \quad \text{and} \quad T(\alpha) = \mathrm{Tr}(\alpha),$$

where N and Tr are the norm and trace of Exercise 3.

VII, §3. SPLITTING FIELDS

In this section we do not assume characteristic 0.

Let E be a finite extension of F. Let σ be an embedding of F, and τ an extension of σ to an embedding of E. We shall also say that τ is **over** σ. If σ is the identity map, then we say that τ is an embedding of E **over** F. Thus τ is an embedding of E over F means that $\tau x = x$ for all $x \in F$. We also say that τ leaves F **fixed**.

If τ is an embedding of E over F, then τ is called a conjugate mapping and the image field τE is called a **conjugate** of E over F. Observe that if $\alpha \in E$, then $\tau\alpha$ is a conjugate of α over F. Thus we apply the word "conjugate" both to elements and to fields. By Corollary 2.4, if F has characteristic 0, then the number of conjugate mappings of E over F is equal to the degree $[E:F]$.

Let $f(t) \in F[t]$ be a polynomial of degree $n \geqq 1$. By a **splitting field** K for f we mean a finite extension of F such that f has a factorization in K into factors of degree 1, that is

$$f(t) = c(t - \alpha_1) \cdots (t - \alpha_n),$$

with $c \in F$ the leading coefficient of f, and $K = F(\alpha_1, \ldots, \alpha_n)$. Thus roughly we may say that a splitting field is the field generated by "all" the roots of f such that f factors into factors of degree 1.

A priori, we could have a factorization as above in some field K, and another factorization

$$f(t) = c(t - \beta_1) \cdots (t - \beta_n)$$

with β_i in some field $L = F(\beta_1, \ldots, \beta_n)$. The question arises whether there is any relation between K and L, and the answer is that these splitting fields must be isomorphic. The next theorems prove both the existence of a splitting field and its uniqueness up to isomorphism.

Theorem 3.1. *Let $f(t) \in F[t]$ be a polynomial of degree $\geqq 1$. Then there exists a splitting field for f.*

Proof. By induction on the degree of f. Let p be an irreducible factor of f. By Theorem 2.2, there exists an extension $E_1 = F(\alpha_1)$ with some root α_1 of p, and hence of f. Let

$$f(t) = (t - \alpha_1)g(t)$$

be a factorization of f in $E_1[t]$. Then $\deg g = \deg f - 1$. Hence by induction, there exists a field $E = E_1(\alpha_2, \ldots, \alpha_n)$ such that

$$g(t) = c(t - \alpha_2) \cdots (t - \alpha_n)$$

with some element $c \in F$. This concludes the proof.

The splitting field of a polynomial is uniquely determined up to isomorphism. More precisely:

Theorem 3.2. *Let $f(t) \in F[t]$ be a polynomial of degree $\geqq 1$. Let K and L be extensions of F which are splitting fields of f. Then there exists an isomorphism*

$$\sigma: K \to L$$

over F.

Proof. Without loss of generality, we may assume that the leading coefficient of f is 1. We shall prove the following more precise statement by induction.

Let $\sigma: F \to \sigma F$ be an isomorphism. Let $f(t) \in F[t]$ be a polynomial with leading coefficient 1, and let $K = F(\alpha_1, \ldots, \alpha_n)$ be a splitting field for f with the factorization

$$f(t) = \prod_{i=1}^{n} (t - \alpha_i).$$

Let L be a splitting field of σf, with $L = (\sigma F)(\beta_1, \ldots, \beta_n)$ and

$$\sigma f(t) = \prod_{i=1}^{n} (t - \beta_i).$$

Then there exists an isomorphism $\tau: K \to L$ extending σ such that, after a permutation of $\beta_1, \ldots, \beta_n$ if necessary, we have $\tau\alpha_i = \beta_i$ for $i = 1, \ldots, n$.

Let $p(t)$ be an irreducible factor of f. By Theorem 2.1, given a root α_1 of p and a root β_1 of σp, there exists an isomorphism

$$\tau_1: F(\alpha_1) \to (\sigma F)(\beta_1)$$

extending σ and mapping α_1 on β_1. We can factor

$$f(t) = (t - \alpha_1)g(t) \text{ over } F(\alpha_1),$$
$$\sigma f(t) = (t - \beta_1)\tau_1 g(t) \text{ over } \sigma F(\beta_1),$$

because $\tau_1\alpha_1 = \beta_1$. By induction, applied to the polynomials g and $\tau_1 g$, we obtain the isomorphism τ, which concludes the proof.

Remark. Although we have collected together theorems concerning finite fields in a subsequent chapter, the reader may wish to look at that chapter immediately to see how finite fields are determined up to isomorphism as splitting fields. These provide nice examples for the considerations of this section.

By an **automorphism** of a field K we shall mean an isomorphism $\sigma: K \to K$ of K with itself. The context always makes it clear that we mean field-isomorphism (and not another kind of isomorphism, e.g. group, or vector space isomorphism).

Let σ be an embedding of a finite extension K of F, over F. Assume that $\sigma(K)$ is contained in K. Then $\sigma(K) = K$.

Indeed, σ induces a linear map of the vector space K over F, and is injective. You should know from linear algebra that σ is then surjective. Indeed, $\dim_F(K) = \dim_F(\sigma K)$. If a subspace of a finite dimensional vector space has the same dimension as the vector space, then the subspace is equal to the whole space. Then it follows that σ is an isomorphism of fields, whence an automorphism of K.

We observe that the set of all automorphisms of a field K is a group. Trivial verification. We shall be concerned with certain subgroups.

Let G be a group of automorphisms of a field K. Let K^G be the set of all elements $x \in K$ such that $\sigma x = x$ for all $\sigma \in G$. Then K^G is a field. Indeed, K^G contains 0 and 1. If x, y are in K^G, then

$$\sigma(x + y) = \sigma x + \sigma y = x + y,$$
$$\sigma(xy) = \sigma(x)\sigma(y) = xy,$$

so $x + y$ and xy are in K^G. Also $\sigma(x^{-1}) = \sigma(x)^{-1} = x^{-1}$, so x^{-1} is in K^G. This proves that K^G is a field, called the **fixed field of** G.

If G is a group of automorphisms of K over a subfield F, then F is contained in the fixed field (by definition), but the fixed field may be bigger than F. For instance, G could consist of the identity alone, in which case its fixed field is K itself.

Example 1. The field of rational numbers has no automorphisms except the identity. Proof?

Example 2. Prove that the field $\mathbf{Q}(\alpha)$ where $\alpha^3 = 2$ has no automorphism except the identity.

Example 3. Let F be a field of characteristic $\neq 2$, and $a \in F$. Assume that a is not a square in F, and let $\alpha^2 = a$. Then $F(\alpha)$ has precisely two automorphisms over F, namely the identity, and the automorphism which maps α on $-\alpha$.

Remark. The isomorphism of Theorem 3.2 between the splitting fields K and L is not uniquely determined. If σ, τ are two isomorphisms of K onto L leaving F fixed, then

$$\sigma\tau^{-1}: L \to L$$

is an automorphism of L over F. We shall study the group of such automorphisms later in this chapter. We emphasize the need for the possible permutation of $\beta_1, \ldots, \beta_n$ in the statement of the result. Furthermore, it will be a problem to determine which permutations of these roots actually may occur.

A finite extension K of F in an algebraically closed field A will be said to be **normal extension** if every embedding of K over F in A is an automorphism of K.

Theorem 3.3. *A finite extension of F is normal if and only if it is the splitting field of a polynomial.*

Proof. Let K be a normal extension of F. Suppose $K = F(\alpha)$ for some element α. Let $p(t)$ be the irreducible polynomial of α over F. For each root α_i of p, there exists a unique embedding σ_i of K over F such that $\sigma_i\alpha = \alpha_i$. Since each embedding is an automorphism, it follows that α_i is contained in K. Hence

$$K = F(\alpha) = F(\alpha_1, \ldots, \alpha_n),$$

and K is the splitting field of p.

If F has characteristic 0, then we are done by Theorem 2.5, because we know $K = F(\alpha)$ for some α. In general, $K = F(\alpha_1, \ldots, \alpha_r)$, and we can argue in the same way with each irreducible polynomial $p_i(t)$ with respect to α_i. We then let $f(t)$ be the product

$$f(t) = p_1(t) \cdots p_r(t).$$

Assuming that K is normal over F, it follows that K is the splitting field of the polynomial $f(t)$, which in this case is not irreducible, but so what?

Conversely, suppose that K is the splitting field of a polynomial $f(t)$, not necessarily irreducible, with roots $\alpha_1,\dots,\alpha_n$. If σ is an embedding of K over F, then $\sigma\alpha_i$ must also be a root of f. Hence σ maps K into itself, and hence σ is an automorphism.

Theorem 3.4. *Let K be a normal extension of F. If $p(t)$ is a polynomial in $F[t]$, and is irreducible over F, and if p has one root in K, then p has all its roots in K.*

Proof. Let α be one root of p in K. Let β be another root. By Theorem 2.1 there exists an embedding σ of $F(\alpha)$ on $F(\beta)$ mapping α on β, and equal to the identity on F. Extend this embedding to K. Since an embedding of K over F is an automorphism, we must have $\sigma\alpha \in K$, and hence $\beta \in K$.

VII, §3. EXERCISES

1. Let F be a field of characteristic p. Let $c \in F$ and let

$$f(t) = t^p - t - c.$$

(a) Show that either all roots of f lie in F or f is irreducible in $F[t]$. [*Hint*: Let α be a root, and $a \in \mathbf{Z}/p\mathbf{Z}$. What can you say about $\alpha + a$? For the irreducibility, suppose there is a factor g such that $1 < \deg g < \deg f$. Look at the coefficient of the term of next to highest degree. Such a coefficient must lie in F.]

(b) Let $F = \mathbf{Z}/p\mathbf{Z}$. Let $c \in F$ and $c \neq 0$. Show that $t^p - t - c$ is irreducible in $F[t]$.

(c) Let F again be any field of characteristic p. Let α be a root of f. Show that $F(\alpha)$ is a splitting field for f.

(d) Assume f irreducible. Prove that there is a group of automorphisms of $F(\alpha)$ over F isomorphic to $\mathbf{Z}/p\mathbf{Z}$ (and so cyclic of order p).

(e) Prove that there are no other automorphisms of $F(\alpha)$ over F besides those found in (d).

2. Let F be a field and let n be a positive integer not divisible by the characteristic of F if this characteristic is $\neq 0$. Let $a \in F$. Let ζ be a primitive n-th root of unity, and α one root of $t^n - a$. Prove that the splitting field of $t^n - a$ is $F(\alpha, \zeta)$.

3. (a) Let p be an odd prime. Let F be a field of characteristic 0, and let $a \in F$, $a \neq 0$. Assume that a is not a p-th power in F. Prove that $t^p - a$ is irreducible in $F[t]$. [*Hint*: Suppose $t^p - a$ factors over F. Look at the constant term of one of the factors, expressed as a product of some roots, and deduce that a is a p-th power in F.]

(b) Again assume that a is not a p-th power in F. Prove that for every positive integer r, $t^{p^r} - a$ is irreducible in $F[t]$. [*Hint*: Use induction. Distinguish the cases whether a root α of $t^{p^r} - a$ is a p-th power in $F(\alpha)$ or not, and take the norm from $F(\alpha)$ to F. The norm was defined in the exercises of §2.]

4. Let F be a field of chacteristic 0, let $a \in F$, $a \neq 0$. Let n be an odd integer $\geqq 3$. Assume that for all prime numbers p such that $p|n$ we have $a \notin F^p$ (where F^p is the set of p-th powers in F). Show that $t^n - a$ is irreducible in $F[t]$. [*Hint*: Write $n = p^r m$ with $p \nmid m$. Assume inductively that $t^m - a$ is irreducible in $F[t]$. Show that α is not a p-th power in $F(\alpha)$ and use induction.]

Remark. When n is even, the analogous result is not quite true because of the factorization of $t^4 + 4$. Essentially this is the only exception, and the general result can be stated as follows.

Theorem. *Let F be a field and n an integer $\geqq 2$. Let $a \in F$, $a \neq 0$. Assume that for all prime numbers p such that $p|n$ we have $a \notin F^p$, and if $4|n$ then $a \notin -4F^4$. Then $t^n - a$ is irreducible in $F[t]$.*

It is more tedious to handle this general case, but you can always have a try at it. The main point is that the prime 2 causes some trouble.

5. Let E be a finite extension of F. Let $E = E_1, E_2, \ldots, E_r$ be all the distinct conjugates of E over F. Prove that the composite

$$K = E_1 E_2 \cdots E_r$$

is the smallest normal extension of F containing E. We can say that this smallest normal extension is the composite of E and all its conjugates over F.

6. Let A be an algebraic extension of a field F of characteristic 0. Assume that every polynomial of degree $\geqq 1$ in $F[t]$ has at least one root in A. Prove that A is algebraically closed. [*Hint*: If not, there is some element α in an extension of A which is algebraic over A but not in A. Show that α is algebraic over F. Let $f(t)$ be the irreducible polynomial of α over F, and let K be the splitting field of f in some algebraic closure of A. Write $K = F(\gamma)$, and let $g(t)$ be the irreducible polynomial of γ over F. Now use the assumption that g has a root in A. *Remark*: the result of this exercise is valid even if we don't assume characteristic 0, but one must then use additional arguments to deal with the possibility that the primitive element theorem cannot always be applied.]

VII, §4. GALOIS THEORY

Throughout this section we assume that all fields have characteristic 0.

In the case of characteristic 0, a normal extension will also be called a **Galois** extension.

Theorem 4.1. *Let K be a Galois extension of F. Let G be the group of automorphisms of K over F. Then F is the fixed field of G.*

Proof. Let F' be the fixed field. Then trivially, $F \subset F'$. Suppose $\alpha \in F'$ and $\alpha \notin F$. Then by Theorem 2.1 there exists an embedding σ_0 of $F(\alpha)$ over F such that $\sigma_0\alpha \neq \alpha$. Extend σ_0 to an embedding σ of K over F by Theorem 2.3. By hypothesis, σ is an automorphism of K over F, and $\sigma\alpha = \sigma_0\alpha \neq \alpha$, thereby contradicting the assumption that $\alpha \in F'$ but $\alpha \notin F$. This proves our theorem.

If K is a Galois extension of F, the group of automorphisms of K over F is called the **Galois group** of K over F and is denoted by $G_{K/F}$. If K is the splitting field of a polynomial $f(t)$ in $F[t]$, then we also say that $G_{K/F}$ is the **Galois group of** f.

Theorem 4.2. *Let K be a Galois extension of F. To each intermediate field E, associate the subgroup $G_{K/E}$ of automorphisms of K leaving E fixed. Then K is Galois over E. The map*

$$E \mapsto G_{K/E}$$

is an injective and surjective map from the set of intermediate fields onto the set of subgroups of G, and E is the fixed field of $G_{K/E}$.

Proof. Every embedding of K over E is an embedding over F, and hence is an automorphism of K. It follows that K is Galois over E. Furthermore, E is the fixed field of $G_{K/E}$ by Theorem 4.1. This shows in particular that the map

$$E \mapsto G_{K/E}$$

is injective, i.e. if $E \neq E'$ then $G_{K/E} \neq G_{K/E'}$. Finally, let H be a subgroup of G. Write $K = F(\alpha)$ with some element α. Let $\{\sigma_1, \ldots, \sigma_r\}$ be the elements of H, and let

$$f(t) = (t - \sigma_1\alpha) \cdots (t - \sigma_r\alpha).$$

For any σ in H, we note that $\{\sigma\sigma_1, \ldots, \sigma\sigma_r\}$ is a permutation of $\{\sigma_1, \ldots, \sigma_r\}$. Hence from the expression

$$\sigma f(t) = (t - \sigma\sigma_1\alpha) \cdots (t - \sigma\sigma_r\alpha) = f(t),$$

we see that f has its coefficients in the fixed field E of H. Furthermore, $K = E(\alpha)$, and α is a root of a polynomial of degree r over E. Hence $[K:E] \leqq r$. But K has r distinct embeddings over E (those of H), and hence by Corollary 2.4, $[K:E] = r$, and $H = G_{K/E}$. This proves our theorem.

Let $f(t) \in F[t]$, and let

$$f(t) = (t - \alpha_1) \cdots (t - \alpha_n).$$

Let $K = F(\alpha_1, \ldots, \alpha_n)$, and let σ be an element of $G_{K/F}$. Then $\{\sigma\alpha_1, \ldots, \sigma\alpha_n\}$ is a permutation of $\{\alpha_1, \ldots, \alpha_n\}$, which we may denote by π_σ. If $\sigma \neq \tau$, then $\pi_\sigma \neq \pi_\tau$, and clearly,

$$\pi_{\sigma\tau} = \pi_\sigma \circ \pi_\tau.$$

Hence we have represented the Galois group $G_{K/F}$ as a group of permutations of the roots of f. More precisely, we have an injective homomorphism of the Galois group $G_{K/F}$ into the symmetric group S_n:

$$G_{K/F} \to S_n \quad \text{given by} \quad \sigma \mapsto \pi_\sigma.$$

Of course, it is not always true that every permutation of $\{\alpha_1, \ldots, \alpha_n\}$ is represented by an element of $G_{K/F}$, even if f is irreducible over F. Cf. the next section for examples. In other words, given a permutation π of $\{\alpha_1, \ldots, \alpha_n\}$ such a permutation is merely a bijective map of the set of roots with itself. It is not always true that there exists an automorphism σ of K over F whose restriction to this set of roots is equal to π. Or, put another way, the permutation π cannot necessarily be extended to an automorphism of K over F. We shall find later conditions when $G_{K/F} \approx S_n$.

The notion of a Galois extension and a Galois group are defined completely algebraically. Hence they behave formally under isomorphisms the way one expects from general algebraic situations. We describe this behavior explicitly.

Let K be a Galois extension of F. Let

$$\lambda : K \to \lambda K$$

be an isomorphism. Then λK is a Galois extension of λF. Indeed, if K is the splitting field of the polynomial f, then λK is the splitting field of λf.

Let G be the Galois group of K over F. Then the map

$$\sigma \mapsto \lambda \circ \sigma \circ \lambda^{-1}$$

gives a homomorphism of G into the Galois group of λK over λF, whose inverse is given by

$$\lambda^{-1} \circ \tau \circ \lambda \leftarrow\!\shortmid \tau.$$

$$\begin{array}{ccc} K & \xrightarrow{\lambda} & \lambda K \\ | & & | \\ F & \longrightarrow & \lambda F \end{array}$$

Hence $G_{\lambda K/\lambda F}$ is isomorphic to $G_{K/F}$ under the above map. We may write

$$G_{\lambda K/\lambda F} = \lambda G_{K/F} \lambda^{-1}.$$

This is a "conjugation", just like conjugations in the theory of groups in Chapter II.

In particular, let E be an intermediate field, say

$$F \subset E \subset K.$$

Let $\lambda: E \to \lambda E$ be an embedding of E in K, which we assume extends to an automorphism of K. Then $\lambda K = K$. Hence

$$G_{K/\lambda E} = \lambda G_{K/E} \lambda^{-1}.$$

Theorem 4.3. *Let K be a Galois extension of a field F, and let E be an intermediate field, $F \subset E \subset K$. Let*

$$H = G_{K/E} \qquad \textit{and} \qquad G = G_{K/F}.$$

Then E is a Galois extension of F if and only if H is normal in G. If this is the case, then the restriction map

$$\sigma \mapsto \mathrm{res}_E\, \sigma \qquad \textit{of} \qquad G \to G_{E/F}$$

induces an isomorphism of G/H onto $G_{E/F}$.

Proof. Assume that $G_{K/E}$ is normal in G. Let λ_0 be an embedding of E over F. It suffices to prove that λ_0 is an automorphism of E. Let λ be an extension of λ_0 to K. Since K is Galois over F, it follows that λ is an automorphism of K over F, and by assumption

$$G_{K/E} = \lambda G_{K/E} \lambda^{-1} = G_{K/\lambda E}.$$

By Theorem 4.2 it follows that $\lambda E = E$, so λ is an automorphism of E, whence E is normal over F.

Conversely suppose E normal over F. Then the restriction map

$$\sigma \mapsto \sigma | E \qquad \text{for} \quad \sigma \in G_{K/F}$$

is a homomorphism $G_{K/F} \to G_{E/F}$ whose kernel is $G_{K/E}$ by definition. Hence $G_{K/E}$ is normal. This homomorphism is surjective, because given $\sigma_0 \in G_{E/F}$ we can extend σ_0 to an embedding σ of K over F, and since K is Galois over F, we actually have $\sigma \in G_{K/F}$ and σ_0 is the restriction of σ to E. This concludes the proof of Theorem 4.3.

A Galois extension K/F is said to be **abelian** if its Galois group is abelian. It is said to be **cyclic** if its Galois group is cyclic.

Corollary 4.4. *Let K/F be an abelian extension. If E is an intermediate field, $F \subset E \subset K$, then E is Galois over F and abelian over F. Similarly, if K/F is a cyclic extension, then E is cyclic over F.*

Proof. This follows at once from the fact that a subgroup of an abelian group is normal and abelian, and a factor group of an abelian group is abelian. Similarly for cyclic subgroups and factor groups.

The next theorem describes what happens to a Galois extension under translation.

Theorem 4.5. *Let K/F be a Galois extension with group G. Let E be an arbitrary extension field of F. Assume that K, E are both contained in some field, and let KE be the composite field. Then KE is Galois over E. The map*

$$G_{KE/E} \to G_{K/F} \quad \text{given by} \quad \sigma \mapsto \mathrm{res}_K(\sigma),$$

i.e. the restriction of an element of $G_{KE/E}$ to K, gives an isomorphism of $G_{KE/E}$ with $G_{K/(K \cap E)}$.

The field diagram for Theorem 4.5 is as follows:

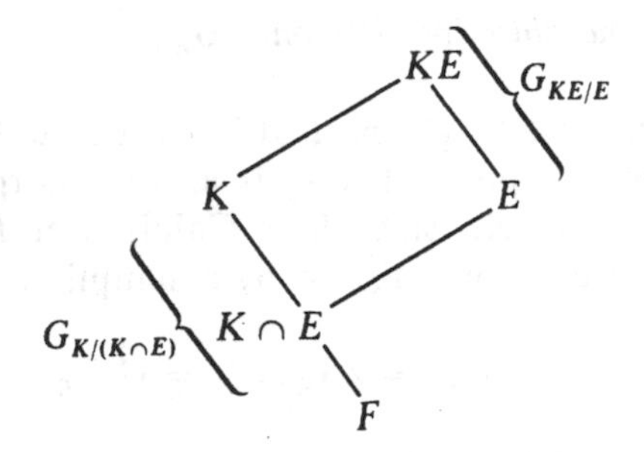

Proof. The extension KE/E is clearly Galois. The restriction maps $G_{KE/E}$ a priori into the set of embeddings of K over F. But since we assume that K is Galois over F, the image of the restriction lies in $G_{K/F}$. The restriction is obviously a homomorphism

$$\mathrm{res}\colon G_{KE/E} \to G_{K/F}.$$

The kernel of the restriction is the identity, for if $\sigma \in G_{KE/E}$ is the identity on K then σ is the identity on KE, because KE is generated by elements of K and elements of E. Thus we get an injective homomorphism of $G_{KE/E}$ into $G_{K/F}$. Since an element of $G_{KE/E}$ is the identity on E, the restriction of this element to K is the identity on $K \cap E$, so the image of the restriction lies in the Galois group of $K/(K \cap E)$. Thus we get an injective homomorphism

$$G_{KE/E} \hookrightarrow G_{K/(K \cap E)}.$$

We must finally prove that this map is surjective. It will suffice to prove that

$$[K : K \cap E] = [KE : E],$$

because if a subgroup of a finite group has the same order as the group, then the subgroup is equal to the group. So write $K = F(\alpha)$ for some generator α. Let

$$f(t) = \prod_{\sigma \in G_{KE/E}} (t - \sigma\alpha),$$

where the product is taken over all elements $\sigma \in G_{KE/E}$. Then the coefficients of $f(t)$ lie in E since they are fixed under every element of $G_{KE/E}$ which permutes the roots of $f(t)$. Furthermore, these coefficients lie in K, because each $\sigma\alpha \in K$ by the hypothesis that K is normal over F. Hence the coefficients of f lie in $K \cap E$. Hence

$$[K : K \cap E] \leqq \deg f = [KE : E] \leqq [K : K \cap E],$$

where this last inequality is by Proposition 1.5 to the effect that the degree does not increase under translation. This proves the theorem.

Corollary 4.6. *Let K/F be Galois and let E be an arbitrary extension of F. Then*

$$[KE : E] \quad \text{divides} \quad [K : F].$$

Furthermore if $K \cap E = F$ then

$$[KE : E] = [K : F].$$

Proof. We have

$$[K : F] = [K : K \cap E][K \cap E : F] = [KE : E][K \cap E : F].$$

Both assertions follow from this relation.

In the example preceding Proposition 5, we have seen that the relations of Corollary 4.6 are not necessarily true when K/F is not Galois.

As a special case of Corollary 4.6, suppose K/F Galois and $K \cap E = F$. Let $K = F(\alpha)$. The relation $[KE:E] = [K:F]$ tells us that the degree of α over F is equal to the degree of α over E. This is a statement about irreducibility, namely the irreducible polynomial of α over F is the same as the irreducible polynomial of α over E. The example preceding Proposition 1.5 shows that if we do not assume $F(\alpha)$ normal over F, even if $F(\alpha) \cap E = F$, it does not necessarily follow that

$$[F(\alpha):F] = [E(\alpha):E].$$

Corollary 4.7. *Let K/F be a Galois extension and let E be a finite extension. If $K \cap E = F$ then*

$$[KE:K] = [E:F].$$

Proof. Exercise 6.

We shall now prove an interesting theorem due to Artin, showing how to get a Galois extension going from the top down rather than from the bottom up as we have done up to now. The technique of proof is interesting in that it is similar to the technique used to prove Theorem 4.5, and also uses aspects from the primitive element theorem, so you see field theory at work. We start with a lemma.

Lemma 4.8. *Let E be an extension field of F such that every element of E is algebraic over F. Assume that there is an integer $n \geqq 1$ such that every element of E is of degree $\leqq n$ over F. Then E is a finite extension and $[E:F] \leqq n$.*

Proof. Let α be an element of E such that the degree $[F(\alpha):F]$ is maximal, say $m \leqq n$. We claim that $F(\alpha) = E$. If this is not true, then there exists an element $\beta \in E$ such that $\beta \notin F(\alpha)$, and by Theorem 2.5, there exists an element $\gamma \in F(\alpha, \beta)$ such that $F(\alpha, \beta) = F(\gamma)$. But from the tower

$$F \subset F(\alpha) \subset F(\alpha, \beta)$$

we see that $[F(\alpha, \beta):F] > m$, whence γ has degree $> m$ over F, a contradiction which proves the lemma.

Theorem 4.9 (Artin). *Let K be a field and let G be a finite group of automorphisms of K, of order n. Let $F = K^G$ be the fixed field. Then K is a Galois extension of F, and its Galois group is G. We have $[K:F] = n$.*

Proof. Let $\alpha \in K$ and let $\sigma_1, \ldots, \sigma_r$ be a maximal set of elements of G such that $\sigma_1\alpha, \ldots, \sigma_r\alpha$ are distinct. If $\tau \in G$ then $(\tau\sigma_1\alpha, \ldots, \tau\sigma_r\alpha)$ differs from

$(\sigma_1\alpha,\ldots,\sigma_r\alpha)$ by a permutation, because τ is injective, and every $\tau\sigma_i\alpha$ is among the set $\{\sigma_1\alpha,\ldots,\sigma_r\alpha\}$; otherwise this set is not maximal. Hence α is a root of the polynomial

$$f(t) = \prod_{i=1}^{r} (t - \sigma_i\alpha),$$

and for any $\tau \in G$, $\tau f = f$. Hence the coefficients of f lie in $K^G = F$. Hence every element α of K is a root of a polynomial of degree $\leqq n$ with coefficients in F. Furthermore, this polynomial splits in linear factors in K. By Lemma 4.8 and Theorem 2.5 we can write $K = F(\gamma)$, and so K is Galois over F. The Galois group of K over F has order $[K:F] \leqq n$ by Corollary 2.4. Since G is a group of automorphisms of K over F, it follows that G is equal to the Galois group, thus proving the theorem.

Remark. Let A be an algebraically closed field, and let G be a non-trivial finite group of automorphisms of A. It is a theorem of Artin that G has order 2, and that essentially, the fixed field is something like the real numbers. For a precise statement, see, for instance, my book *Algebra*.

For those who have read the section on Sylow groups, we shall now give an application of Galois theory.

Theorem 4.10. *The complex numbers are algebraically closed.*

Proof. The only facts about the real numbers used in the proof are:

1. Every polynomial of odd degree with real coefficients has a real root.
2. Every positive real number is the square of a real number.

We first remark that every complex number has a square root in the complex numbers. If you are willing to use the polar form, then the formula for the square root is easy. Indeed, if $z = re^{i\theta}$ then

$$r^{1/2}e^{i\theta/2}$$

is a square root of z. You can also write down a formula for the square root of $x + iy$ $(x, y \in \mathbf{R})$ directly in terms of x and y, using only the fact that a positive real number is the square of a real number. Do this as an exercise.

Now let E be a finite extension of $\mathbf{R}$ containing $\mathbf{C}$. Let K be a Galois extension of $\mathbf{R}$ containing E. Let $G = G_{K/\mathbf{R}}$, and let H be a 2-Sylow subgroup of G. Let F be the fixed field. Since $[F:\mathbf{R}] = (G:H)$ it follows that $[F:\mathbf{R}]$ is odd. We can write $F = \mathbf{R}(\alpha)$ for some element α by Theorem 2.5. Let $f(t)$ be the irreducible polynomial of α over $\mathbf{R}$. Then

$\deg f$ is odd, whence f has a root in $\mathbf{R}$, whence f has degree 1, whence $F = \mathbf{R}$. Hence G is a 2-group. Let $G_0 = G_{K/\mathbf{C}}$. If $G_0 \neq \{1\}$, then being a 2-group, G_0 has a subgroup H_0 of index 2 by Theorem 9.1 of Chapter II. The fixed field of H_0 is then an extension of $\mathbf{C}$ of degree 2. But every such extension is generated by the root of a polynomial $t^2 - \beta$, for some $\beta \in \mathbf{C}$. This contradicts the fact that every complex number is the square of a complex number, and concludes the proof of the theorem.

Remark. The above proof is a variation by Artin of a proof by Gauss.

VII, §4. EXERCISES

1. By a **primitive** n-th root of unity, one means a number ζ whose period is exactly n. For instance, $e^{2\pi i/n}$ is a primitive n-th root of unity. Show that every other primitive n-th root of unity is equal to a power $e^{2\pi ir/n}$ where r is an integer > 0 and relatively prime to n.

2. Let F be a field, and $K = F(\zeta)$, where ζ is a primitive n-th root of unity. Show that K is Galois over F, and that its Galois group is commutative. [*Hint*: For each embedding σ over F, note that $\sigma\zeta = \zeta^{r(\sigma)}$ with some integer $r(\sigma)$.] If τ is another embedding, what is $\tau\sigma\zeta$, and $\sigma\tau\zeta$?

3. (a) Let K_1, K_2 be two Galois extensions of a field F. Say $K_1 = F(\alpha_1)$ and $K_2 = F(\alpha_2)$. Let $K = F(\alpha_1, \alpha_2)$. Show that K is Galois over F. Let G be its Galois group. Map G into the direct product $G_{K_1/F} \times G_{K_2/F}$ by associating with each σ in G the pair (σ_1, σ_2), where σ_1 is the restriction of σ to K_1, and σ_2 is the restriction of σ to K_2. Show that this mapping is an injective homomorphism. If $K_1 \cap K_2 = F$, show that the map is an isomorphism.

   (b) More generally, let $K_1, \ldots, K_r$ be finite extensions of a field F contained in some field. Denote by $K_1 \cdots K_r$ the smallest field containing $K_1, \ldots, K_r$. Thus if $K_i = F(\alpha_i)$ then $K_1 \cdots K_r = F(\alpha_1, \ldots, \alpha_r)$. Let $K = K_1 \cdots K_r$. We call K the **composite field**. Suppose that $K_1, \ldots, K_r$ are finite Galois extensions of F. Show that K is a Galois extension of F. Show that the map

$$\sigma \mapsto (\mathrm{res}_{K_1} \sigma, \ldots, \mathrm{res}_{K_r} \sigma)$$

   is an injective homomorphism of $G_{K/F}$ into $G_{K_1/F} \times \cdots \times G_{K_r/F}$. Finally, assume that for each i,

$$(K_1 \cdots K_i) \cap K_{i+1} = F.$$

   Show that the above map is an isomorphism of $G_{K/F}$ with the product. [This follows from (a) by induction.]

4. (a) Let K be an abelian extension of F and let E be an arbitrary extension of F. Prove that KE is abelian over E.

(b) Let K_1, K_2 be abelian extensions of F. Prove that K_1K_2 is abelian over F.

(c) Let K be a Galois extension of F. Prove that there exists a maximal abelian subextension of K. In other words, there exists a subfield K' of K containing F such that K' is abelian, and if E is a subfield of K abelian over F then $E \subset K'$.

(d) Prove that $G_{K/K'}$ is the commutator group of $G_{K/F}$, in other words, $G_{K/K'}$ is the group generated by all elements

$$\sigma\tau\sigma^{-1}\tau^{-1} \quad \text{with} \quad \sigma, \tau \in G_{K/F}.$$

5. (a) Let K be a cyclic extension of F and let E be an arbitrary extension of F. Prove that KE is cyclic over E.
   (b) Let K_1, K_2 be cyclic extensions of F. Is K_1K_2 cyclic over F? Proof?

6. Let K be Galois over F and let E be finite over F. Assume that $K \cap E = F$. Prove that $[KE:K] = [E:F]$.

7. Let F be a field containing $i = \sqrt{-1}$. Let K be a splitting field of the polynomial $t^4 - a$ with $a \in F$. Show that the Galois group of K over F is a subgroup of a cyclic group of order 4. If $t^4 - a$ is irreducible over F, show that this Galois group is cyclic of order 4. If α is a root of $t^4 - a$, express all the other roots in terms of α and i.

8. More generally, let F be a field containing all n-th roots of unity. Let K be a splitting field of the equation $t^n - a = 0$ with $a \in F$. Show that K is Galois over F, with a Galois group which is a subgroup of a cyclic group of order n. If $t^n - a$ is irreducible, prove that the Galois group is cyclic of order n.

9. Show that the Galois group of the polynomial $t^4 - 2$ over the rational numbers has order 8, and contains a cyclic subgroup of order 4. [*Hint*: Prove first that the polynomial is irreducible over $\mathbf{Q}$. Then, if α is a real fourth root of 2, consider $K = \mathbf{Q}(\alpha, i)$.]

10. Give an example of extension fields $F \subset E \subset K$ such that E/F is Galois, K/E is Galois, but K/F is not Galois.

11. Let K/F be a Galois extension whose Galois group is the symmetric group on 3 elements. Prove that K does not contain a cyclic extension of F of degree 3. How many non-cyclic extensions of degree 3 does K contain?

12. (a) Let K be the splitting field of $t^4 - 2$ over the rationals. Prove that K does not contain a cyclic extension of $\mathbf{Q}$ of degree 4.
    (b) Let K be a Galois extension of F with group G of order 8, generated by two elements σ, τ such that $\sigma^4 = 1$, $\tau^2 = 1$ and $\tau\sigma\tau = \sigma^3$. Prove that K does not contain a subfield cyclic over F of degree 4.

13. Let K be Galois over F. Suppose the Galois group is generated by two elements σ, τ such that $\sigma^m = 1$, $\tau^n = 1$, $\tau\sigma\tau^{-1} = \sigma^r$ where $r - 1 > 0$ is prime to m. Assume that $[K:F] = mn$. Prove that the maximal subfield K' of K which is abelian over F has degree n over F.

14. Let S_n be the symmetric group of permutations of $\{1,\dots,n\}$.
    (a) Show that S_n is generated by the transpositions (12), (13), ..., $(1n)$. [*Hint*: Use conjugations.]
    (b) Show that S_n is generated by the transpositions (12), (23), (34), ..., $(n-1, n)$.
    (c) Show that S_n is generated by the cycles (12) and $(123\cdots n)$.
    (d) Let p be a prime number. Let $i \neq 1$ be an integer with $1 < i \leqq p$. Show that S_p is generated by $(1i)$ and $(123\cdots p)$.
    (e) Let $\sigma \in S_p$ be a permutation of period p. Show that S_p is generated by σ and any transposition.

15. Let $f(t)$ be an irreducible polynomial of degree p over the rationals, where p is an odd prime. Suppose that f has $p-2$ real roots and two complex roots which are not real. Prove that the Galois group of f is S_p. [Use Exercise 14.] Exercise 8 of Chapter IV, §5 showed how to construct such polynomials.

The following two exercises are due to Artin.

16. Let F be a field of characteristic 0, contained in its algebraic closure A. Let $\alpha \in A$ and suppose $\alpha \notin F$, but every finite extension E of F with $E \neq F$ contains α. In other words, F is a maximal subfield of A not containing α. Prove that every finite extension of F is cyclic.

17. Again let F_0 be a field of characteristic 0, contained in its algebraic closure A. Let σ be an automorphism of A over F_0 and let F be the fixed field. Prove that every finite extension of F is cyclic.

The next five exercises will show you how to determine the Galois groups of certain abelian extensions.

18. Assume that the field F contains all the n-th roots of unity. Let B be a subgroup of F^* containing F^{*n}. [Recall that F^* is the multiplicative group of F, so F^{*n} is the group of n-th powers of elements in F^*.] We assume that the factor group B/F^{*n} is finitely generated.
    (a) Let $K = F(B^{1/n})$, i.e. K is the smallest field containing F and all n-th roots of all elements in B. If $b_1,\dots,b_r$ are distinct coset representatives of B/F^{*n}, show that $K = F(b_1^{1/n},\dots,b_r^{1/n})$, so K is actually finite over F.
    (b) Prove that K is Galois over F.
    (c) Given $b \in B$ and $\sigma \in G_{K/F}$, let $\beta \in K$ be such that $\beta^n = b$, and define the **Kummer symbol**

$$\langle b, \sigma \rangle = \sigma\beta/\beta.$$

    Prove that $\langle b, \sigma \rangle$ is an n-th root of unity independent of the choice of β such that $\beta^n = b$.
    (d) Prove that the symbol $\langle b, \sigma \rangle$ is **bimultiplicative**, in other words:

$$\langle ab, \sigma \rangle = \langle a, \sigma \rangle \langle b, \sigma \rangle \quad \text{and} \quad \langle b, \sigma\tau \rangle = \langle b, \sigma \rangle \langle b, \tau \rangle.$$

(e) Let $b \in B$ be such that $\langle b, \sigma \rangle = 1$ for all $\sigma \in G_{K/F}$. Prove that $b \in F^{*n}$.
(f) Let $\sigma \in G_{K/F}$ be such that $\langle b, \sigma \rangle = 1$ for all $b \in B$. Prove that $\sigma = \text{id}$.

Remark. As in Exercise 18, assume that the n-th roots of unity lie in F. Let $a_1, \dots, a_s \in F^*$. There is in most people a strong feeling that the Galois group of $F(a_1^{1/n}, \dots, a_s^{1/n})$ should be determined only by multiplicative relations between $a_1, \dots, a_s$. Exercise 18 is the main step in formulating this idea precisely. The next exercises show you how to complete this idea. Note that Exercises 19, 20, 21 concern only finite abelian groups, and not fields. The application to field theory comes in Exercise 22.

19. (a) Let A be a cyclic group of order n. Let C be a cyclic group of order n. Show that the group of homomorphisms of A into C is cyclic of order n. We let $\text{Hom}(A, C)$ denote the group of homomorphisms of A into C.
(b) Let A be a cyclic group of order d, and assume $d|n$. Let again C be a cyclic group of order n. Show that $\text{Hom}(A, C)$ is cyclic of order d.

20. Let $A = A_1 \times \cdots \times A_r$ be a product of cyclic groups of orders $d_1, \dots, d_r$ respectively, and assume that $d_i | n$ for all i. Prove that

$$\text{Hom}(A, C) \approx \prod_{i=1}^{r} \text{Hom}(A_i, C),$$

and hence that $\text{Hom}(A, C)$ is isomorphic to A, by using Exercise 19(b).

21. Let A, B be two finite abelian groups such that $x^n = 1$ for all $x \in A$ and all $x \in B$. Suppose given a bimultiplicative mapping

$$A \times B \to C \qquad \text{denoted by} \qquad (a, b) \mapsto \langle a, b \rangle$$

into a cyclic group C of order n. **Bimultiplicative** means that for all $a, a' \in A$ and $b, b' \in B$ we have

$$\langle aa', b \rangle = \langle a, b \rangle \langle a', b \rangle \qquad \text{and} \qquad \langle a, bb' \rangle = \langle a, b \rangle \langle a, b' \rangle.$$

Define *a* **perpendicular to** *b* to mean that $\langle a, b \rangle = 1$. *Assume* that if $a \in A$ is perpendicular to all elements of B then $a = 1$, and also if $b \in B$ is perpendicular to all elements of A, then $b = 1$. *Prove* that there is a natural isomorphism

$$A \approx \text{Hom}(B, C),$$

and hence by Exercise 20, that there is some isomorphism $A \approx B$.

22. In Exercise 18, prove that $G_{K/F} \approx B/F^{*n}$.

VII, §5. QUADRATIC AND CUBIC EXTENSIONS

In this section we continue to assume that all fields have characteristic 0.

Quadratic polynomials

Let F be a field. Any irreducible polynomial $t^2 + bt + c$ over F has a splitting field $F(\alpha)$, with

$$\alpha = \frac{-b \pm \sqrt{b^2 - 4c}}{2}.$$

Thus $F(\alpha)$ is Galois over F, and its Galois group is cyclic of order 2. If we let $d = b^2 - 4c$, then $F(\alpha) = F(\sqrt{d})$. Conversely, the polynomial $t^2 - d$ is irreducible over F if and only if d is not a square in F.

Cubic polynomials

Consider now the cubic case. Let

$$f(t) = (t - \alpha_1)(t - \alpha_2)(t - \alpha_3) \in F[t]$$

be a cubic polynomial with coefficients in F. The roots may not be in F, however. We let the **discriminant** of f be defined as

$$D = [(\alpha_2 - \alpha_1)(\alpha_3 - \alpha_1)(\alpha_3 - \alpha_2)]^2.$$

Any automorphism of $F(\alpha_1, \alpha_2, \alpha_3)$ leaves D fixed because it changes the product

$$\Delta = (\alpha_2 - \alpha_1)(\alpha_3 - \alpha_1)(\alpha_3 - \alpha_2)$$

at most by a sign. This is a special case of the proof of Chapter II, §6.

Let K be the splitting field, so

$$K = F(\alpha_1, \alpha_2, \alpha_3).$$

Given $\sigma \in G_{K/F}$ we let

$$\sigma\Delta = \varepsilon(\sigma)\Delta \qquad \text{with} \quad \varepsilon(\sigma) = 1 \text{ or } -1.$$

Then the map

$$\sigma \mapsto \varepsilon(\sigma)$$

is immediately verified to be a homomorphism of $G_{K/F}$ into the cyclic group with two elements. The kernel of this homomorphism consists of those σ which induce an even permutation of the roots. In any case we have an injective homomorphism

$$G_{K/F} \to S_3$$

which to each σ associates the corresponding permutation of the roots. Since S_3 has order 6, it follows that $G_{K/F}$ has order dividing 6, so $G_{K/F}$ has order 1, 2, 3, or 6. We let A_3 be the alternating group, i.e. the subgroup of even permutations in S_3. One possible question is whether $G_{K/F}$ is isomorphic to A_3 or to S_3.

Note that $D = \Delta^2$ and $D \in F$. This follows from the general theory of the discriminant, which is a symmetric function of the roots, and therefore a polynomial in the coefficients of f with integer coefficients by Chapter IV, Theorem 8.1. Since the coefficients of f are in F, the discriminant is in F. We can also see this result from Galois theory, since D is fixed under the Galois group $G_{K/F}$.

Theorem 5.1. *Let f be an irreducible polynomial over F, of degree 3. Let K be its splitting field and $G = G_{K/F}$. Then G is isomorphic to S_3 if and only if the discriminant D of f is not a square in F. If D is a square in F, then K has degree 3 over F and $G_{K/F}$ is cyclic of degree 3.*

Proof. Suppose D is a square in F. Since the square root of D is uniquely determined up to a sign, it follows that $\Delta \in F$. Hence Δ is fixed under the Galois group. This implies that $G_{K/F}$ is isomorphic to a subgroup of the alternating group which has order 3. Since $[K:F] = 3$ because f is assumed irreducible, we conclude that $G_{K/F}$ is isomorphic to A_3, and therefore $G_{K/F}$ is cyclic of order 3.

Suppose D is not a square in F. Then

$$[F(\Delta):F] = 2.$$

Since by irreducibility $[F(\alpha):F] = 3$, it follows that $[F(\alpha, \Delta):F] = 6$, whence $G_{K/F}$ has order at least 6. But $G_{K/F}$ is isomorphic to a subgroup of S_3, so $G_{K/F}$ has order exactly 6 and is isomorphic to S_3. This concludes the proof.

Actually the proofs of the two cases in Theorem 5.1 has also given us a proof of the following result.

Theorem 5.2. *Let $f(t) = t^3 + \cdots$ be a polynomial of degree 3, irreducible over the field F. Let D be the discriminant and let α be a root. Then the splitting field of f is $K = F(\sqrt{D}, \alpha)$.*

After replacing the variable t by $t - c$ with a suitable constant $c \in F$ if necessary a cubic polynomial with leading coefficient 1 in $F[t]$ can be brought into the form

$$f(t) = t^3 + at + b = (t - \alpha_1)(t - \alpha_2)(t - \alpha_3)$$

with $a, b \in F$. The roots may or may not be in F. If f has no root in F, then f is irreducible. We find

$$\alpha_1 + \alpha_2 + \alpha_3 = 0, \qquad \alpha_1\alpha_2 + \alpha_1\alpha_3 + \alpha_2\alpha_3 = a, \qquad -\alpha_1\alpha_2\alpha_3 = b.$$

As in Chapter IV, §8 the discriminant has the form

$$\boxed{D = -4a^3 - 27b^2.}$$

Theorems 5.1, 5.2 and this formula now give you the tools to determine the Galois group of a cubic polynomial, provided you have a means of determining its irreducibility.

We emphasize that before doing anything else, one must *always* determine the irreducibility of f.

Example. We consider the polynomial $f(t) = t^3 - 3t + 1$. It has no integral root, and hence is irreducible over $\mathbf{Q}$. Its discriminant is

$$D = -4a^3 - 27b^2 = 3^4.$$

The discriminant is a square in $\mathbf{Q}$, and hence the Galois group of f over $\mathbf{Q}$ is cyclic of order 3. The splitting field is $\mathbf{Q}(\alpha)$ for any root α.

VII, §5. EXERCISES

1. Determine the Galois groups of the following polynomials over the rational numbers.
   (a) $t^2 - t + 1$ (b) $t^2 - 4$ (c) $t^2 + t + 1$ (d) $t^2 - 27$

2. Let $f(t)$ be a polynomial of degree 3, irreducible over the field F. Prove that the splitting field K of f contains at most one subfield of degree 2 over F namely $F(\sqrt{D})$, where D is the discriminant. If D is a square in F, then K does not contain any subfield of degree 2 over F.

3. Determine the Galois groups of the following polynomials over the rational numbers. Find the discriminants.
   (a) $t^3 - 3t + 1$ (b) $t^3 + 3$ (c) $t^2 - 5$
   (d) $t^3 - a$ where a is rational, $\neq 0$, and is not a cube of a rational number
   (e) $t^3 - 5t + 7$ (f) $t^3 + 2t + 2$ (g) $t^3 - t - 1$

4. Determine the Galois groups of the following polynomials over the indicated field.

   (a) $t^3 - 10$ over $\mathbf{Q}(\sqrt{2})$ (b) $t^3 - 10$ over $\mathbf{Q}$
   (c) $t^3 - t - 1$ over $\mathbf{Q}(\sqrt{-23})$ (d) $t^3 - 10$ over $\mathbf{Q}(\sqrt{-3})$
   (e) $t^3 - 2$ over $\mathbf{Q}(\sqrt{-3})$ (f) $t^3 - 9$ over $\mathbf{Q}(\sqrt{-3})$
   (g) $t^2 - 5$ over $\mathbf{Q}(\sqrt{-5})$ (h) $t^2 + 5$ over $\mathbf{Q}(\sqrt{-5})$

5. Let F be a field and let

$$f(t) = t^3 + a_2t^2 + a_1t + a_0 \quad \text{and} \quad g(t) = t^2 - c$$

   be irreducible polynomials over F. Let D be the discriminant of f. Assume that

$$[F(D^{1/2}):F] = 2 \quad \text{and} \quad F(D^{1/2}) \neq F(c^{1/2}).$$

   Let α be a root of f and β a root of g in an algebraic closure. Prove:
   (a) The splitting field of $f(t)g(t)$ over F has degree 12.
   (b) Let $\gamma = \alpha + \beta$. Then $[F(\gamma):F] = 6$.

6. Let f, g be irreducible polynomials of degree 3 over the field F. Let D_f, D_g be their discriminants. Assume that D_f is not a square in F but D_g is a square in F.
   (a) Prove that the splitting field of fg over F has degree 18.
   (b) Let S_3 be the symmetric group on 3 elements, and let C_3 be a cyclic group of order 3. Prove that the Galois group of fg over F is isomorphic to $S_3 \times C_3$.

7. Let $f(t) = t^3 + at + b$. Let α be a root, and let β be a number such that

$$\alpha = \beta - \frac{a}{3\beta}.$$

   Show that such a β can be found if $a \neq 0$. Show that

$$\beta^3 = -b/2 + \sqrt{-D/108}.$$

   In this way we get an expression of α in terms of radicals.

For the next exercises, recall the following result, or prove it if you have not already done so.

Let $a, b \in F$ and suppose $F(\sqrt{a})$ and $F(\sqrt{b})$ have degree 2 over F. Then $F(\sqrt{a}) = F(\sqrt{b})$ if and only if there exists $c \in F$ such that $a = c^2b$.

8. Let $f(t) = t^4 + 30t^2 + 45$. Let α be a root of f. Prove that $\mathbf{Q}(\alpha)$ is cyclic of degree 4 over $\mathbf{Q}$. [*Hint*: Note that to solve the equation, you can apply the quadratic formula twice.]
9. Let $f(t) = t^4 + 4t^2 + 2$.
   (a) Prove that $f(t)$ is irreducible over $\mathbf{Q}$.
   (b) Prove that the Galois group over $\mathbf{Q}$ is cyclic.
10. Let $K = \mathbf{Q}(\sqrt{2}, \sqrt{3}, \alpha)$ where $\alpha^2 = (9 - 5\sqrt{3})(2 - \sqrt{2})$.
    (a) Prove that K is a Galois extension of $\mathbf{Q}$.
    (b) Prove that $G_{K/\mathbf{Q}}$ is not cyclic but contains a unique element of order 2.
11. Let $[E:F] = 4$. Prove that E contains a subfield L with $[L:F] = 2$ if and only if $E = F(\alpha)$ where α is a root of an irreducible polynomial in $F[t]$ of the form $t^4 + bt^2 + c$.

VII, §6. SOLVABILITY BY RADICALS

We continue to assume that all fields are of characteristic 0.

A Galois extension whose Galois group is abelian is said to be an **abelian extension**. Let K be a Galois extension of F, $K = F(\alpha)$. Let σ, τ be automorphisms of K over F. To verify that $\sigma\tau = \tau\sigma$ it suffices to verify that $\sigma\tau\alpha = \tau\sigma\alpha$. Indeed, any element of K can be written in the form

$$x = a_0 + a_1\alpha + \cdots + a_{d-1}\alpha^{d-1}$$

if d is the degree of α over F. Since $\sigma\tau a_i = \tau\sigma a_i$ for all i, it follows that if in addition $\sigma\tau\alpha = \tau\sigma\alpha$, then $\sigma\tau\alpha^i = \tau\sigma\alpha^i$ for all i, whence $\tau\sigma x = \sigma\tau x$. We shall describe two important cases.

Theorem 6.1. *Let F be a field and n a positive integer. Let ζ be a primitive n-th root of unity, that is $\zeta^n = 1$, and every n-th root of unity can be written in the form ζ^r for some r, $0 \leqq r < n$. Let $K = F(\zeta)$. Then K is Galois and abelian over F.*

Proof. Let σ be an embedding of K over F. Then

$$(\sigma\zeta)^n = \sigma(\zeta^n) = 1.$$

Hence $\sigma\zeta$ is also an n-th root of unity, and there exists an integer r such that $\sigma\zeta = \zeta^r$. In particular, K is Galois over F. Furthermore, if τ is another automorphism of K over F, then $\tau\zeta = \zeta^s$ for some s, and

$$\sigma\tau\zeta = \sigma(\zeta^s) = \sigma(\zeta)^s = \zeta^{rs} = \tau\sigma\zeta.$$

Hence $\sigma\tau = \tau\sigma$, and the Galois group is abelian, as was to be shown.

Theorem 6.2. *Let F be a field and assume that the n-th roots of unity lie in F. Let $a \in F$. Let α be a root of the polynomial $t^n - a$, so $\alpha^n = a$, and let $K = F(\alpha)$. Then K is abelian over F, and in fact K is cyclic over F, that is $G_{K/F}$ is cyclic.*

Proof. Let σ be an embedding of K over F. Then

$$(\sigma\alpha)^n = \sigma(\alpha^n) = \sigma a = a.$$

Hence $\sigma\alpha$ is also a root of $t^n - a$, and

$$(\sigma\alpha/\alpha)^n = 1.$$

Hence there exists an n-th root of unity ζ_σ (not necessarily primitive) such that

$$\sigma\alpha = \zeta_\sigma\alpha.$$

(Note that we index ζ_σ by σ to denote its dependence on σ.) In particular, K is Galois over F. Let τ be any automorphism of K over F. Then similarly there is a root of unity ζ_τ such that

$$\tau\alpha = \zeta_\tau\alpha.$$

In addition,

$$\sigma(\tau(\alpha)) = \sigma(\zeta_\tau\alpha) = \zeta_\tau\sigma(\alpha) = \zeta_\tau\zeta_\sigma\alpha,$$

because the roots of unity are in F, and so are left fixed by σ. Therefore the association

$$\sigma \mapsto \zeta_\sigma$$

is a homomorphism of the Galois group $G_{K/F}$ into the cyclic group of n-th roots of unity. If $\zeta_\sigma = 1$ then σ is the identity on α, and therefore the identity on K. Hence the kernel of this homomorphism is 1. Therefore $G_{K/F}$ is isomorphic to a subgroup of the cyclic group of n-th roots of unity, and is therefore cyclic. This concludes the proof.

Remark. It may of course happen that in Theorem 6.2 the Galois group $G_{K/F}$ is cyclic of order less than n. For instance, a could be a d-th power in F with $d|n$. However, if $t^n - a$ is irreducible, then $G_{K/F}$ has order n. Indeed, we know that $[K:F] \geqq n$ by irreducibility, so $G_{K/F}$ has order at least n, whence $G_{K/F}$ has order exactly n since it is a subgroup of a cyclic group of order n.

For a concrete example, let $F = \mathbf{Q}(\zeta)$ where ζ is a primitive cube root of unity, so $F = \mathbf{Q}(\sqrt{-3})$. Let α be a root of $t^9 - 27$. Then

$$[F(\alpha):F] = 3.$$

Let F be a field and f a polynomial of degree $n \geqq 1$ over F. We shall say that f is **solvable by radicals** if its splitting field is contained in a finite extension K which admits a sequence of subfields

$$F = F_0 \subset F_1 \subset F_2 \subset \cdots \subset F_m = K$$

such that:

(a) K is Galois over F.
(b) $F_1 = F(\zeta)$ for some primitive n-th root of unity ζ.
(c) For each i with $1 \leqq i \leqq m-1$, the field F_{i+1} can be written in the form $F_{i+1} = F_i(\alpha_{i+1})$, where α_{i+1} is a root of some polynomial

$$t^{d_i} - a_i = 0,$$

where d_i divides n, and a_i is an element of F_i.

Observe that if d divides n, then $\zeta^{n/d}$ is a primitive d-th root of unity (proof?) and hence by Theorems 6.1 and 6.2 the extension F_{i+1} of F_i is abelian. We also have seen that F_1 is abelian over F. Thus K is decomposed into a tower of abelian extensions. Let G_i be the Galois group of K over F_i. Then we obtain a corresponding sequence of groups

$$G \supset G_1 \supset G_2 \supset \cdots \supset G_m = \{e\}$$

such that G_{i+1} is normal in G_i, and the factor group G_i/G_{i+1} is abelian by Theorem 6.2.

Theorem 6.3. *If f is solvable by radicals, then its Galois group is solvable.*

Proof. Let L be the splitting field of f, and suppose that L is contained in a Galois extension K as above. By the definition of a solvable group, it follows that $G_{K/F}$ is solvable. But $G_{L/F}$ is a factor group of $G_{K/F}$, and so $G_{L/F}$ is solvable. This proves the theorem.

Remark. In the definition of solvability by radicals, we built in from the start the hypothesis that K is Galois over F. In the exercises, you can develop a proof of a slightly stronger result, without making this assumption. The reason for making the assumption was to exhibit clearly and briefly the essential part of the argument, which from solvability by radicals implies the solvability of the Galois group. The steps given in the exercises will be of a more routine type.

The converse is also true: if the Galois group of an extension is solvable, then the extension is solvable by radicals. This takes somewhat more arguments to prove, and you can look it up in a more advanced text, e.g. my *Algebra*.

It was a famous problem once to determine whether every polynomial is solvable by radicals. To show that this is not the case, it will suffice to exhibit a polynomial whose Galois group is the symmetric group S_5 (or S_n for $n \geqq 5$), according to Theorem 6.3 of Chapter II, §6. This is easily done:

Theorem 6.4. *Let $x_1,\dots,x_n$ be algebraically independent over a field F_0, and let*

$$f(t) = \prod_{i=1}^{n} (t - x_i) = t^n - s_1 t^{n-1} + \cdots + (-1)^n s_n,$$

where

$$s_1 = x_1 + \cdots + x_n, \dots, s_n = x_1 \cdots x_n$$

are the coefficients of f. Let $F = F_0(s_1,\dots,s_n)$. Let $K = F(x_1,\dots,x_n)$. Then K is Galois over F, with Galois group S_n.

Proof. Certainly K is a Galois extension of F because

$$K = F(x_1,\dots,x_n)$$

is the splitting field of f. Let $G = G_{K/F}$. By general Galois theory, we know that there is an injective homomorphism of G into S_n:

$$\sigma \mapsto \pi_\sigma$$

which to each automorphism σ of K over F associates a permutation of the roots. So we have to prove that given a permutation π of the roots $x_1,\dots,x_n$ there exists an automorphism of K over F which restricts to this permutation. But $(\pi(x_1),\dots,\pi(x_n))$ being a permutation of the roots, the elements $\pi(x_1),\dots,\pi(x_n)$ are algebraically independent, and by the basic fact of polynomials in several variables, there is an isomorphism

$$F_0[x_1,\dots,x_n] \xrightarrow{\approx} F_0[\pi(x_1),\dots,\pi(x_n)]$$

which extends to an isomorphism of the quotient fields

$$F_0(x_1,\dots,x_n) \xrightarrow{\approx} F_0(\pi(x_1),\dots,\pi(x_n)),$$

sending x_i on $\pi(x_i)$ for $i = 1,\dots,n$. This isomorphism is therefore an automorphism of $F_0(x_1,\dots,x_n)$, which leaves the ring of symmetric polynomials $F_0[s_1,\dots,s_n]$ fixed, and therefore leaves the quotient field $F_0(s_1,\dots,s_n)$ fixed. This proves that the map $\sigma \mapsto \pi_\sigma$ is surjective on S_n, and concludes the proof of the theorem.

In the next section, we shall show that one can always select n complex numbers algebraically independent over $\mathbf{Q}$.

In some sense, in exhibiting a Galois extension whose Galois group is S_n we have "cheated". We really would like to see a Galois extension of the rational numbers $\mathbf{Q}$ whose Galois group is S_n. This is somewhat more difficult to achieve. It can be shown by techniques beyond this book that for a lot of special values $\bar{s}_1,\dots,\bar{s}_n$ in the rational numbers, the polynomial

$$\bar{f}(t) = t^n - \bar{s}_1 t^{n-1} + \cdots + (-1)^n \bar{s}_n$$

has the symmetric group as its Galois group.

Also the polynomial $t^5 - t - 1$ has the symmetric group as its Galois group. You can refer to a more advanced algebra book to see how to prove such statements.

Remark. Radicals are merely the simplest way of expressing irrational numbers. One can ask much more general questions, for instance along the following lines. Let α be an **algebraic number**, that is the root of a polynomial of degree $\geqq 1$ with rational coefficients. Let us start with the question: Is there a root of unity ζ such that $\alpha \in \mathbf{Q}(\zeta)$? We suppose $\mathbf{Q}(\zeta)$ is embedded in the complex numbers. We can write

$$\zeta = e^{2\pi i r/n}$$

where r, n are positive integers. If we define the function

$$f(z) = e^{2\pi i z},$$

then our question amounts to whether $\alpha \in \mathbf{Q}(f(a))$ for some rational number a. Now the question can be generalized, by taking for f an arbitrary classical function, not just the exponential function. In analysis courses, you may have heard of the Bessel function, the gamma function, the zeta function, various other functions, such as solutions of differential equations, whatever. The question then runs almost immediately into unsolved problems, and leads into a mixture of algebra, number theory, and analysis.

VII, §6. EXERCISES

1. Let K_1, K_2 be Galois extensions of F whose Galois groups are solvable. Prove that the Galois groups of K_1K_2 and $K_1 \cap K_2$ are solvable.

2. Let K be a Galois extension of F whose Galois group is solvable. Let E be any extension of F. Prove that KE/E has solvable Galois group.

3. By a **radical tower** over a field F we mean a sequence of finite extensions

$$F = F_0 \subset F_1 \subset \cdots \subset F_r,$$

having the property that there exist positive integers d_i, elements $a_i \in F_i$ and α_i with $\alpha_i^{d_i} = a_i$ such that

$$F_{i+1} = F_i(\alpha_i).$$

We say that E is **contained in a radical tower** if there exists a radical tower as above such that $E \subset F_r$.

Let E be a finite extension of F. Prove:

(a) If E is contained in a radical tower and E' is a conjugate of E over F, then E' is contained in a radical tower.

(b) Suppose E is contained in a radical tower. Let L be an extension of F. Then EL is contained in a radical tower of L.

(c) If E_1, E_2 are finite extensions of F, each one contained in a radical tower, then the composite E_1E_2 is contained in a radical tower.

(d) If E is contained in a radical tower, then the smallest normal extension of F containing E is contained in a radical tower.

(e) If $F_0 \subset \cdots \subset F_r$ is a radical tower, let K be the smallest normal extension of F_0 containing F_r. Then K has a radical tower over F.

(f) Let E be a finite extension of F and suppose E is contained in a radical tower. Show that there exists a radical tower

$$F \subset E_0 \subset E_1 \subset \cdots \subset E_m$$

such that:

E_m is Galois over F and $E \subset E_m$;
$E_0 = F(\zeta)$ where ζ is a primitive n-th root of unity;
For each i, $E_{i+1} = E_i(\alpha_i)$ where $\alpha_i^{d_i} = a_i \in E_i$ and $d_i | n$.

Thus if E is contained in a radical tower, then E is solvable by radicals in the sense given in the text. The property of being contained in a radical tower is closer to the naive notion of an algebraic element being expressible in terms of radicals, and so we gave the development of this exercise to show that this naive notion is equivalent to the notion given in the text.

VII, §7. INFINITE EXTENSIONS

We begin with some cardinality statements concerning fields. We use only denumerable or finite sets in the present situation, and all that we need about such sets is the definition of Chapter X, §3 and the following:

If D is denumerable, then a finite product $D \times \cdots \times D$ is denumerable.

A denumerable union of denumerable sets is denumerable.

An infinite subset of a denumerable set is denumerable.

If D is denumerable, and $D \to S$ is a surjective map onto some set S which is not finite, then S is denumerable.

The reader will find simple self-contained proofs in Chapter X (cf. Theorem 3.2 and its corollaries), and for denumerable sets, these statements are nothing but simple exercises.

Let F be a field, and E an extension of F. We shall say that E is **algebraic** over F if every element of E is algebraic over F. Let A be an algebraically closed field containing F. Let F^{a} be the subset of A consisting of all elements which are algebraic over F. The superscript "a" denotes "**algebraic closure**" of F in A. Then F^{a} is a field, because we have seen that whenever α, β are algebraic, then $\alpha + \beta$ and $\alpha\beta$ are algebraic, being contained in the finite extension $F(\alpha, \beta)$ of F.

Theorem 7.1. *Let F be a denumerable field. Then F^{a} is denumerable.*

Proof. We proceed stepwise. Let P_n be the set of irreducible polynomials of degree $n \geqq 1$ with coefficients in F and leading coefficient 1. To each polynomial $f \in P_n$,

$$f(t) = t^n + a_{n-1}t^{n-1} + \cdots + a_0,$$

we associate its coefficients $(a_{n-1}, \ldots, a_0)$. We thus obtain an injection of P_n into $F \times \cdots \times F = F^n$, whence we conclude that P_n is denumerable.

Next, for each $f \in P_n$, we let $\alpha_{f,1}, \ldots, \alpha_{f,n}$ be its roots, in a fixed order. Let $J_n = \{1, \ldots, n\}$, and let

$$P_n \times \{1, \ldots, n\} \to A$$

be the map of $P_n \times J_n$ into A such that

$$(f, i) \mapsto \alpha_{f,i}$$

for $i = 1, \ldots, n$ and $f \in P_n$. Then this map is a surjection of $P_n \times J_n$ onto the set of numbers of degree n over F, and hence this set is denumerable. Taking the union over all $n = 1, 2, \ldots$ we conclude that the set of all numbers algebraic over F is denumerable. This proves our theorem.

Theorem 7.2. *Let F be a denumerable field. Then the field of rational functions $F(t)$ is denumerable.*

Proof. It will suffice to prove that the ring of polynomials $F[t]$ is denumerable, because we have a surjective map

$$F[t] \times F[t]_0 \to F(t),$$

where $F[t]_0$ denotes the set of non-zero elements of $F[t]$. The map is of course $(a, b) \mapsto a/b$. For each n, let P_n be the set of polynomials of degree $\leqq n$ with coefficients in F. Then P_n is denumerable, and hence $F[t]$ is denumerable, being the denumerable union of P_0, P_1, $P_2, \ldots$ together with the single element 0.

Note. The fact that **R** (and hence **C**) is not denumerable will be proved in Chapter IX, Corollary 4.4.

Corollary 7.3. *Given an integer $n \geqq 1$, there exist n algebraically independent complex numbers over* **Q**.

Proof. The field $\mathbf{Q}^a$ is denumerable, and **C** is not. Hence there exists $x_1 \in \mathbf{C}$ which is transcendental over $\mathbf{Q}^a$. Let $F_1 = \mathbf{Q}^a(x_1)$. Then F_1 is denumerable. Proceeding inductively, we let x_2 be transcendental over F_1^a, and so on, to find our desired elements $x_1, x_2, \ldots, x_n$.

The complex numbers form a convenient algebraically closed field of characteristic 0 for many applications.

Theorem 7.4. *Let F be a field. Then there exists an algebraic closure of F, that is, there exists a field A algebraic over F such that A is algebraically closed.*

Proof. In general, we face a problem which is set-theoretic in nature. In Exercise 9 of Chapter X, §3 you will see how to carry out the proof in general. We give the proof here in the most important special case, when F is denumerable. All the basic features of the proof already occur in this case.

As in Theorem 7.1 we give an enumeration of all polynomials of degree $\geqq 1$ over F, say $\{f_1, f_2, f_3, \ldots\}$. By induction we can find a splitting field K_1 of f_1 over F, then a splitting field K_2 of $f_1 f_2$ over F containing K_1 as subfield; then a splitting field of $f_1 f_2 f_3$ containing K_2; and so on. In general, we let K_{n+1} be a splitting field of $f_1 f_2 \cdots f_{n+1}$ containing K_n. We take the union

$$A = \bigcup_{n=1}^{\infty} K_n.$$

Observe first that A is a field. Two elements of A lie in some K_n, so their sum, product and quotient (by a non-zero element) are defined in K_n, and since K_n is a subfield of K_m for $m > n$, this sum, product and quotient do not depend on the choice of n. Furthermore, A is algebraic over F because every element of A lies in some K_n, which is finite over F. We claim that A is algebraically closed. Let α be algebraic over A. Then α is the root of a polynomial $f(t) \in A[t]$, and f actually has coefficients in some field K_n, so α is algebraic over K_n. Then $K_n(\alpha)$ is algebraic, so finite over F, so α lies in K_m for some m, whence $\alpha \in A$. This concludes the proof.

Next, we deal with the uniqueness of an algebraic closure.

Theorem 7.5. *Let A, B be algebraic closures of F. Then there exists an isomorphism $\sigma: A \to B$ over F (that is, σ is the identity on F).*

Proof. Again in general we meet a set-theoretic difficulty which disappears in the denumerable case, so we give the proof when F is denumerable. By an argument similar to the one used in the previous theorem, we write A as a union

$$A = \bigcup_{n=1}^{\infty} K_n$$

where K_n is finite normal over F and $K_n \subset K_{n+1}$ for all n. By the embedding theorem for finite extensions, there exists an embedding

$$\sigma_1: K_1 \to B$$

which is the identity on F. By induction, suppose we have obtained an embedding

$$\sigma_n: K_n \to B$$

over F. By the embedding theorem, there exists an embedding

$$\sigma_{n+1}: K_{n+1} \to B$$

which is an extension of σ_n. Then we can define σ on A as follows. Given an element $x \in A$, there is some n such that $x \in K_n$. The element $\sigma_n x$ of B does not depend on the choice of n, because of the compatibility condition that if $m > n$ then the restriction of σ_m to K_n is σ_n. We define σx to be $\sigma_n x$. Then σ is an embedding of A into B which restricts to σ_n on K_n. It will now suffice to prove:

Let A, B be algebraic closures of F and let $\sigma: A \to B$ be an embedding over F. Then $B = \sigma A$, so σ is an isomorphism.

Proof. Since A is algebraically closed, it follows that σA is algebraically closed, and $\sigma A \subset B$. Since B is algebraic over F, it follows that B is algebraic over σA. Let $y \in B$. Then y is algebraic over σA, and therefore lies in σA, so $B = \sigma A$, as was to be shown. This concludes the proof of Theorem 7.5.

Remark. The element σ in the above proof was defined in a way which could be called a limit of a sequence of embeddings σ_n. The study of such limits would constitute a further chapter in field theory. We shall not go into this matter except by giving some examples as exercises. The study of such limits is also analogous to the considerations of Chapter IX, which deals with completions.

Just as in the finite case, we can speak of the group of automorphisms $\mathrm{Aut}(A/F)$, and this group is also called a **Galois group** for the possibly infinite extension A of F.

More generally, let

$$K = \bigcup_{n=1}^{\infty} K_n$$

be a union of finite Galois extensions K_n of F, such that $K_n \subset K_{n+1}$ for all n. Then we let

$$G_{K/F} = \mathrm{Aut}(K/F)$$

be the group of automorphisms of K over F. Each such automorphism restricts to an embedding of K_n, which must be an automorphism of K_n. By the extension theorem and an inductive definition as in the proof of the uniqueness of the algebraic closure, we conclude that the restriction homomorphism

$$\mathrm{res}\colon G_{K/F} \to G_{K_n/F}$$

is surjective. Its kernel is G_{K/K_n}. A major problem is to determine $G_{K/F}$ for various infinite extensions, and especially when $K = \mathbf{Q}^{\mathrm{a}}$ is the algebraic closure of the rational numbers. In an exercise, you can see an example for a tower of abelian extensions. The point is that automorphisms in $G_{K/F}$ are limits of sequences $\{\sigma_n\}$ of automorphisms of the finite Galois extensions K_n/F. Thus the consideration of infinite extensions leads to a study of limits of sequences of elements in Galois groups, and more general types of groups. In the next chapter we shall meet two basic examples: the extensions of finite fields, and extensions generated by roots of unity.

VII, §7. EXERCISES

1. Let $\{G_n\}$ be a sequence of multiplicative groups and for each n suppose given a surjective homomorphism

$$h_{n+1}: G_{n+1} \to G_n.$$

Let G be the set of all sequences

$$(s_1, s_2, s_3, \ldots, s_n, \ldots) \quad \text{with} \quad s_n \in G_n$$

satisfying the condition $h_n s_n = s_{n-1}$. Define multiplication of such sequences componentwise. Prove that G is a group, which is called the **projective limit** of $\{G_n\}$. If the groups G_n are additive groups, then G is an additive group in a similar way.

Examples. Let p be a prime number. Let $\mathbf{Z}(p^n) = \mathbf{Z}/p^n\mathbf{Z}$, and let

$$h_{n+1}: \mathbf{Z}/p^{n+1}\mathbf{Z} \to \mathbf{Z}/p^n\mathbf{Z}$$

be the natural homomorphism. The projective limit is called the group of ***p*-adic integers**, and is denoted by $\mathbf{Z}_p$.

2. (a) Using the fact that h_{n+1} is actually a surjective ring homomorphism, show that $\mathbf{Z}_p$ is a ring in a natural way.
(b) Prove that $\mathbf{Z}_p$ has a unique prime ideal which is $p\mathbf{Z}_p$.

Let again p be a prime number. Instead of $\mathbf{Z}(p^n)$ as above, consider the group of units in the ring $\mathbf{Z}/p^n\mathbf{Z}$, so let $G_n = (\mathbf{Z}/p^n\mathbf{Z})^*$. We can define

$$h_{n+1}^*: (\mathbf{Z}/p^{n+1}\mathbf{Z})^* \to (\mathbf{Z}/p^n\mathbf{Z})^*$$

to be the restriction of h_{n+1}, and you will see immediately that h_{n+1}^* is surjective. Cf. the exercises of Chapter II, §7. The projective limit of $\{(\mathbf{Z}/p^n\mathbf{Z})^*\}$ is called the group of ***p*-adic units**, and is denoted by $\mathbf{Z}_p^*$. Thus a p-adic unit consists of a sequence

$$(u_1, u_2, u_3, \ldots, u_n, \ldots)$$

where each $u_n \in (\mathbf{Z}/p^n\mathbf{Z})^*$ and

$$u_{n+1} \equiv u_n \bmod p^n.$$

Each element u_n can be represented by a positive integer prime to p, and is well-defined mod p^n.

In Chapter VIII, §5 you will find an application of the projective limit of the groups $(\mathbf{Z}/p^n\mathbf{Z})^*$ to the Galois theory of roots of unity. If you read Theorem 5.1 of Chapter VIII, you can do right away Exercise 12 of VIII, §5 following the above discussion. However, you can now do the following exercise, which depends only on the notions and results which have already been dealt with.

3. Let F be a field and let $\{K_n\}$ be a sequence of finite Galois extensions such that $K_n \subset K_{n+1}$. Let

$$K = \bigcup_{n=1}^{\infty} K_n \quad \text{and} \quad G_n = \text{Gal}(K_n/F).$$

By finite Galois theory, the restriction homomorphism $G_{n+1} \to G_n$ is surjective. Define a natural map

$$\lim G_n \to \text{Aut}(K/F),$$

and prove that your map is an isomorphism. The limit is the projective limit defined in Exercise 1.

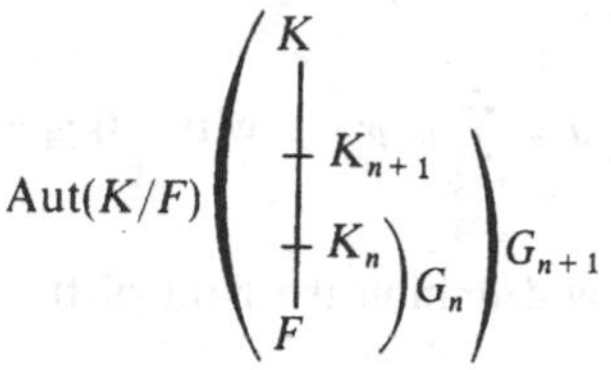

K
K_{n+1}
$\text{Aut}(K/F)$ K_n G_n G_{n+1}
F

The next exercises give another approach to the limits which we have just considered, in another context which will relate to the context of Chapter IX. You may wish to postpone these exercises until you read Chapter IX.

4. Let G be a group. Let $\mathcal{F}$ be the family of all subgroups of finite index. Let $\{x_n\}$ be a sequence in G. We define this sequence to be **Cauchy** if given $H \in \mathcal{F}$ there exists n_0 such that for $m, n \geqq n_0$ we have $x_n x_m^{-1} \in H$. If $\{x_n\}$, $\{y_n\}$ are two sequences, define their product to be the sequence $\{x_n y_n\}$.
   (a) Show that the set of Cauchy sequences forms a group.
   (b) Define $\{x_n\}$ to be a **null sequence** if given $H \in \mathcal{F}$ there exists n_0 such that for $n \geqq n_0$ we have $x_n \in H$. Show that the null sequences form a normal subgroup.

   The factor group of all Cauchy sequences modulo the null sequences is called the **completion** of G. Note that we have not assumed G to be commutative.
   (c) Prove that the map which sends an element $x \in G$ on the class of the sequence $(x, x, x, \ldots)$ modulo null sequences, is a homomorphism of G into the completion, whose kernel is the intersection

$$\bigcap_{H \in \mathcal{F}} H.$$

   It may be useful to you to refer to the exercises of Chapter II, §4, where you should have proved that a subgroup H of finite index always contains a normal subgroup of finite index.

5. Instead of the family $\mathcal{F}$ of all subgroups of finite index, let p be a prime number, and let $\mathcal{F}$ be the family of all normal subgroups whose index is a power of p. Again define Cauchy sequences and null sequences and prove the

analogous statements of (a), (b), (c) in Exercise 4. This time, the completion is called the p-**adic completion**.

6. Let $G = \mathbf{Z}$ be the additive group of integers, and let $\mathcal{F}$ be the family of subgroups $p^n\mathbf{Z}$ where p is a prime number. Let R_p be the completion of $\mathbf{Z}$ in the sense of Exercise 5.
   (a) If $\{x_n\}$ and $\{y_n\}$ are Cauchy sequences, show that $\{x_n y_n\}$ is a Cauchy sequence, and prove that R_p is a ring.
   (b) Prove that the map which to each $x \in \mathbf{Z}$ associates the class of the sequence $(x, x, x, \dots)$ modulo null sequences gives an embedding of $\mathbf{Z}$ into R_p.
   (c) Prove that R_p has a unique prime ideal, which is pR_p.
   (d) Prove that every ideal of R_p has the form $p^m R_p$ for some integer m.

7. Let $\{x_n\}$ be a sequence in R_p and let $x \in R_p$. Define $x = \lim x_n$ if given a positive integer r, there exists n_0 such that for all $n \geqq n_0$ we have $x - x_n \in p^r R_p$. Show that every element $a \in R_p$ has a unique expression

$$a = \sum_{i=0}^{\infty} m_i p^i \qquad \text{with} \quad 0 \leqq m_i \leqq p - 1.$$

The infinite sum is by definition the limit of the partial sums

$$\lim_{N \to \infty} (m_0 + m_1 p + \cdots + m_N p^N).$$

8. Let $G = \mathbf{Z}$ be the additive group of integers, and let $\mathcal{F}$ be the family of subgroups $p^n\mathbf{Z}$ where p is a prime number. Let R_p be the completion of $\mathbf{Z}$ in the sense of Exercises 5, 6, and let $\mathbf{Z}_p$ be the projective limit of $\mathbf{Z}/p^n\mathbf{Z}$ in the sense of Exercise 1. Prove that there is an isomorphism $\mathbf{Z}_p \to R_p$. [In practice, one does not distinguish between $\mathbf{Z}_p$ and R_p, and one uses $\mathbf{Z}_p$ to denote the completion.]

CHAPTER VIII

Finite Fields

It is worth while to consider separately finite fields, which exhibit some very interesting features. We preferred to do Galois theory in characteristic 0 first, in order not to obscure the basic ideas by the special phenomena which can occur when finite fields are involved. On the other hand, finite fields occur so frequently that we now deal with them more systematically.

VIII, §1. GENERAL STRUCTURE

Let F be a finite field with q elements. Let e be the unit element of F. As with any ring, there is a unique ring homomorphism

$$\mathbf{Z} \to F$$

such that

$$n \mapsto ne = \underbrace{e + e + \cdots + e}_{n \text{ times}}.$$

The kernel is an ideal of $\mathbf{Z}$, which we know is principal, and cannot be 0 since the image is finite. Let p be the smallest positive integer in that ideal, so p generates the ideal. Then we have an isomorphism

$$\mathbf{Z}/p\mathbf{Z} \to \mathbf{F}_p$$

between $\mathbf{Z}/p\mathbf{Z}$ and its image in F. Denote this image by $\mathbf{F}_p$. Since $\mathbf{F}_p$ is a subfield of F, it has no divisors of 0, and consequently the ideal $p\mathbf{Z}$ is prime. Hence p is a prime number, uniquely determined by the field F.

We call p the **characteristic** of F, and the subfield $\mathbf{F}_p$ is called the **prime field**.

As an exercise, prove that $\mathbf{Z}/p\mathbf{Z}$ has no automorphisms other than the identity. We then identify $\mathbf{Z}/p\mathbf{Z}$ with its image $\mathbf{F}_p$. This is possible in only one way. We write 1 instead of e.

Theorem 1.1. *The number of elements of F is equal to a power of p.*

Proof. We may view F as a vector space over $\mathbf{F}_p$. Since F has only a finite number of elements, it follows that the dimension of F over $\mathbf{F}_p$ is finite. Let this dimension be n. If $\{w_1, w_2, \ldots, w_n\}$ is a basis, then every element of F has a unique expression

$$x = a_1 w_1 + \cdots + a_n w_n,$$

with elements a_i in $\mathbf{F}_p$. Since the choice of these a_i is arbitrary, it follows that there are p^n possible elements in F, thus proving that $q = p^n$, as desired.

The multiplicative group F^* of non-zero elements has $q - 1$ elements, and they all satisfy the equation

$$x^{q-1} - 1 = 0 \quad \text{if} \quad x \in F^*.$$

Hence all elements of F satisfy the equation

$$x^q - x = 0.$$

(Of course, the only other element is 0.)

In Chapter IV, we discussed polynomials over arbitrary fields. Let us consider the polynomial

$$f(t) = t^q - t$$

over the finite field $\mathbf{F}_p$. It has q distinct roots in the field F, namely all elements of F. The proof that a polynomial of degree n has at most n roots applies as well to the present case. Hence if K is another finite field containing F, then $t^q - t$ cannot have any roots in K other than the elements of F.

If we use the definition of a splitting field as in the previous chapter, we then find:

Theorem 1.2. *The finite field F with q elements is the splitting field of the polynomial $t^q - t$ over the field $\mathbf{Z}/p\mathbf{Z}$.*

By Theorem 3.2 of Chapter VII, two finite fields with the same number of elements are isomorphic. We denote the field with q elements by $\mathbf{F}_q$. In particular, consider the polynomial

$$t^p - t.$$

It has p roots, namely the elements of the prime field. Therefore the elements of $\mathbf{F}_p$ are precisely the roots of $t^p - t$ in $\mathbf{F}_q$.

In the previous chapter, we had used an algebraically closed field right from the start as a matter of convenience. Let us do the same thing here, and postpone to a last section the discussion of the existence of such a field containing our field F. Thus:

We assume that all of our fields are contained in an algebraically closed field A, containing $\mathbf{F}_p = \mathbf{Z}/p\mathbf{Z}$.

In Theorem 1.2 we started with a finite field F with q elements. We may ask for the converse: Given $q = p^n$ equal to a power of p, what is the nature of the splitting field of $t^q - t$ over the field $\mathbf{F}_p = \mathbf{Z}/p\mathbf{Z}$?

Theorem 1.3. *Given $q = p^n$ the set of elements $x \in A$ such that $x^q = x$ is a finite field with q elements.*

Proof. We first make some remarks on binomial coefficients. In the ordinary binomial expansion

$$(x + y)^p = \sum_{i=0}^{p} \binom{p}{i} x^i y^{p-i},$$

we see from the expression

$$\binom{p}{i} = \frac{p!}{i!\,(p-i)!}$$

that all binomial coefficients are divisible by p except the first and the last. Hence in the finite field, all the coefficients are 0 except the first and the last, and we obtain the basic formula:

For any elements $x, y \in A$, we have

$$(x + y)^p = x^p + y^p.$$

By induction, we then see that for any positive integer m, we have

$$(x + y)^{p^m} = x^{p^m} + y^{p^m}.$$

Let K be the set of elements $x \in A$ such that

$$x^{p^n} = x.$$

It is then easily verified that K is a field. Indeed, the above formula shows that this set is closed under addition. It is closed under multiplication, since

$$(xy)^{p^n} = x^{p^n}y^{p^n}$$

in any commutative ring. If $x \neq 0$ and $x^{p^n} = x$, then we see at once that

$$(x^{-1})^{p^n} = x^{-1}.$$

Let $f(t) = t^q - t$. Then K contains all the roots of f, and is then obviously the smallest field containing $\mathbf{F}_p$ and all roots of f. Consequently, K is the splitting field of f.

The theory of unique factorization in Chapter IV, §3 applies to polynomials over any field, and in particular, the derivative criterion of Theorem 3.6 applies. Observe here how peculiar the derivative is. We have

$$f'(t) = qt^{q-1} - 1 = -1,$$

because $q = 0$ in $\mathbf{F}_p = \mathbf{Z}/p\mathbf{Z}$. Consequently the polynomial $f(t)$ has no multiple roots. Since it has at most q roots in A, we conclude that it has exactly q roots in A. Hence K has exactly q elements. This concludes the proof of Theorem 1.3.

VIII, §1. EXERCISES

1. Let F be a finite field with q elements. Let $f(t) \in F[t]$ be irreducible.
   (a) Prove that $f(t)$ divides $t^{q^n} - t$ if and only if $\deg f$ divides n.
   (b) Show that

   $$t^{q^n} - t = \prod_{d \mid n} \prod_{f_d \text{ irr}} f_d(t)$$

   where the product on the inside is over all irreducible polynomials of degree d with leading coefficient 1.
   (c) Let $\psi(d)$ be the number of irreducible polynomials over F of degree d. Show that

   $$q^n = \sum_{d \mid n} d\psi(d).$$

   (d) Let μ be the Moebius function. Prove that

   $$n\psi(n) = \sum_{d \mid n} \mu(d)q^{n/d}.$$

   Dividing by n yields an explicit formula for the number of irreducible polynomials of degree n, and leading coefficient 1 over F.

VIII, §2. THE FROBENIUS AUTOMORPHISM

Theorem 2.1. *Let F be a finite field with q elements. The mapping*

$$\varphi\colon x \mapsto x^p$$

is an automorphism of F.

Proof. We already know that the map

$$x \mapsto x^p$$

is a ring-homomorphism of F into itself. Its kernel is 0, and since F is finite, the map has to be a bijection, so an automorphism.

The automorphism φ is called the **Frobenius automorphism** of F (relative to the prime field). It generates a cyclic group, which is finite, because F has only a finite number of elements. Let $q = p^n$. Note that

$$\varphi^n = \text{id}.$$

Indeed, for any positive integer m,

$$\varphi^m x = x^{p^m}$$

for all x in F. Hence the period of φ divides n, because

$$\varphi^n x = x^{p^n} = x^q = x.$$

Theorem 2.2. *The period of φ is exactly n.*

Proof. Suppose the period is $m < n$. Then every element x of F satisfies the equation

$$x^{p^m} - x = 0.$$

But we have remarked in the preceding section that the polynomial

$$t^{p^m} - t$$

has at most p^m roots. Hence we cannot have $m < n$, as desired.

Theorem 2.3. *Suppose $\mathbf{F}_{p^m}$ is a subfield of $\mathbf{F}_{p^n}$. Then $m|n$. Conversely, if $m|n$ then $\mathbf{F}_{p^m}$ is a subfield of $\mathbf{F}_{p^n}$.*

Proof. Let F be a subfield of $\mathbf{F}_q$, where $q = p^n$. Then F contains the prime field $\mathbf{F}_p$ so F has p^m elements for some m. We view $\mathbf{F}_q$ as a vector space over F, say of dimension d. Then after representing elements of $\mathbf{F}_q$ as linear combinations of basis elements with coefficients in F we see that $\mathbf{F}_q$ has $(p^m)^d = p^{md}$ elements, whence $n = md$ and $m|n$.

Conversely, let $m|n$, $n = md$. Let F be the fixed field of φ^m. Then F is the set of all elements $x \in \mathbf{F}_q$ such that

$$x^{p^m} = x.$$

By Theorem 1.2, this is precisely the field $\mathbf{F}_{p^m}$. But

$$\varphi^{md} = \varphi^n,$$

so $\mathbf{F}_{p^m}$ is fixed under φ^n. Since $\mathbf{F}_{p^n}$ is the fixed field of φ^n, it follows that

$$\mathbf{F}_{p^m} \subset \mathbf{F}_{p^n}.$$

This concludes the proof of the theorem.

If $n = md$, we see that the order of φ^m is precisely equal to d. Thus φ^m generates a cyclic group of automorphisms of $\mathbf{F}_q$, whose fixed field is $\mathbf{F}_{p^m}$. The order of this cyclic group is exactly equal to the degree

$$d = [\mathbf{F}_{p^n} : \mathbf{F}_{p^m}].$$

In the next section, we shall prove that the multiplicative group of $\mathbf{F}_q$ is cyclic. Using this, we now prove:

Theorem 2.4. *The only automorphisms of* $\mathbf{F}_q$ *are the powers of the Frobenius automorphism* $1, \varphi, \ldots, \varphi^{n-1}$.

Proof. Let $\mathbf{F}_q = \mathbf{F}_p(\alpha)$ for some element α (this is what we are assuming now). Then α is a root of a polynomial of degree n, if $q = p^n$. Therefore by Theorem 2.1 of Chapter VII there are at most n automorphisms of F_q over the prime field $\mathbf{F}_p$. Since $1, \varphi, \ldots, \varphi^{n-1}$ constitute n distinct automorphisms, there cannot be any others. This concludes the proof.

The above theorems carry out the Galois theory for finite fields. We have obtained a bijection between subfields of $\mathbf{F}_q$ and subgroups of the group of automorphisms of $\mathbf{F}_q$, each subfield being the fixed field of a subgroup.

VIII, §3. THE PRIMITIVE ELEMENTS

Let F be a finite field with $q = p^n$ elements. In this section, we prove more than the fact that $F = \mathbf{F}_p(\alpha)$ for some α.

Theorem 3.1. *The multiplicative group F^* of F is cyclic.*

Proof. In Theorem 1.10 of Chapter IV we gave a proof based on the structure theorem for abelian groups. Here we give a proof based on a similar idea but self contained. We start with a remark. Let A be a finite abelian group, written additively. Let a be an element of A, of period d, and let b be an element of period d', with $(d, d') = 1$. Then

$$a + b$$

has period dd'. The proof is easy, and is left as an exercise.

Lemma 3.2. *Let z be an element of A having maximal period, that is, whose period d is $\geqq$ the period of any other element of A. Then the period of any element of A divides d.*

Proof. Suppose w has period m not dividing d. Let l be a prime number such that a power l^k divides the period of w, but does not divide d. Write

$$m = l^k m', \qquad d = l^\nu d',$$

where m', d' are not divisible by l. Let

$$a = m'w \qquad \text{and} \qquad b = l^\nu z.$$

Then a has period l^k and b has period d'. Then

$$a + b$$

has period $l^k d' > d$, a contradiction.

We apply these remarks to the multiplicative group F^*, having $q - 1$ elements. Let α be an element of F^* having maximal period d. Then α is a root of the polynomial

$$t^d - 1.$$

By the lemma, all the powers

$$1, \alpha, \alpha^2, \dots, \alpha^{d-1}$$

are distinct, and are roots of this polynomial. Hence they constitute all the distinct roots of the polynomial $t^d - 1$. Suppose that α does not generate F^*, so there is another element x in F^* which is not a power of α. By the lemma, this element x is also a root of $t^d - 1$. This implies that $t^d - 1$ has more than d roots, a contradiction which proves the theorem.

Example. Consider the prime field $\mathbf{F}_p = \mathbf{Z}/p\mathbf{Z}$ itself. The theorem asserts the existence of an element $\alpha \in \mathbf{F}_p^*$ such that every element of $\mathbf{F}_p^*$ is an integral power of α. In terms of congruences, this means that there exists an integer a such that every integer x prime to p satisfies a relation

$$x \equiv a^\nu \mod p,$$

for some positive integer ν. Such integer a is called a **primitive root** mod p in the classical literature.

VIII, §3. EXERCISES

1. Find the smallest positive integer which is a primitive root mod p in each case: $p = 3$, $p = 5$, $p = 7$, $p = 11$, $p = 13$.

2. Make a list of all the primes $\leqq 100$ for which 2 is a primitive root. Do you think there are infinitely many such primes? The answer (yes) was conjectured by Artin, together with a density; cf. his collected works.

3. If α is a cyclic generator of F^*, where F is a finite field, show that α^p is also a cyclic generator.

4. Let p be a prime $\geqq 3$. An integer $a \not\equiv 0 \mod p$ is called a **quadratic residue** mod p if there exists an integer x such that

$$a \equiv x^2 \mod p.$$

It is called a **quadratic non-residue** if there is no such integer x. Show that the number of quadratic residues is equal to the number of quadratic non-residues. [*Hint*: Consider the map $x \mapsto x^2$ on $\mathbf{F}_p^*$.]

VIII, §4. SPLITTING FIELD AND ALGEBRAIC CLOSURE

In Chapter VII, §7 we gave a general method for constructing an algebraic closure. Here in the case of finite fields, we can express the proof more simply. Let g be polynomial of degree $\geqq 1$ over the finite field F.

We have shown in Chapter VII, Theorem 3.1 how to construct a splitting field for g.

For each positive integer n, let K_n be the splitting field of the polynomial

$$t^{p^n} - t$$

over the prime field $\mathbf{F}_p$. By Theorem 3.2 of Chapter VII, given two splitting fields E and E' of a polynomial f, there is an isomorphism

$$\sigma: E' \to E$$

leaving F fixed. In particular, if $m|n$, there is an embedding

$$K_m \to K_n,$$

because any root of $t^{p^m} - t$ is also a root of $t^{p^n} - t$. If we then consider the union

$$A = \bigcup K_n$$

for $n = 1, 2, \ldots$, then this union is easily shown to be algebraically closed. Indeed, let f be a polynomial of degree $\geqq 1$ with coefficients in A. Then f has coefficients in some K_m, so f splits into factors of degree 1 in K_n for some n. This concludes the proof.

VIII, §5. IRREDUCIBILITY OF THE CYCLOTOMIC POLYNOMIALS OVER Q

In Chapter VII, §6 we considered an extension $F(\zeta)$, where ζ is a primitive n-th root of unity. When $F = \mathbf{Q}$ is the rational numbers, the Galois group can be determined completely, and we shall now do so by using finite fields, although we use essentially nothing of what precedes in this chapter, only the flavor.

Let σ be an automorphism of $F(\zeta)$ over F. Then there is some integer $r(\sigma)$ prime to n such that

$$\sigma\zeta = \zeta^{r(\sigma)},$$

and this integer mod n is uniquely determined by σ. Thus we get a map

$$G_{F(\zeta)/F} \to (\mathbf{Z}/n\mathbf{Z})^* \quad \text{by} \quad \sigma \mapsto r(\sigma).$$

This map is injective. It is a homomorphism, because

$$\sigma(\tau(\zeta)) = \sigma(\zeta^{r(\tau)}) = (\sigma\zeta)^{r(\tau)} = \zeta^{r(\sigma)r(\tau)},$$

whence $r(\sigma\tau) = r(\sigma)r(\tau)$. In this way, we can view $G_{F(\zeta)/F}$ as embedded in the multiplicative group $(\mathbf{Z}/n\mathbf{Z})^*$.

So far, we have not used any special property of the rational numbers or the integers. We shall now do so.

Let $f(t) \in \mathbf{Z}[t]$ be a polynomial with integer coefficients. Let p be a prime number. Then we can reduce f mod p. If

$$f(t) = a_n t^n + \cdots + a_0$$

with $a_0, \ldots, a_n \in \mathbf{Z}$, then we let its **reduction** mod p be

$$\bar{f}(t) = \bar{a}_n t^n + \cdots + \bar{a}_0,$$

where $\bar{a}_i$ is the reduction mod p of a_i. The map $f \mapsto \bar{f}$ is a homomorphism

$$\mathbf{Z}[t] \to (\mathbf{Z}/p\mathbf{Z})[t] = \mathbf{F}_p[t].$$

This is easily verified by using the definition of addition and multiplication of polynomials as in Chapter IV, §5.

Theorem 5.1. *The map $\sigma \mapsto r(\sigma)$ is an isomorphism*

$$G_{\mathbf{Q}(\zeta)/\mathbf{Q}} \to (\mathbf{Z}/n\mathbf{Z})^*.$$

Proof. Let m be a positive integer prime to n. Then the map $\zeta \mapsto \zeta^m$ can be decomposed as a composite of maps

$$\zeta \mapsto \zeta^p,$$

where p ranges over primes dividing m. Thus it will suffice to prove: if p is a prime number and $p \nmid n$, and if $f(t)$ is the irreducible polynomial of ζ over $\mathbf{Q}$, then ζ^p is also a root of $f(t)$. Since the roots of $t^n - 1$ are all the n-th roots of unity, primitive or not, it follows that $f(t)$ divides $t^n - 1$, so there is a polynomial $h(t) \in \mathbf{Q}[t]$ with leading coefficient 1 such that

$$t^n - 1 = f(t)h(t).$$

By the Gauss lemma, it follows that f, h have integral coefficients.

Suppose ζ^p is not a root of f. Then ζ^p is a root of h, and ζ itself is a root of $h(t^p)$. Hence $f(t)$ divides $h(t^p)$, and we can write

$$h(t^p) = f(t)g(t).$$

Since $f(t)$ has integral coefficients and leading coefficient 1, we see that g has integral coefficients, again by the Gauss lemma. Since $a^p \equiv a \pmod{p}$ for any integer a, we conclude that

$$h(t^p) \equiv h(t)^p \pmod{p},$$

and hence

$$h(t)^p \equiv f(t)g(t) \pmod{p}.$$

In particular, if we denote by $\bar{f}$ and $\bar{h}$ the polynomial over $\mathbf{Z}/p\mathbf{Z}$ obtained by reducing f and h respectively mod p, we see that $\bar{f}$ and $\bar{h}$ are not relatively prime, i.e. have a factor in common. But $t^n - \bar{1} = \bar{f}(t)\bar{h}(t)$, and hence $t^n - \bar{1}$ has multiple roots. This is impossible, as one sees by taking the derivative, and our theorem is proved.

As a consequence of Theorem 5.1, it follows that the cyclotomic polynomials of Chapter IV, §3, Exercise 13 are irreducible over $\mathbf{Q}$. Prove this as an exercise.

VIII, §5. EXERCISES

1. Let F be a finite extension of the rationals. Show that there is only a finite number of roots of unity in F.

2. (a) Determine which roots of unity lie in the following fields: $\mathbf{Q}(i)$, $\mathbf{Q}(\sqrt{-2})$, $\mathbf{Q}(\sqrt{2})$, $\mathbf{Q}(\sqrt{-3})$, $\mathbf{Q}(\sqrt{3})$, $\mathbf{Q}(\sqrt{-5})$.
   (b) Let ζ be a primitive n-th root of unity. For which n is

$$[\mathbf{Q}(\zeta):\mathbf{Q}] = 2?$$

   Prove your assertion, of course.

3. Let F be a field of characteristic 0, and n an odd integer $\geqq 1$. Let ζ be a primitive n-th root of unity in F. Show that F also contains a primitive $2n$-th root of unity.

4. Given a prime number p and a positive integer m, show that there exists a cyclic extension of $\mathbf{Q}$ of degree p^m. [*Hint*: Use the exercises of Chapter II, §7.]

5. Let n be a positive integer. Prove that there exist infinitely many cyclic extensions of $\mathbf{Q}$ of degree n which are independent. That is, if $K_1,\dots,K_r$ are such extensions then

$$K_i \cap (K_1K_2\cdots K_{i-1}) = \mathbf{Q} \quad \text{for all} \quad i = 2,\dots,r.$$

   (For this exercise, you may assume that there exist infinitely many primes p such that $p \equiv 1 \bmod n$. Again, use Chapter II, §7 and its exercises.)

6. Let G be a finite abelian group. Prove that there exists a Galois extension of $\mathbf{Q}$ whose Galois group is G. In fact, prove that there exist infinitely many such extensions. [You may use previous exercises.]

7. Let $\alpha^3 = 2$ and $\beta^5 = 7$. Let $\gamma = \alpha + \beta$. Prove that $\mathbf{Q}(\alpha, \beta) = \mathbf{Q}(\gamma)$ and that $[\mathbf{Q}(\alpha, \beta):\mathbf{Q}] = 15$.

In the next two exercises, you will see a non-abelian linear group appearing as a Galois group.

8. Describe the splitting field of $t^5 - 7$ over the rationals. What is its degree? Show that the Galois group is generated by two elements σ, τ satisfying the relation

$$\sigma^5 = 1, \qquad \tau^4 = 1, \qquad \tau\sigma\tau^{-1} = \sigma^2.$$

9. Let p be an odd prime and let a be a rational number which is not a p-th power in $\mathbf{Q}$. Let K be the splitting field of $t^p - a$ over the rationals.
   (a) Prove that $[K:\mathbf{Q}] = p(p-1)$. [Cf. Exercise 3 of §3.]
   (b) Let α be a root of $t^p - a$. Let ζ be a primitive p-th root of unity. Let $\sigma \in G_{K/\mathbf{Q}}$. Prove that there exists some integer $b = b(\sigma)$, uniquely determined mod p, such that

$$\sigma(\alpha) = \zeta^b\alpha.$$

   (c) Show that there exists some integer $d = d(\sigma)$ prime to p, uniquely determined mod p, such that

$$\sigma(\zeta) = \zeta^d.$$

   (d) Let G be the subgroup of $GL_2(\mathbf{Z}/p\mathbf{Z})$ consisting of all matrices

$$\begin{pmatrix} 1 & 0 \\ b & d \end{pmatrix} \quad \text{with} \quad b \in \mathbf{Z}/p\mathbf{Z} \quad \text{and} \quad d \in (\mathbf{Z}/p\mathbf{Z})^*.$$

   Prove that the association

$$\sigma \mapsto M(\sigma) = \begin{pmatrix} 1 & 0 \\ b(\sigma) & d(\sigma) \end{pmatrix}$$

   is an isomorphism of $G_{K/\mathbf{Q}}$ with G.
   (e) Let r be a primitive root mod p, i.e. a positive integer prime to p which generates the cyclic group $(\mathbf{Z}/p\mathbf{Z})^*$. Show that there exist elements ρ, $\tau \in G_{K/\mathbf{Q}}$ which generate $G_{K/\mathbf{Q}}$, and satisfy the relations:

$$\rho^p = 1, \qquad \tau^{p-1} = 1, \qquad \tau\rho\tau^{-1} = \rho^r.$$

   (f) Let F be a subfield of K which is abelian over $\mathbf{Q}$. Prove that $F \subset \mathbf{Q}(\zeta)$.

10. Let p, q be distinct odd primes. Let a, b be rational numbers such that a is not a p-th power in $\mathbf{Q}$ and b is not a q-th power in $\mathbf{Q}$. Let $f(t) = t^p - a$ and $g(t) = t^q - b$. Let K_1 be the splitting field of $f(t)$ and K_2 the splitting field of

$g(t)$. Prove that $K_1 \cap K_2 = \mathbf{Q}$. It follows (from what?) that if K is the splitting field of $f(t)g(t)$, then

$$G_{K/\mathbf{Q}} \approx G_{K_1/\mathbf{Q}} \times G_{K_2/\mathbf{Q}}.$$

11. Generalize Exercise 10 as much as you can.

12. Refer back to the exercises of Chapter VII, §7. Let p be a prime number. Let $\boldsymbol{\mu}(p^n)$ denote the group of p^n-th roots of unity. Let

$$\boldsymbol{\mu}(p^\infty) = \bigcup_{n=1}^{\infty} \boldsymbol{\mu}(p^n).$$

Let $K_n = \mathbf{Q}(\boldsymbol{\mu}(p^n))$ be the splitting field of $t^{p^n} - 1$ over the rationals, and let

$$K_\infty = \bigcup_{n=1}^{\infty} K_n.$$

Prove:

Theorem. *Let $\mathbf{Z}_p^*$ be the projective limit of the groups $(\mathbf{Z}/p^n\mathbf{Z})^*$, as in Chapter VII, §7, Exercise 2. There is an isomorphism*

$$\mathbf{Z}_p^* \xrightarrow{\approx} \operatorname{Aut}(K_\infty/\mathbf{Q}),$$

which can be obtained as follows.

(a) Given $a \in \mathbf{Z}_p^*$ prove that there exists an automorphism $\sigma_a \in \operatorname{Aut}(K_\infty/\mathbf{Q})$ having the following property. Let $\zeta \in \boldsymbol{\mu}(p^\infty)$. Choose n such that $\zeta \in \boldsymbol{\mu}(p^n)$. Let $u \in \mathbf{Z}$ be such that $u \equiv a \bmod p^n$. Then $\sigma_a \zeta = \zeta^u$.

(b) Prove that the map $a \mapsto \sigma_a$ is an injective homomorphism of $\mathbf{Z}_p^*$ into $\operatorname{Aut}(K_\infty/\mathbf{Q})$.

(c) Given $\sigma \in \operatorname{Aut}(K_\infty/\mathbf{Q})$, prove that there exists $a \in \mathbf{Z}_p^*$ such that $\sigma = \sigma_a$.

VIII, §6. WHERE DOES IT ALL GO? OR RATHER, WHERE DOES SOME OF IT GO?

You have now learned some facts about Galois groups, and I thought you should get some idea of the type of questions which remain unsolved. If you find the going too hard, then sleep on it. If you still find it too hard, then you can obviously skip this entire section without affecting your understanding of any other part of the book. I want this section to be stimulating, not scary.

One fundamental question is whether given a finite group G, there exists a Galois extension K of $\mathbf{Q}$ whose Galois group is G. This problem has been realized explicitly for at least a century. Emmy Noether thought of one possibility: construct a Galois extension of an extension

$\mathbf{Q}(u_1,\ldots,u_n)$ with the given Galois group, and then specialize the parameters $u_1,\ldots,u_n$ to rational numbers. Here $u_1,\ldots,u_n$ are independent variables. You have seen how one can construct an extension $\mathbf{Q}(x_1,\ldots,x_n)$ of $\mathbf{Q}(s_1,\ldots,s_n)$ where $s_1,\ldots,s_n$ are the elementary symmetric functions of $x_1,\ldots,x_n$. The Galois group is the symmetric group S_n. Let G be a subgroup of S_n. It was an open question for a long time whether the fixed field of G can be written in the form $\mathbf{Q}(u_1,\ldots,u_n)$. Swan showed that this was impossible in general, even if G was a cyclic group (*Inventiones Math.*, 1969).

In the nineteenth century already, number theorists realized the difference between abelian and non-abelian extensions. Kronecker stated and gave what are today considered as incomplete arguments that every finite abelian extension of $\mathbf{Q}$ is contained in some extension $\mathbf{Q}(\zeta)$ where ζ is a root of unity. Such extensions are called **cyclotomic**. A complete proof was given by Weber at the end of the nineteenth century.

If F is a finite extension of $\mathbf{Q}$, the situation is harder to describe, but I shall give one significant example exhibiting the flavor. The field F contains a subring R_F, called the ring of **algebraic integers** in F, consisting of all elements $\alpha \in F$ such that the irreducible polynomial of α with rational coefficients and leading coefficient 1 has in fact all its coefficients contained in the integers $\mathbf{Z}$. It can be shown that the set of all such elements is a ring R_F, and that F is its quotient field.

Let P be a prime ideal of R_F. It is easy to show that $P \cap \mathbf{Z} = (p)$ is generated by a prime number p. Furthermore, R_F/P is a finite field with q elements. Let K be a finite Galois extension of F. It can be shown that there exists a prime ideal Q of R_K such that $Q \cap R_F = P$. Furthermore, there exists an element $\sigma_Q \in G = \mathrm{Gal}(K/F)$ such that $\sigma_Q(Q) = Q$ and for all $\alpha \in R_K$ we have

$$\sigma_Q\alpha \equiv \alpha^q \mod Q.$$

We call σ_Q a **Frobenius element** in the Galois group G associated with Q. In fact, it can be shown that for all but a finite number of Q, two such elements are conjugate to each other in G. We denote any of them by σ_P. If G is abelian, then there is only one element σ_Q in the Galois group.

Theorem. *There exists a unique abelian extension K of F having the following property: If P_1, P_2 are prime ideals of R_F, then $\sigma_{P_1} = \sigma_{P_2}$ if and only if there is an element α of K such that $\alpha P_1 = P_2$.*

In a similar but more complicated manner, one can characterize all abelian extensions of F. This theory is known as **class field theory**, developed by Kronecker, Weber, Hilbert, Takagi, and Artin. The main statement concerning the Frobenius automorphism is Artin's **Reciprocity Law**. You can find an account of class field theory in books on algebraic number theory. Although some fundamental results are known, by no means all results are known.

The non-abelian case is much more difficult. I shall indicate briefly one special case which gives some of the flavor of what's going on. The problem is to do for non-abelian extensions what Artin did for abelian extensions in "class field theory". Artin went as far as saying that the problem was not to give proofs but to formulate what was to be proved. The insight of Langlands and others in the sixties showed that actually Artin was mistaken: The problem lies in both. Shimura made several computations in this direction involving "modular forms", see, for instance, *A reciprocity law in non-solvable extensions*, J. reine angew. Math. **221**, 1966. Langlands gave a number of conjectures, relating Galois groups with "automorphic forms", which showed that the answer lay in deeper theories, whose formulations, let alone their proofs, were difficult. Great progress was made in the seventies by Serre and Deligne, who proved a first case of Langland's conjectures, *Annales Ecole Normale Supérieure*, 1974.

The study of non-abelian Galois groups occurs via their linear "representations". For instance let l be a prime number. We can ask whether $GL_n(\mathbf{F}_l)$, or $GL_2(\mathbf{F}_l)$, or $PGL_2(\mathbf{F}_l)$ occurs as a Galois group over $\mathbf{Q}$, and "how". The problem is to find natural objects on which a Galois group operates as a linear map, such that we get in a natural way an isomorphism of this Galois group with one of the above linear groups. The theories which indicate in which direction to find such objects are much beyond the level of this course. Again I pick a special case to give the flavor.

Let K be a finite Galois extension of the rational numbers, with Galois group $G = \mathrm{Gal}(K/\mathbf{Q})$. Let

$$\rho\colon G \to GL_2(\mathbf{F}_l)$$

be a homomorphism of G into the group of 2×2 matrices over the finite field $\mathbf{F}_l$ for some prime l. Such a homomorphism is called a **representation** of G. Recall from elementary linear algebra that if

$$M = \begin{pmatrix} a & b \\ c & d \end{pmatrix}$$

is a 2×2 matrix, then its **trace** is defined to be the sum of the diagonal elements, that is

$$\mathrm{tr}\, M = a + d.$$

Thus we can take the trace and determinant

$$\operatorname{tr} \rho(\sigma) \quad \text{and} \quad \det \rho(\sigma),$$

which are elements of the field $\mathbf{F}_l$ itself.

Consider the infinite product

$$\Delta = \Delta(z) = z \prod_{n=1}^{\infty} (1 - z^n)^{24}$$

$$= \sum_{n=1}^{\infty} a_n z^n.$$

The coefficients a_n are integers, and $a_1 = 1$.

Theorem. *For each prime l there exists a unique Galois extension K of $\mathbf{Q}$, with Galois group G, and an injective homomorphism*

$$\rho: G \to \mathrm{GL}_2(\mathbf{F}_l)$$

having the following property: For all but a finite number of primes p, if a_p is the coefficient of z^p in $\Delta(z)$, then we have

$$\operatorname{tr} \rho(\sigma_p) \equiv a_p \bmod l \quad \text{and} \quad \det \rho(\sigma_p) \equiv p^{11} \bmod l.$$

Furthermore, for all but a finite number of primes l (which can be explicitly determined), the image $\rho(G)$ in $\mathrm{GL}_2(\mathbf{F}_l)$ consists of those matrices $M \in \mathrm{GL}_2(\mathbf{F}_l)$ such that $\det M$ is an eleventh power in $\mathbf{F}_l^*$.

The above theorem was conjectured by Serre in 1968, Séminaire de Théorie des Nombres, Delange-Pisot-Poitou. A proof of the existence as in the first statement was given by Deligne, Séminaire Bourbaki, 1968-1969. The second statement, describing how big the Galois group actually is in the group of matrices $\mathrm{GL}_2(\mathbf{F}_l)$ is due to Serre and Swinnerton-Dyer, Bourbaki Seminar, 1972, see also Swinnerton-Dyer's article in Springer Lecture Notes 350 on "Modular Functions of One Variable III".

Of course, the product and series for $\Delta(z)$ have been pulled out of nowhere. To explain the somewhere which makes such product and series natural would take another book.

Still another type of question about $G_{\mathbf{Q}}$

The above theorems involve the arithmetic of prime ideals and Frobenius automorphisms. There is still another possibility for describing Galois groups. Let F be a field, F^a an algebraic closure, and let

$$G_F = \mathrm{Gal}(F^a/F)$$

be the group of automorphisms of its algebraic closure, leaving F fixed. The question is what does $G_{\mathbf{Q}}$ look like?

Artin showed that the only elements of finite order in $G_{\mathbf{Q}}$ are complex conjugation (for any imbedding of $\mathbf{Q}^a$ in $\mathbf{C}$), and all conjugates of complex conjugation in $G_{\mathbf{Q}}$ (*Abhandlung Math. Seminar Hamburg*, 1924).

The structure of $G_{\mathbf{Q}}$ is complicated. But I shall now develop some notions leading to a conjecture of Shafarevich which gives some idea of one possible formulation for part of the answer.

Let G be any group. Let $\mathscr{F}$ be the family of all subgroups of finite index. Let $\{x_n\}$ be a sequence in G. We say that $\{x_n\}$ is a **Cauchy sequence** if given a subgroup $H \in \mathscr{F}$ there exists n_0 such that for $m, n \geqq n_0$ we have $x_n x_m^{-1} \in H$. A sequence is said to be **null** if given $H \in \mathscr{F}$ there exists n_0 such that for all $n \geqq n_0$ we have $x_n \in H$. As an exercise (see Exercise 4 of Chapter VII, §7), show that the Cauchy sequences form a group, the null sequences form a normal subgroup. The factor group is called the **completion of** G with respect to the subgroups of finite index, and this completion is denoted by $\bar{G}$.

Let $X = \{x_1, x_2, \ldots\}$ be a denumerable sequence of symbols. It can be shown that there is a group G, called the **free group** on X, having the following property. Every element of G can be written in the form

$$x_{i_1}^{m_1} \cdots x_{i_r}^{m_r},$$

where $m_1, \ldots, m_r$ are integers, $x_{i_1}, \ldots, x_{i_r}$ are elements of the sequence X, and $i_j \neq i_{j+1}$ for $j = 1, \ldots, r-1$. Furthermore, such an element is equal to 1 if and only if $m_1 = \cdots = m_r = 0$. We also call G the **free group on a countable set of symbols.**

The following conjecture is due to Shafarevich, following work of Iwasawa (*Annals of Mathematics*, 1953) and Shafarevich (*Izvestia Akademia Nauk*, 1954).

Conjecture. *Let F_0 be the union of all fields $\mathbf{Q}(\zeta)$, where ζ ranges over all roots of unity. Let F be a finite extension of F_0. Then G_F is isomorphic to the completion $\bar{G}$, where G is the free group on a countable set of symbols.*

For an excellent historical account and other matters, see Matzat's paper, *Jahresbericht Deutsche Math. Vereinigung*, 1986–1987.

CHAPTER IX

The Real and Complex Numbers

IX, §1. ORDERING OF RINGS

Let R be an integral ring. By an **ordering** of R one means a subset P of R satisfying the following conditions:

ORD 1. *For every $x \in R$ we have $x \in P$, or $x = 0$, or $-x \in P$, and these three possibilities are mutually exclusive.*

ORD 2. *If $x, y \in P$ then $x + y \in P$ and $xy \in P$.*

We also say that R is **ordered** by P, and call P the set of **positive** elements.

Let us assume that R is ordered by P. Since $1 \neq 0$, and $1 = 1^2 = (-1)^2$ we see that 1 is an element of P, i.e. 1 is positive. By **ORD 2** and induction, it follows that $1 + \cdots + 1$ (sum taken n times) is positive. An element $x \in R$ such that $x \neq 0$ and $x \notin P$ is called **negative**. If x, y are negative elements of R, then xy is positive (because $-x \in P$, $-y \in P$, and hence $(-x)(-y) = xy \in P$). If x is positive and y is negative, then xy is negative, because $-y$ is positive, and hence $x(-y) = -xy$ is positive. For any $x \in R$, $x \neq 0$, we see that x^2 is positive.

Suppose that R is a field. If x is positive and $x \neq 0$ then $xx^{-1} = 1$, and hence by the preceding remarks, it follows that x^{-1} is also positive.

Let R be an arbitrary ordered integral ring again, and let R' be a subring. Let P be the set of positive elements in R, and let $P' = P \cap R'$. Then it is clear that P' defines an ordering on R', which is called the **induced ordering**.

More generally, let R' and R be ordered rings, and let P', P be their sets of positive elements respectively. Let $f\colon R' \to R$ be an embedding (i.e. an injective homomorphism). We shall say that f is **order-preserving** if for every $x \in R'$ such that $x \in P'$ we have $f(x) \in P$. This is equivalent to saying that $f^{-1}(P) = P'$ [where $f^{-1}(P)$ is the set of all $x \in R'$ such that $f(x) \in P$].

Let $x, y \in R$. We define $x < y$ (or $y > x$) to mean that $y - x \in P$. Thus to say that $x > 0$ is equivalent to saying that $x \in P$; and to say that $x < 0$ is equivalent to saying that x is negative, or $-x$ is positive. One verifies easily the usual relations for inequalities, namely for $x, y, z \in R$:

IN 1. $x < y$ *and* $y < z$ *implies* $x < z$.

IN 2. $x < y$ *and* $z > 0$ *implies* $xz < yz$.

IN 3. $x < y$ *implies* $x + z < y + z$.

If R is a field, then

IN 4. $x < y$ *and* $x, y > 0$ *implies* $1/y < 1/x$.

As an example, we shall prove **IN 2**. We have $y - x \in P$ and $z \in P$, so that by **ORD 2**, $(y - x)z \in P$. But $(y - x)z = yz - xz$, so that by definition, $xz < yz$. As another example, to prove **IN 4**, we multiply the inequality $x < y$ by x^{-1} and y^{-1} to find the assertion of **IN 4**. The others are left as exercises.

If $x, y \in R$ we define $x \leqq y$ to mean that $x < y$ or $x = y$. Then one verifies at once that **IN 1, 2, 3** hold if we replace throughout the $<$ sign by $\leqq$. Furthermore, one also verifies at once that if $x \leqq y$ and $y \leqq x$ then $x = y$.

In the next theorem, we see how an ordering on an integral ring can be extended to an ordering of its quotient field.

Theorem 1.1. *Let R be an integral ring, ordered by P. Let K be its quotient field. Let P_K be the set of elements of K which can be written in the form a/b with $a, b \in R$, $b > 0$ and $a > 0$. Then P_K defines an ordering on K extending P.*

Proof. Let $x \in K$, $x \neq 0$. Multiplying a numerator and denominator of x by -1 if necessary, we can write x in the form $x = a/b$ with $a, b \in R$ and $b > 0$. If $a > 0$ then $x \in P_K$. If $-a > 0$ then $-x = -a/b \in P_K$. We cannot have both x and $-x \in P_K$, for otherwise, we could write

$$x = a/b \quad \text{and} \quad -x = c/d$$

with $a, b, c, d \in R$ and $a, b, c, d > 0$. Then

$$-a/b = c/d,$$

whence $-ad = bc$. But $bc \in P$ and $ad \in P$, a contradiction. This proves that P_K satisfies **ORD 1**. Next, let $x, y \in P_K$ and write

$$x = a/b \qquad \text{and} \qquad y = c/d$$

with $a, b, c, d \in R$ and $a, b, c, d > 0$. Then $xy = ac/bd \in P_K$. Also

$$x + y = \frac{ad + bc}{bd}$$

lies in P_K. This proves that P_K satisfies **ORD 2**. If $a \in R$, $a > 0$, then $a = a/1 \in P_K$ so $P \subset P_K$. This proves our theorem.

Theorem 1.1 shows in particular how one extends the usual ordering on the ring of integers **Z** to the field of rational numbers **Q**. How one defines the integers, and the ordering on them, will be discussed in an appendix.

Let R be an ordered ring as before. If $x \in R$, we define

$$|x| = \begin{cases} x & \text{if } x \geqq 0, \\ -x & \text{if } x < 0. \end{cases}$$

We then have the following characterization of the function $x \mapsto |x|$, which is called the **absolute value**:

For every $x \in R$, $|x|$ is the unique element $z \in R$ such that $z \geqq 0$ and $z^2 = x^2$.

To prove this, observe first that certainly $|x|^2 = x^2$, and $|x| \geqq 0$ for all $x \in R$. On the other hand, given $a \in R$, $a > 0$ there exist at most two elements $z \in R$ such that $z^2 = a$ because the polynomial $t^2 - a$ has at most two roots. If $w^2 = a$ then $w \neq 0$ and $(-w)^2 = w^2 = a$ also. Hence there is at most one positive element $z \in R$ such that $z^2 = a$. This proves our assertion.

We define the symbol $\sqrt{a}$ for $a \geqq 0$ in R to be the element $z \geqq 0$ in R such that $z^2 = a$, if such z exists. Otherwise, $\sqrt{a}$ is not defined. It is now easy to see that if $a, b \geqq 0$ and $\sqrt{a}, \sqrt{b}$ exist, then $\sqrt{ab}$ exists and

$$\sqrt{ab} = \sqrt{a}\sqrt{b}.$$

Indeed, if $z, w \geqq 0$ and $z^2 = a$, $w^2 = b$, then $(zw)^2 = z^2w^2 = ab$. Thus we may express the definition of the absolute value by means of the expression $|x| = \sqrt{x^2}$.

The absolute value satisfies the following rules:

AV 1. *For all* $x \in R$, *we have* $|x| \geqq 0$, *and* $|x| > 0$ *if* $x \neq 0$.

AV 2. $|xy| = |x||y|$ *for all* $x, y \in R$.

AV 3. $|x + y| \leqq |x| + |y|$ *for all* $x, y \in R$.

The first one is obvious. As to **AV 2**, we have

$$|xy| = \sqrt{(xy)^2} = \sqrt{x^2y^2} = \sqrt{x^2}\sqrt{y^2} = |x||y|.$$

For **AV 3**, we have

$$\begin{aligned} |x + y|^2 = (x + y)^2 &= x^2 + xy + xy + y^2 \\ &\leqq |x|^2 + 2|xy| + |y|^2 \\ &= |x|^2 + 2|x||y| + |y|^2 \\ &= (|x| + |y|)^2. \end{aligned}$$

Taking square roots yields what we want. (We have used implicitly two properties of inequalities, cf. Exercise 1.)

IX, §1. EXERCISES

1. Let R be an ordered integral ring.
   (a) Prove that $x \leqq |x|$ for all $x \in R$.
   (b) If $a, b \geqq 0$ and $a \leqq b$, and if $\sqrt{a}, \sqrt{b}$ exist, show that $\sqrt{a} \leqq \sqrt{b}$.
2. Let K be an ordered field. Let P be the set of polynomials

   $$f(t) = a_n t^n + \cdots + a_0$$

   over K, with $a_n > 0$. Show that P defines an ordering on $K[t]$.
3. Let R be an ordered integral ring. If $x, y \in R$, prove that $|-x| = |x|$,

   $$|x - y| \geqq |x| - |y|,$$

   and also

   $$|x + y| \geqq |x| - |y|.$$

   Also prove that $|x| \leqq |x + y| + |y|$.
4. Let K be an ordered field and $f: \mathbf{Q} \to K$ the embedding of the rational numbers into K. Show that f is necessarily order preserving.

IX, §2. PRELIMINARIES

Let K be an ordered field. From Exercise 4 of the preceding section, we know that the embedding of $\mathbf{Q}$ in K is order preserving. We shall identify $\mathbf{Q}$ as a subfield of K.

We recall formally a definition. Let S be a set. A **sequence** of elements of S is simply a mapping

$$\mathbf{Z}^+ \to S$$

from the positive integers into S. One usually denotes a sequence with the notation

$$\{x_1, x_2, \ldots\}$$

or

$$\{x_n\}_{n \geqq 1}$$

or simply

$$\{x_n\}$$

if there is no danger of confusing this with the set consisting of the single element x_n.

A sequence $\{x_n\}$ in K is said to be a **Cauchy sequence** if given an element $\epsilon > 0$ in K, there exists a positive integer N such that for all integers m, $n \geqq N$ we have

$$|x_n - x_m| \leqq \epsilon.$$

(For simplicity, we agree to let N, n, m denote positive integers unless otherwise specified. We also agree that ϵ denotes elements of K.)

To avoid the use of excessively many symbols, we shall say that a certain statement S concerning positive integers holds for all **sufficiently large** integers if there exists N such that the statement $S(n)$ holds for all $n \geqq N$. It is clear that if $S_1, \ldots, S_r$ is a finite number of statements, each holding for all sufficiently large integers, then they are valid simultaneously for all sufficiently large integers. Indeed, if

$$S_1(n) \text{ is valid for } n \geqq N_1, \ldots, S_r(n) \text{ is valid for } n \geqq N_r,$$

we let N be the maximum of $N_1, \ldots, N_r$ and see that each $S_i(n)$ is valid for $n \geqq N$.

We shall say that a statement holds for **arbitrarily large** integers if given N, the statement holds for some $n \geqq N$.

A sequence $\{x_n\}$ in K is said to **converge** if there exists an element $x \in K$ such that, given $\epsilon > 0$ we have

$$|x - x_n| \leqq \epsilon$$

for all sufficiently large n.

An ordered field in which every Cauchy sequence converges is said to be **complete**. We observe that the number x above, if it exists, is uniquely determined, for if $y \in K$ is such that

$$|y - x_n| \leqq \epsilon$$

for all sufficiently large n, then

$$|x - y| \leqq |x - x_n + x_n - y| \leqq |x - x_n| + |x_n - y| \leqq 2\epsilon.$$

This is true for every $\epsilon > 0$ in K, and it follows that $x - y = 0$, that is $x = y$. We call this number x the **limit of the sequence** $\{x_n\}$.

An ordered field K will be said to be **archimedean** if given $x \in K$ there exists a positive integer n such that $x \leqq n$. It then follows that given $\epsilon > 0$ in K, we can find an integer $m > 0$ such that $1/\epsilon < m$, whence $1/m < \epsilon$.

It is easy to see that the field of rational numbers is not complete. For instance, one can construct Cauchy sequences of rationals whose square approaches 2, but such that the sequence has no limit in **Q** (otherwise, $\sqrt{2}$ would be rational). In the next section, we shall construct an archimedean complete field, which will be called the real numbers. Here, we prove one property of such fields, which is taken as the starting point of analysis.

Let S be a subset of K. By an **upper bound** for S one means an element $z \in K$ such that $x \leqq z$ for all $x \in S$. By a **least upper bound** of S one means an element $w \in K$ such that w is an upper bound, and such that, if z is an upper bound, then $w \leqq z$. If w_1, w_2 are least upper bounds of S, then $w_1 \leqq w_2$ and $w_2 \leqq w_1$ so that $w_1 = w_2$: A least upper bound is uniquely determined.

Theorem 2.1. *Let K be a complete archimedean ordered field. Then every non-empty subset S of K which has an upper bound also has a least upper bound.*

Proof. For each positive integer n we consider the set T_n consisting of all integers y such that for all $x \in S$, we have $nx \leqq y$ (and consequently, $x \leqq y/n$). Then T_n is bounded from below by any element nx (with $x \in S$), and is not empty because if b is an upper bound for S, then any integer y such that $nb \leqq y$ will be in T_n. (We use the archimedean property.)

Let y_n be the smallest element of T_n. Then there exists an element x_n of S such that

$$y_n - 1 < nx_n \leqq y_n$$

(otherwise, y_n is not the smallest element of T_n). Hence

$$\frac{y_n}{n} - \frac{1}{n} < x_n \leqq \frac{y_n}{n}.$$

Let $z_n = y_n/n$. We contend that the sequence $\{z_n\}$ is Cauchy. To prove this, let m, n be positive integers and say $y_n/n \leqq y_m/m$. We contend that

$$\frac{y_m}{m} - \frac{1}{m} < \frac{y_n}{n} \leqq \frac{y_m}{m}.$$

Otherwise

$$\frac{y_n}{n} \leqq \frac{y_m}{m} - \frac{1}{m}$$

and

$$\frac{y_m}{m} - \frac{1}{m}$$

is an upper bound for S, which is not true because x_m is bigger. This proves our contention, from which we see that

$$\left|\frac{y_n}{n} - \frac{y_m}{m}\right| \leqq \frac{1}{m}.$$

For m, n sufficiently large, this is arbitrarily small, and we have proved that our sequence $\{z_n\}$ is Cauchy.

Let w be its limit. We first prove that w is an upper bound for S. Suppose there exists $x \in S$ such that $w < x$. There exists an n such that

$$|z_n - w| \leqq \frac{x - w}{2}.$$

Then

$$\begin{aligned} x - z_n = x - w + w - z_n &\geqq x - w - |w - z_n| \\ &\geqq x - w - \frac{x - w}{2} \\ &\geqq \frac{x - w}{2} > 0, \end{aligned}$$

so $x > z_n$, contradicting the fact that z_n is an upper bound for S.

We now show that w is a least upper bound for S. Let $u < w$. There exists some n such that

$$|z_n - x_n| \leqq \frac{1}{n} < \frac{w-u}{4}.$$

(Just select n sufficiently large.) We can also select n sufficiently large so that

$$|z_n - w| \leqq \frac{w-u}{4}$$

since w is the limit of $\{z_n\}$. Now

$$\begin{aligned} x_n - u &= w - u + x_n - z_n + z_n - w \\ &\geqq w - u - |x_n - z_n| - |z_n - w| \\ &\geqq w - u - \frac{w-u}{4} - \frac{w-u}{4} \\ &\geqq \frac{w-u}{2} > 0, \end{aligned}$$

whence $u < x_n$. Hence u is not an upper bound. This proves that w is the least upper bound, and concludes the proof of the theorem.

IX, §3. CONSTRUCTION OF THE REAL NUMBERS

We start with the rational numbers $\mathbf{Q}$ and their ordering obtained from the ordering of the integers $\mathbf{Z}$ as in Theorem 1.1 of §1. We wish to define the real numbers. In elementary school, one uses the real numbers as infinite decimals, like

$$\sqrt{2} = 1.414\ldots.$$

Such an infinite decimal is nothing but a sequence of rational numbers, namely

$$1,\ 1.4,\ 1.41,\ 1.414,\ldots$$

and it should be noted that there exist other sequences which "approach" $\sqrt{2}$. If one wishes to *define* $\sqrt{2}$, it is then reasonable to take as definition an equivalence class of sequences of rational numbers, under a suitable concept of equivalence. We shall do this for all real numbers.

We start with our ordered field **Q** and Cauchy sequences in **Q**. Let $\gamma = \{c_n\}$ be a sequence of rational numbers. We say that γ is a **null sequence** if given a rational number $\epsilon > 0$ we have

$$|c_n| \leqq \epsilon$$

for all sufficiently large n. Unless otherwise specified we deal with rational ϵ in what follows, and our sequences are sequences of rational numbers.

If $\alpha = \{a_n\}$ and $\beta = \{b_n\}$ are sequences of rational numbers, we define $\alpha + \beta$ to be the sequence $\{a_n + b_n\}$, i.e. the sequence whose n-th term is $a_n + b_n$. We define the product $\alpha\beta$ to be the sequence whose n-th term is $a_n b_n$. Thus the set of sequences of rational numbers is nothing but the ring of all mappings of $\mathbf{Z}^+$ into **Q**. We shall see in a moment that the Cauchy sequences form a subring.

Lemma 3.1. *Let $\alpha = \{a_n\}$ be a Cauchy sequence. There exists a positive rational number B such that $|a_n| \leqq B$ for all n.*

Proof. Given 1 there exists N such that for all $n \geqq N$ we have

$$|a_n - a_N| \leqq 1.$$

Then for all $n \geqq N$,

$$|a_n| \leqq |a_N| + 1.$$

We let B be the maximum of $|a_1|, |a_2|, \ldots, |a_{N-1}|, |a_N| + 1$.

Lemma 3.2. *The Cauchy sequences form a commutative ring.*

Proof. Let $\alpha = \{a_n\}$ and $\beta = \{b_n\}$ be Cauchy sequences. Given $\epsilon > 0$, we have

$$|a_n - a_m| \leqq \frac{\epsilon}{2}$$

for all m, n sufficiently large, and also

$$|b_n - b_m| \leqq \frac{\epsilon}{2}$$

for all m, n sufficiently large. Hence for all m, n sufficiently large, we have

$$\begin{aligned} |a_n + b_n - (a_m + b_m)| &= |a_n - a_m + b_n - b_m| \\ &\leqq |a_n - a_m| + |b_n - b_m| \\ &\leqq \frac{\epsilon}{2} + \frac{\epsilon}{2} = \epsilon. \end{aligned}$$

Hence the sum $\alpha + \beta$ is a Cauchy sequence. One sees at once that

$$-\alpha = \{-a_n\}$$

is a Cauchy sequence. As for the product, we have

$$\begin{aligned}|a_nb_n - a_mb_m| &= |a_nb_n - a_nb_m + a_nb_m - a_mb_m| \\ &\leqq |a_n||b_n - b_m| + |a_n - a_m||b_m|.\end{aligned}$$

By Lemma 3.1, there exists $B_1 > 0$ such that $|a_n| \leqq B_1$ for all n, and $B_2 > 0$ such that $|b_n| \leqq B_2$ for all n. Let $B = \max(B_1, B_2)$. For all m, n sufficiently large, we have

$$|a_n - a_m| \leqq \frac{\epsilon}{2B} \quad \text{and} \quad |b_n - b_m| \leqq \frac{\epsilon}{2B},$$

and consequently

$$|a_nb_n - a_mb_m| \leqq \frac{\epsilon}{2} + \frac{\epsilon}{2} = \epsilon.$$

So the product $\alpha\beta$ is a Cauchy sequence. It is clear that the sequence $e = \{1, 1, 1, \ldots\}$ is a Cauchy sequence. Hence Cauchy sequences form a ring, and a subring of the ring of all mappings of $\mathbf{Z}^+$ into $\mathbf{Q}$. This ring is obviously commutative.

Lemma 3.3. *The null sequences form an ideal in the ring of Cauchy sequences.*

Proof. Let $\beta = \{b_n\}$ and $\gamma = \{c_n\}$ be null sequences. Given $\epsilon > 0$, for all n sufficiently large we have

$$|b_n| \leqq \frac{\epsilon}{2} \quad \text{and} \quad |c_n| \leqq \frac{\epsilon}{2}.$$

Hence for all n sufficiently large, we have

$$|b_n + c_n| \leqq \epsilon$$

so $\beta + \gamma$ is a null sequence. It is clear that $-\beta$ is a null sequence.

By Lemma 3.1, given a Cauchy sequence $\alpha = \{a_n\}$, there exists a rational number $B > 0$ such that $|a_n| \leqq B$ for all n. For all n sufficiently large, we have

$$|b_n| \leqq \frac{\epsilon}{B},$$

whence

$$|a_nb_n| \leqq B\frac{\epsilon}{B} = \epsilon,$$

so that $\alpha\beta$ is a null sequence. This proves that null sequences form an ideal, as desired.

Let R be the ring of Cauchy sequences and M the ideal of null sequences. We then have the notion of congruence, that is if α, $\beta \in R$, we had defined $\alpha \equiv \beta \pmod{M}$ to mean $\alpha - \beta \in M$, or in other words $\alpha = \beta + \gamma$ for some null sequence γ. We define a **real number** to be a congruence class of Cauchy sequences. As we know from constructing arbitrary factor rings, the set of such congruence classes is itself a ring, denoted by R/M, but which we shall also denote by $\mathbf{R}$. The congruence class of the sequence α will be denoted by $\bar{\alpha}$ for the moment. Then by definition,

$$\overline{\alpha + \beta} = \bar{\alpha} + \bar{\beta}, \qquad \overline{\alpha\beta} = \bar{\alpha}\bar{\beta}.$$

The unit element of $\mathbf{R}$ is the class of the Cauchy sequence $\{1, 1, 1, \ldots\}$.

Theorem 3.4. *The ring $R/M = \mathbf{R}$ of real numbers is in fact a field.*

Proof. We must prove that if α is a Cauchy sequence, and is not a null sequence, then there exists a Cauchy sequence β such that $\alpha\beta \equiv e \pmod{M}$, where $e = \{1, 1, 1, \ldots\}$. We need a lemma on null sequences.

Lemma 3.5. *Let α be a Cauchy sequence, and not a null sequence. Then there exist N_0 and a rational number $c > 0$ such that $|a_n| \geqq c$ for all $n \geqq N_0$.*

Proof. Suppose otherwise. Let $\alpha = \{a_n\}$. Then given $\epsilon > 0$, there exists an infinite sequence $n_1 < n_2 < \cdots$ of positive integers such that

$$|a_{n_i}| < \frac{\epsilon}{3}$$

for each $i = 1, 2, \ldots$. By definition, there exists N such that for m, $n \geqq N$ we have

$$|a_n - a_m| \leqq \frac{\epsilon}{3}.$$

Let $n_i \geqq N$. We have for $m \geqq N$,

$$|a_m| \leqq |a_m - a_{n_i}| + |a_{n_i}| \leqq \frac{2\epsilon}{3},$$

and for m, $n \geqq N$,

$$|a_n| \leqq |a_m| + \frac{\epsilon}{3} \leqq \epsilon.$$

This shows that α is a null sequence, contrary to hypothesis, and proves our lemma.

We return to the proof of the theorem. By Lemma 3.5, there exists N_0 such that for $n \geqq N_0$ we have $a_n \neq 0$. Let $\beta = \{b_n\}$ be the sequence such that $b_n = 1$ if $n < N_0$, and $b_n = a_n^{-1}$ if $n \geqq N_0$. Then $\beta\alpha$ differs from e only in a finite number of terms, and so $\beta\alpha - e$ is certainly a null sequence. There remains to prove that β is a Cauchy sequence. By Lemma 3.5, we can select N_0 such that for all $n \geqq N_0$ we have $|a_n| \geqq c > 0$. It follows that

$$\frac{1}{|a_n|} \leqq \frac{1}{c}.$$

Given $\epsilon > 0$, there exists N (which we can take $\geqq N_0$) such that for all $m, n \geqq N$ we have

$$|a_n - a_m| \leqq \epsilon c^2.$$

Then for $m, n \geqq N$ we get

$$\left|\frac{1}{a_n} - \frac{1}{a_m}\right| = \left|\frac{a_m - a_n}{a_m a_n}\right| \leqq \frac{\epsilon c^2}{c^2} = \epsilon,$$

thereby proving that β is a Cauchy sequence, and concluding the proof of our theorem.

We have constructed the field of real numbers.

Observe that we have a natural ring-homomorphism of **Q** into **R**, obtained by mapping each rational number a on the class of the Cauchy sequence $\{a, a, a, \ldots\}$. This is a composite of two homomorphisms, first the map

$$a \mapsto \{a, a, a, \ldots\}$$

of **Q** into the ring of Cauchy sequences, followed by the map

$$R \to R/M.$$

Since this is not the zero homomorphism, it follows that it is an isomorphism of **Q** onto its image.

The next lemma is designed for the purpose of defining an ordering on the real numbers.

Lemma 3.6. *Let $\alpha = \{a_n\}$ be a Cauchy sequence. Exactly one of the following possibilities holds:*

(1) *α is a null sequence.*

(2) *There exists a rational* $c > 0$ *such that for all* n *sufficiently large,* $a_n \geqq c$.

(3) *There exists a rational* $c < 0$ *such that for all* n *sufficiently large,* $a_n \leqq c$.

Proof. It is clear that if α satisfies one of the three possibilities, then it cannot satisfy any other, i.e. the possibilities are mutually exclusive. What we must show is that at least one of the possibilities holds. Suppose that α is not a null sequence. By Lemma 3.5 there exists N_0 and a rational number $c > 0$ such that $|a_n| \geqq c$ for all $n \geqq N_0$. Thus $a_n \geqq c$ if a_n is positive, and $-a_n \geqq c$ if a_n is negative. Suppose that there exist arbitrarily large integers n such that a_n is positive, and arbitrarily large integers m such that a_m is negative. Then for such m, n we have

$$a_n - a_m \geqq 2c > 0,$$

thereby contradicting the fact that α is a Cauchy sequence. This proves that (2) or (3) must hold, and concludes the proof of the lemma.

Lemma 3.7. *Let* $\alpha = \{a_n\}$ *be a Cauchy sequence and let* $\beta = \{b_n\}$ *be a null sequence. If* α *satisfies property* (2) *of Lemma* 3.6 *then* $\alpha + \beta$ *also satisfies this property, and if* α *satisfies property* (3) *of Lemma* 3.6, *then* $\alpha + \beta$ *also satisfies property* (3).

Proof. Suppose that α satisfies property (2). For all n sufficiently large, by definition of a null sequence, we have $|b_n| \leqq c/2$. Hence for sufficiently large n,

$$a_n + b_n \geqq |a_n| - |b_n| \geqq c/2.$$

A similar argument proves the analogue for property (3). This proves the lemma.

We may now define an ordering on the real numbers. We let P be the set of real numbers which can be represented by a Cauchy sequence α having property (2), and prove that P defines an ordering.

Let α be a Cauchy sequence representing a real number. If α is not null and does not satisfy (2), then $-\alpha$ obviously satisfies (2). By Lemma 3.7, every Cauchy sequence representing the same real number as α also satisfies (2). Hence P satisfies condition **ORD 1**.

Let $\alpha = \{a_n\}$ and $\beta = \{b_n\}$ be Cauchy sequences representing real numbers in P, and satisfying (2). There exists $c_1 > 0$ such that $a_n \geqq c_1$ for all sufficiently large n, and there exists $c_2 > 0$ such that $b_n \geqq c_2$ for all sufficiently large n. Hence $a_n + b_n \geqq c_1 + c_2 > 0$ for sufficiently large n, thereby proving that $\alpha + \beta$ is also in P. Furthermore,

$$a_n b_n \geqq c_1 c_2 > 0$$

for all sufficiently large n, so that $\overline{\alpha\beta}$ is in P. This proves that P defines an ordering on the real numbers.

We recall that we had obtained an isomorphism of **Q** onto a subfield of **R**, given by the map

$$a \mapsto \overline{\{a, a, a, \ldots\}}.$$

In view of Exercise 4, §1 this map is order preserving, but this is also easily seen directly from our definitions. For a while, we shall not identify a with its image in **R**, and we denote by $\bar{a}$ the class of the Cauchy sequence $\{a, a, a, \ldots\}$.

Theorem 3.8. *The ordering of* **R** *is archimedean.*

Proof. Let $\bar{\alpha}$ be a real number, represented by a Cauchy sequence $\alpha = \{a_n\}$. By Lemma 3.1, we can find a rational number r such that $a_n \leqq r$ for all n, and multiplying r by a positive denominator, we see that there exists an integer b such that $a_n \leqq b$ for all n. Then $\bar{b} - \bar{\alpha}$ is represented by the sequence $\{b - a_n\}$ and $b - a_n \geqq 0$ for all n. By definition, it follows that

$$\bar{b} - \bar{\alpha} \geqq 0,$$

whence $\bar{\alpha} \leqq \bar{b}$, as desired.

The following lemma gives us a criterion for inequalities between real numbers in terms of Cauchy sequences.

Lemma 3.9. *Let $\gamma = \{c_n\}$ be a Cauchy sequence of rational numbers, and let c be a rational number > 0. If $|c_n| \leqq c$ for all n sufficiently large, then $|\bar{\gamma}| \leqq \bar{c}$.*

Proof. If $\bar{\gamma} = 0$, our assertion is trivial. Suppose $\bar{\gamma} \neq 0$, and say $\bar{\gamma} > 0$. Then $|\bar{\gamma}| = \bar{\gamma}$, and thus we must show that $\bar{c} - \bar{\gamma} \geqq 0$. But for all n sufficiently large, we have

$$c - c_n \geqq 0.$$

Since $\bar{c} - \bar{\gamma} = \overline{\{c - c_n\}}$, it follows from our definition of the ordering in **R** that $\bar{c} - \bar{\gamma} \geqq 0$. The case when $\bar{\gamma} < 0$ is proved by considering $-\bar{\gamma}$.

Given a real number $\epsilon > 0$, by Theorem 3.8 there exists a rational number $\epsilon_1 > 0$ such that $0 < \overline{\epsilon_1} < \epsilon$. Hence in the definition of limit, when we are given the ϵ, it does not matter whether we take it real or rational.

Lemma 3.10. *Let $\alpha = \{a_n\}$ be a Cauchy sequence of rational numbers. Then $\bar{\alpha}$ is the limit of the sequence $\{\bar{a}_n\}$.*

Proof. Given a *rational* number $\epsilon > 0$, there exists N such that, for m, $n \geqq N$ we have

$$|a_n - a_m| \leqq \epsilon.$$

Then for all $m \geqq N$ we have by Lemma 3.9, for all $n \geqq N$,

$$|\bar{\alpha} - \bar{a}_m| = |\overline{\{a_n - a_m\}}| \leqq \bar{\epsilon}.$$

This proves our assertion.

Theorem 3.11. *The field of real numbers is complete.*

Proof. Let $\{A_n\}$ be a Cauchy sequence of real numbers. For each n, by Lemma 3.10, we can find a rational number a_n such that

$$|A_n - \bar{a}_n| \leqq \frac{1}{n}.$$

(Strictly speaking, we still should write $1/\bar{n}$ on the right-hand side!) Furthermore, by definition, given $\epsilon > 0$ there exists N such that for all m, $n \geqq N$ we have

$$|A_n - A_m| \leqq \frac{\epsilon}{3}.$$

Let N_1 be an integer $\geqq N$, and such that $1/N_1 \leq \epsilon/3$. Then for all m, $n \geqq N_1$ we get

$$\begin{aligned}
|\bar{a}_n - \bar{a}_m| &= |\bar{a}_n - A_n + A_n - A_m + A_m - \bar{a}_m| \\
&\leqq |\bar{a}_n - A_n| + |A_n - A_m| + |A_m - \bar{a}_m| \\
&\leqq \frac{\epsilon}{3} + \frac{\epsilon}{3} + \frac{\epsilon}{3} = \epsilon.
\end{aligned}$$

This proves that $\{\bar{a}_n\}$ is a Cauchy sequence of rational numbers. Let A be its limit. For all n, we have

$$|A_n - A| \leqq |A_n - \bar{a}_n| + |\bar{a}_n - A|.$$

If we take n sufficiently large, we see that A is also the limit of the sequence $\{A_n\}$, thereby proving our theorem.

The procedure we have followed to construct a complete field from the rational numbers can be generalized in many contexts, and occurs often in mathematics. The exercises will show you some number-theoretic contexts, as in the construction of p-adic fields for prime numbers p. But in analysis, the construction is also applied to vector spaces, not necessarily fields. For instance, let V be the real vector space of all continuous functions on $\mathbf{R}$, vanishing outside some bounded interval. On V we can define a norm as follows. Let $f \in V$. We define

$$\|f\|_1 = \int_{-\infty}^{\infty} |f(x)|\, dx.$$

This norm satisfies properties analogous to **AV 1**, **AV 2** and **AV 3**, namely:

N 1. *Let* $f \in V$. *Then* $\|f\|_1 \geqq 0$, *and* $= 0$ *if and only if* $f = 0$.

N 2. *If* $c \in \mathbf{R}$ *and* $f \in V$, *then* $\|cf\|_1 = |c|\,\|f\|_1$.

N 3. *If* $f, g \in \mathbf{R}$, *then* $\|f + g\|_1 \leqq \|f\|_1 + \|g\|_1$.

One can then define Cauchy sequences, null sequences, and one can construct the factor space of Cauchy sequences modulo null sequences, to obtain a vector space $\bar{V}$. The norm can be extended to $\bar{V}$, and one can then prove the theorem that $\bar{V}$ is complete. In analysis, one analyzes this completion, which is in some sense the largest vector space of functions whose absolute value is integrable over $\mathbf{R}$.

In the above context of vector spaces, there is no question of ordering the elements of V, so the whole part of the construction of the real numbers having to do with ordering simply drops out of consideration. Similarly, in the exercises having to do with completions, there will be no ordering properties. The completions will be constructed only with the ring of Cauchy sequences and the maximal ideal of null sequences.

IX, §3. EXERCISES

1. Prove that every positive real number has a positive square root in $\mathbf{R}$. Since the polynomial $t^2 - a$ has at most two roots in a field, and since for any root α, the number $-\alpha$ is also a root, it follows that for every $a \in \mathbf{R}$, $a \geqq 0$, there exists a unique $\alpha \in \mathbf{R}$, $\alpha \geqq 0$ such that $\alpha^2 = a$. [*Hint*: For the above proof, let α be the least upper bound of the set of rational numbers b such that $b^2 \leqq a$.]

2. Show that every automorphism of the real numbers is the identity. [*Hint*: Show first that an automorphism is order preserving.]

3. Let p be a prime number. If x is a non-zero rational number, written in the form $x = p^r a/b$ where r is an integer, a, b are integers not divisible by p, we define

$$|x|_p = 1/p^r.$$

Define $|0|_p = 0$. Show that for all rational x, y we have

$$|xy|_p = |x|_p |y|_p \quad \text{and} \quad |x + y|_p \leqq |x|_p + |y|_p.$$

In fact, prove the stronger property

$$|x + y|_p \leqq \max(|x|_p, |y|_p).$$

4. Let F be a field. By an **absolute value** on F we mean a real-valued function $x \mapsto |x|$ satisfying the following properties:

AV 1. *We have* $|x| \geqq 0$, *and* $= 0$ *if and only if* $x = 0$.
AV 2. *For all* $x, y \in F$ *we have* $|xy| = |x||y|$.
AV 3. $|x + y| \leqq |x| + |y|$.

(a) Define the notion of **Cauchy sequence** and **null sequence** with respect to an absolute value v on a field F. Define what it means for a field to be **complete** with respect to v.
(b) For an absolute value as above, prove that the Cauchy sequences form a ring, the null sequences form a maximal ideal, and the residue class ring is a field. Show that the absolute value can be extended to this field, and that this field is complete.
(c) Let $F \subset E$ be a subfield, and suppose E has an absolute value extending an absolute value on F. We define F to be **dense** in E if given $\varepsilon > 0$, and an element $\alpha \in E$, there exists $a \in F$ such that $|\alpha - a| < \varepsilon$. Prove:

There exists a field E which contains F as a subfield, such that E has an absolute value extending the absolute value on F, such that F is dense in E, and E is complete.

Such a field E is called a **completion** of F.
(d) Prove the uniqueness of a completion, in the following sense. Let E, E' be completions of F. Then there exists an isomorphism

$$\sigma: E \to E'$$

which restricts to the identity on F and such that σ preserves the absolute value, that is for all $\alpha \in E$ we have

$$|\sigma\alpha| = |\alpha|.$$

The completion of a field F with respect to an absolute value v is usually denoted by F_v.

5. An absolute value is called **non-archimedean** if instead of **AV 3** it satisfies the stronger property

$$|x + y| \leqq \max(|x|, |y|).$$

The function $| \ |_p$ on **Q** is called the ***p*-adic absolute value**, and is non-archimedean. Suppose $| \ |$ is a non-archimedean absolute value on a field F. Prove that given $x \in F$, $x \neq 0$, there exists a positive number r such that if $|y - x| < r$ then $|y| = |x|$.

6. Let $| \ |$ be a non-archimedean absolute value on a field F. Let R be the subset of elements $x \in F$ such that $|x| \leqq 1$.
   (a) Show that R is a ring, and that for every $x \in F$ we have $x \in R$ or $x^{-1} \in R$.
   (b) Let M be the subset of elements $x \in R$ such that $|x| < 1$. Show that M is a maximal ideal.

IX, §4. DECIMAL EXPANSIONS

Theorem 4.1. *Let d be an integer $\geqq 2$, and let m be an integer $\geqq 0$. Then m can be written in a unique way in the form*

$$(1) \qquad m = a_0 + a_1 d + \cdots + a_n d^n$$

with integers a_i such that $0 \leqq a_i < d$.

Proof. This is easily seen from the Euclidean algorithm, and we shall give the proof. For the existence, if $m < d$, we let $a_0 = m$ and $a_i = 0$ for $i > 0$. If $m \geqq d$ we write

$$m = qd + a_0$$

with $0 \leqq a_0 < d$, using the Euclidean algorithm. Then $q < m$, and by induction, there exist integers a_i ($0 \leqq a_i < d$ and $i \geqq 1$) such that

$$q = a_1 + a_2 d + \cdots + a_k d^k.$$

Substituting this value for q yields what we want. As for uniqueness, suppose that

$$(2) \qquad m = b_0 + b_1 d + \cdots + b_n d^n$$

with integers b_i satisfying $0 \leqq b_i < d$. (We can use the same n simply by adding terms with coefficients $b_i = 0$ or $a_i = 0$ if necessary.) Say $a_0 \leqq b_0$. Then $b_0 - a_0 \geqq 0$, and $b_0 - a_0 < d$. On the other hand, $b_0 - a_0 = de$ for some integer e [as one sees subtracting (2) from (1)]. Hence $b_0 - a_0 = 0$

and $b_0 = a_0$. Assume by induction that we have shown $a_i = b_i$ for $0 \leqq i \leqq s$ and some $s < n$. Then

$$a_{s+1}d^{s+1} + \cdots + a_n d^n = b_{s+1}d^{s+1} + \cdots + b_n d^n.$$

Dividing both sides by d^{s+1}, we obtain

$$a_{s+1} + \cdots + a_n d^{n-s-1} = b_{s+1} + \cdots + b_n d^{n-s-1}.$$

By what we have just seen, it follows that $a_{s+1} = b_{s+1}$, and thus we have proved uniqueness by induction, as desired.

Let x be a positive real number, and d an integer $\geqq 2$. Then x has a unique expression of the form

$$x = m + \alpha,$$

where $0 \leqq \alpha < 1$. Indeed, we let m be the largest integer $\leqq x$. Then $x < m + 1$, and hence $0 \leqq x - m < 1$. We shall now describe a d-decimal expansion for real numbers between 0 and 1.

Theorem 4.2. *Let x be a real number, $0 \leqq x < 1$. Let d be an integer $\geqq 2$. For each positive integer n there is a unique expression*

$$x = \frac{a_1}{d} + \frac{a_2}{d^2} + \cdots + \frac{a_n}{d^n} + \alpha_n \tag{3}$$

with integers a_i satisfying $0 \leqq a_i < d$ and $0 \leqq \alpha_n < 1/d^n$.

Proof. Let m be the largest integer $\leqq d^n x$. Then $m \geqq 0$ and

$$d^n x = m + \alpha_n$$

with some number α_n such that $0 \leqq \alpha_n < 1$. We apply Theorem 4.1 to m, and then divide by d^n to obtain the desired expression. Conversely, given such an expression (3), we multiply it by d^n and apply the uniqueness part of Theorem 4.1 to obtain the uniqueness of (3). This proves our theorem.

When $d = 10$, the numbers $a_1, a_2, \ldots$ in Theorem 4.2 are precisely those of the decimal expansion of x, which is written

$$x = 0.a_1 a_2 a_3 \ldots$$

since time immemorial.

Conversely:

Theorem 4.3. *Let d be an integer $\geqq 2$. Let $a_1, a_2, \ldots$ be a sequence of integers, $0 \leqq a_i < d$ for all i, and assume that given a positive integer N there exists some $n \geqq N$ such that $a_n \neq d - 1$. Then there exists a real number x such that for each $n \geqq 1$ we have*

$$x = \frac{a_1}{d} + \frac{a_2}{d^2} + \cdots + \frac{a_n}{d^n} + \alpha_n,$$

where α_n is a number with $0 \leqq \alpha_n < 1/d^n$.

Proof. We shall use freely some simple properties of limits and infinite sums, treated in any standard beginning course of analysis. Let

$$y_n = \frac{a_1}{d} + \cdots + \frac{a_n}{d^n}.$$

Then the sequence $y_1, y_2, \ldots$ is increasing, and easily shown to be bounded from above. Let x be its least upper bound. Then x is a limit of the sequence, and

$$x = y_n + \alpha_n,$$

where

$$\alpha_n = \sum_{\nu=n+1}^{\infty} \frac{a_\nu}{d^\nu}.$$

Let

$$\beta_n = \sum_{\nu=n+1}^{\infty} \frac{d-1}{d^\nu}.$$

By hypothesis, we have $\alpha_n < \beta_n$ because there is some a_ν with $\nu \geqq n + 1$ such that $a_\nu \neq d - 1$. On the other hand,

$$\beta_n = \frac{d-1}{d^{n+1}} \sum_{\nu=0}^{\infty} \frac{1}{d^\nu} = \frac{d-1}{d^{n+1}} \frac{1}{1 - \frac{1}{d}} = \frac{1}{d^n}.$$

Hence $0 \leqq \alpha_n < 1/d^n$, as was to be proved.

Corollary 4.4. *The real numbers are not denumerable.*

Proof. Consider the subset of real numbers consisting of all decimal sequences

$$0.a_1a_2\ldots$$

with $1 \leqq a_i \leqq 8$, taking $d = 10$ in Theorems 4.2 and 4.3. It will suffice to prove that this subset is not denumerable. Suppose it is, and let

$$\alpha_1 = 0.a_{11}a_{12}a_{13}\ldots,$$
$$\alpha_2 = 0.a_{21}a_{22}a_{23}\ldots,$$
$$\alpha_3 = 0.a_{31}a_{32}a_{33}\ldots,$$
$$\ldots$$

be an enumeration of this subset. For each positive integer n, let b_n be an integer with $1 \leqq b_n \leqq 8$ such that $b_n \neq a_{nn}$. Let

$$\beta = 0.b_1b_2b_3\ldots b_n\ldots.$$

Then β is not equal to α_n for all n. This proves that there cannot be an enumeration of the real numbers. (*Note*: The simple facts concerning the terminology of denumerable sets used in this proof will be dealt with systematically in the next chapter.)

IX, §5. THE COMPLEX NUMBERS

Our purpose in this section is to identify the real numbers with a subfield of some field in which the equation $t^2 = -1$ has a root. As is usual in these matters, we define the bigger field in a way designed to make this equation obvious, and must then prove all desired properties.

We define a **complex number** to be a pair (x, y) of real numbers. We define addition componentwise. If $z = (x, y)$ we define multiplication of z by a real number a to be

$$az = (ax, ay).$$

Thus the set of complex numbers, denoted by $\mathbf{C}$, is nothing so far but $\mathbf{R}^2$, and can be viewed already as a vector space over $\mathbf{R}$. We let $e = (1, 0)$ and $i = (0, 1)$. Then every complex number can be expressed in a unique way as a sum $xe + yi$ with $x, y \in \mathbf{R}$. We must still define the multiplication of complex numbers. If $z = xe + yi$ and $w = ue + vi$ are complex numbers with $x, y, u, v \in \mathbf{R}$ we **define**

$$zw = (xu - yv)e + (xv + yu)i.$$

Observe at once that $ez = ze = z$ for all $z \in \mathbf{C}$, and $i^2 = -e$. We now contend that $\mathbf{C}$ is a field. We already know that it is an additive (abelian) group. If $z_1 = x_1e + y_1i$, $z_2 = x_2e + y_2i$, and $z_3 = x_3e + y_3i$, then

$$\begin{aligned}(z_1z_2)z_3 &= ((x_1x_2 - y_1y_2)e + (y_1x_2 + x_1y_2)i)(x_3e + y_3i)\\ &= (x_1x_2x_3 - y_1y_2x_3 - y_1x_2y_3 - x_1y_2y_3)e\\ &\quad + (y_1x_2x_3 + x_1y_2x_3 + x_1x_2y_3 - y_1y_2y_3)i.\end{aligned}$$

A similar computation of $z_1(z_2z_3)$ shows that one gets the same value as for $(z_1z_2)z_3$. Furthermore, letting $w = ue + vi$ again, we have

$$\begin{aligned}w(z_1 + z_2) &= (ue + vi)((x_1 + x_2)e + (y_1 + y_2)i)\\ &= (u(x_1 + x_2) - v(y_1 + y_2))e + (v(x_1 + x_2) + u(y_1 + y_2))i\\ &= (ux_1 - vy_1 + ux_2 - vy_2)e + (vx_1 + uy_1 + vx_2 + uy_2)i.\end{aligned}$$

Computing $wz_1 + wz_2$ directly shows that one gets the same thing as $w(z_1 + z_2)$. We also have obviously $wz = zw$ for all $w, z \in \mathbf{C}$, and hence $(z_1 + z_2)w = z_1w + z_2w$. This proves that the complex numbers form a commutative ring.

The map $x \mapsto (x, 0)$ is immediately verified to be an injective homomorphism of $\mathbf{R}$ into $\mathbf{C}$, and from now on, we identify $\mathbf{R}$ with its image in $\mathbf{C}$, that is we write x instead of xe for $x \in \mathbf{R}$.

If $z = x + iy$ is a complex number, we define its **complex conjugate**

$$\bar{z} = x - iy.$$

Then from our multiplication rule, we see that

$$z\bar{z} = x^2 + y^2.$$

If $z \neq 0$, then at least one of the real numbers x or y is not equal to 0, and one sees that

$$\lambda = \frac{\bar{z}}{x^2 + y^2}$$

is such that $z\lambda = \lambda z = 1$, because

$$z\frac{\bar{z}}{x^2 + y^2} = \frac{z\bar{z}}{x^2 + y^2} = 1.$$

Hence every non-zero element of $\mathbf{C}$ has an inverse, and consequently $\mathbf{C}$ is a field, which contains $\mathbf{R}$ as a subfield [taking into account our identification of x with $(x, 0)$].

We define the **absolute value** of a complex number $z = x + iy$ to be

$$|z| = \sqrt{x^2 + y^2}$$

and in terms of the absolute value, we can write the inverse of a non-zero complex number z in the form

$$z^{-1} = \frac{\bar{z}}{|z|^2}.$$

If z, w are complex numbers, it is easily shown that

$$|z + w| \leqq |z| + |w| \qquad \text{and} \qquad |zw| = |z||w|.$$

Furthermore, $\overline{z + w} = \bar{z} + \bar{w}$ and $\overline{zw} = \bar{z}\bar{w}$. We leave these properties as exercises. We have thus brought the theory of complex numbers to the point where the analysts take over.

In particular, using the exponential function, one proves that every positive real number r has a real n-th root, and that in fact any complex number w has an n-th root, for any positive integer n. This is done by using the polar form,

$$w = re^{i\theta}$$

with real θ, in which case $r^{1/n}e^{i\theta/n}$ is such an n-th root.

Aside from this fact, we shall use that a continuous real-valued function on a closed, bounded set of complex numbers has a maximum. All of this is proved in elementary courses in analysis.

Using these facts, we shall now prove that:

Theorem 5.1. *The complex numbers are algebraically closed, in other words, every polynomial* $f \in \mathbf{C}[t]$ *of degree* $\geqq 1$ *has a root in* $\mathbf{C}$.

We may write

$$f(t) = a_n t^n + a_{n-1}t^{n-1} + \cdots + a_0$$

with $a_n \neq 0$. For every real $R > 0$, the function $|f|$ such that

$$t \mapsto |f(t)|$$

is continuous on the closed disc of radius R, and hence has a minimum value on this disc. On the other hand, from the expression

$$f(t) = a_n t^n\left(1 + \frac{a_{n-1}}{a_n t} + \cdots + \frac{a_0}{a_n t^n}\right)$$

we see that when $|t|$ becomes large, then $|f(t)|$ also becomes large, i.e. given $C > 0$ there exists $R > 0$ such that if $|t| > R$ then $|f(t)| > C$. Consequently, there exists a positive number R_0 such that, if z_0 is a minimum point of $|f|$ on the closed disc of radius R_0, then

$$|f(t)| \geqq |f(z_0)|$$

for all complex numbers t. In other words, z_0 is an absolute minimum for $|f|$. We shall prove that $f(z_0) = 0$.

We express f in the form

$$f(t) = c_0 + c_1(t - z_0) + \cdots + c_n(t - z_0)^n$$

with constants c_i. If $f(z_0) \neq 0$, then $c_0 = f(z_0) \neq 0$. Let $z = t - z_0$, and let m be the smallest integer > 0 such that $c_m \neq 0$. This integer m exists because f is assumed to have degree $\geqq 1$. Then we can write

$$f(t) = f_1(z) = c_0 + c_m z^m + z^{m+1} g(z)$$

for some polynomial g, and some polynomial f_1 (obtained from f by changing the variable). Let z_1 be a complex number such that

$$z_1^m = -c_0/c_m,$$

and consider values of z of type

$$z = \lambda z_1,$$

where λ is real, $0 \leqq \lambda \leqq 1$. We have

$$\begin{aligned} f(t) = f_1(\lambda z_1) &= c_0 - \lambda^m c_0 + \lambda^{m+1} z_1^{m+1} g(\lambda z_1) \\ &= c_0[1 - \lambda^m + \lambda^{m+1} z_1^{m+1} c_0^{-1} g(\lambda z_1)]. \end{aligned}$$

There exists a number $C > 0$ such that for all λ with $0 \leqq \lambda \leqq 1$ we have $|z_1^{m+1} c_0^{-1} g(\lambda z_1)| \leqq C$, and hence

$$|f_1(\lambda z_1)| \leqq |c_0|(1 - \lambda^m + C\lambda^{m+1}).$$

If we can now prove that for sufficiently small λ with $0 < \lambda < 1$ we have

$$0 < 1 - \lambda^m + C\lambda^{m+1} < 1,$$

then for such λ we get $|f_1(\lambda z_1)| < |c_0|$, thereby contradicting the hypothesis that $|f(z_0)| \leqq |f(t)|$ for all complex numbers t. The left inequality is of course obvious since $0 < \lambda < 1$. The right inequality amounts to $C\lambda^{m+1} < \lambda^m$, or equivalently $C\lambda < 1$, which is certainly satisfied for sufficiently small λ. This concludes the proof.

Remark. The idea of the proof is quite simple. We have our polynomial

$$f_1(z) = c_0 + c_m z^m + z^{m+1}g(z),$$

and $c_m \neq 0$. If $g = 0$, we simply adjust $c_m z^m$ so as to subtract a term in the same direction as c_0, to shrink it towards the origin. This is done by extracting the suitable m-th root as above. Since $g \neq 0$ in general, we have to do a slight amount of analytic juggling to show that the third term is very small compared to $c_m z^m$, and that it does not disturb the general idea of the proof in an essential way.

IX, §5. EXERCISES

1. Assuming the result just proved about the complex numbers, prove that every irreducible polynomial over the real numbers has degree 1 or 2. [*Hint*: Split the polynomial over the complex numbers and pair off complex conjugate roots.]

2. Prove that an irreducible polynomial of degree 2 over **R**, with leading coefficient 1, can be written in the form

$$(t - a)^2 + b^2$$

with $a, b \in \mathbf{R}$, $b > 0$.

CHAPTER X

Sets

X, §1. MORE TERMINOLOGY

This chapter is the most abstract of the book, and is the one dealing with objects having the least structure, namely just sets. The remarkable thing is that interesting facts can be proved with so little at hand.

We shall first define some terminology. Let S and I be sets. By a **family of elements of** S, **indexed by** I, one means simply a map $f: I \to S$. However, when we speak of a family, we write $f(i)$ as f_i, and also use the notation $\{f_i\}_{i \in I}$ to denote the family.

Example 1. Let S be the set consisting of the single element 3. Let $I = \{1, \ldots, n\}$ be the set of integers from 1 to n. A family of elements of S, indexed by I, can then be written $\{a_i\}_{i=1,\ldots,n}$ with each $a_i = 3$. Note that a family is different from a subset. The same element of S may receive distinct indices.

A family of elements of a set S indexed by positive integers, or non-negative integers, is also called a **sequence**.

Example 2. A sequence of real numbers is written frequently in the form

$$\{x_1, x_2, \ldots\} \quad \text{or} \quad \{x_n\}_{n \geqq 1}$$

and stands for the map $f: \mathbf{Z}^+ \to \mathbf{R}$ such that $f(i) = x_i$. As before, note that a sequence can have all its elements equal to each other, that is

$$\{1, 1, 1, \ldots\}$$

is a sequence of integers, with $x_i = 1$ for each $i \in \mathbf{Z}^+$.

We define a **family of sets indexed by a set** I in the same manner, that is, a family of sets indexed by I is an assignment

$$i \mapsto S_i$$

which to each $i \in I$ associates a set S_i. The sets S_i may or may not have elements in common, and it is conceivable that they may all be equal. As before, we write the family $\{S_i\}_{i \in I}$.

We can define the intersection and union of families of sets, just as for the intersection and union of a finite number of sets. Thus, if $\{S_i\}_{i \in I}$ is a family of sets, we define the **intersection** of this family to be the set

$$\bigcap_{i \in I} S_i$$

consisting of all elements x which lie in all S_i. We define the **union**

$$\bigcup_{i \in I} S_i$$

to be the set consisting of all x such that x lies in some S_i.

If S, S' are sets, we define $S \times S'$ to be the set of all pairs (x, y) with $x \in S$ and $y \in S'$. We can define finite products in a similar way. If $S_1, S_2, \ldots$ is a sequence of sets, we define the product

$$\prod_{i=1}^{\infty} S_i$$

to be the set of all sequences $(x_1, x_2, \ldots)$ with $x_i \in S_i$. Similarly, if I is an indexing set, and $\{S_i\}_{i \in I}$ a family of sets, we define the product

$$\prod_{i \in I} S_i$$

to be the set of all families $\{x_i\}_{i \in I}$ with $x_i \in S_i$.

Let X, Y, Z be sets. We have the formula

$$(X \cup Y) \times Z = (X \times Z) \cup (Y \times Z).$$

To prove this, let $(w, z) \in (X \cup Y) \times Z$ with $w \in X \cup Y$ and $z \in Z$. Then $w \in X$ or $w \in Y$. Say $w \in X$. Then $(w, z) \in X \times Z$. Thus

$$(X \cup Y) \times Z \subset (X \times Z) \cup (Y \times Z).$$

Conversely, $X \times Z$ is contained in $(X \cup Y) \times Z$ and so is $Y \times Z$. Hence their union is contained in $(X \cup Y) \times Z$, thereby proving our assertion.

We say that two sets X, Y are **disjoint** if their intersection is empty. We say that a union $X \cup Y$ is **disjoint** if X and Y are disjoint. Note that if X, Y are disjoint, then $(X \times Z)$ and $(Y \times Z)$ are disjoint.

We can take products with unions of arbitrary families. For instance, if $\{X_i\}_{i \in I}$ is a family of sets, then

$$\left(\bigcup_{i \in I} X_i\right) \times Z = \bigcup_{i \in I} (X_i \times Z).$$

If the family $\{X_i\}_{i \in I}$ is disjoint (that is $X_i \cap X_j$ is empty if $i \neq j$ for $i, j \in I$), then the sets $X_i \times Z$ are also disjoint.

We have similar formulas for intersections. For instance,

$$(X \cap Y) \times Z = (X \times Z) \cap (Y \times Z).$$

We leave the proof to the reader.

Let X be a set and Y a subset. The **complement** of Y in X, denoted by $\mathscr{C}_X Y$, or $X - Y$, is the set of all elements $x \in X$ such that $x \notin Y$. If Y, Z are subsets of X, then we have the following formulas:

$$\mathscr{C}_X(Y \cup Z) = \mathscr{C}_X Y \cap \mathscr{C}_X Z,$$

$$\mathscr{C}_X(Y \cap Z) = \mathscr{C}_X Y \cup \mathscr{C}_X Z.$$

These are essentially reformulations of definitions. For instance, suppose $x \in X$ and $x \notin (Y \cup Z)$. Then $x \notin Y$ and $x \notin Z$. Hence $x \in \mathscr{C}_X Y \cap \mathscr{C}_X Z$. Conversely, if $x \in \mathscr{C}_X Y \cap \mathscr{C}_X Z$, then x lies neither in Y nor Z, and hence $x \in \mathscr{C}_X(Y \cup Z)$. This proves the first formula. We leave the second to the reader. Exercise: Formulate these formulas for the complement of the union of a family of sets, and the complement of the intersection of a family of sets.

Let A, B be sets and $f: A \to B$ a mapping. If Y is a subset of B, we define $f^{-1}(Y)$ to be the set of all $x \in A$ such that $f(x) \in Y$. It may be that $f^{-1}(Y)$ is empty, of course. We call $f^{-1}(Y)$ the **inverse image of** Y (under f). If f is injective, and Y consists of one element y, then $f^{-1}(\{y\})$ either is empty, or has precisely one element. We shall give certain simple properties of the inverse image as exercises.

X, §1. EXERCISES

1. If $f: A \to B$ is a map, and Y, Z are subsets of B, prove the following formulas:

$$f^{-1}(Y \cup Z) = f^{-1}(Y) \cup f^{-1}(Z),$$

$$f^{-1}(Y \cap Z) = f^{-1}(Y) \cap f^{-1}(Z).$$

2. Formulate and prove the analogous properties of Exercise 1 for families of subsets, e.g. if $\{Y_i\}_{i\in I}$ is a family of subsets of B, show that

$$f^{-1}\left(\bigcup_{i\in I} Y_i\right) = \bigcup_{i\in I} f^{-1}(Y_i).$$

3. Let $f: A \to B$ be a surjective map. Show that there exists an injective map of B into A.

X, §2. ZORN'S LEMMA

In order to deal efficiently with infinitely many sets simultaneously, one needs a special axiom. To state it, we need some more terminology.

Let S be a set. A **partial ordering** (also called an **ordering**) of S is a relation, written $x \leqq y$, among some pairs of elements of S, having the following properties.

PO 1. *We have* $x \leqq x$.

PO 2. *If* $x \leqq y$ *and* $y \leqq z$ *then* $x \leqq z$.

PO 3. *If* $x \leqq y$ *and* $y \leqq x$ *then* $x = y$.

Note that we don't require that the relation $x \leqq y$ or $y \leqq x$ hold for every pair of elements (x, y) of S. Some pairs may not be comparable. We sometimes write $y \geqq x$ for $x \leqq y$.

Example 1. Let G be a group. Let S be the set of subgroups. If H, H' are subgroups of G, we define

$$H \leqq H'$$

if H is a subgroup of H'. One verifies immediately that this relation defines a partial ordering on S. Given two subgroups H, H' of G, we do not necessarily have $H \leqq H'$ or $H' \leqq H$.

Example 2. Let R be a ring, and let S be the set of left ideals of R. We define a partial ordering in S in a way similar to the above, namely if L, L' are left ideals of R, we define

$$L \leqq L'$$

if $L \subset L'$.

Example 3. Let X be a set, and S the set of subsets of X. If Y, Z are subsets of X, we define $Y \leqq Z$ if Y is a subset of Z. This defines a partial ordering on S.

In all these examples, the relation of partial ordering is said to be that of inclusion.

In a partially ordered set, if $x \leqq y$ and $x \neq y$ we then write $x < y$.

Remark. We have not defined the word "relation". This can be done in terms of sets as follows. We define a **relation** between pairs of elements of a set A to be a subset R of the product $A \times A$. If $x, y \in A$ and $(x, y) \in R$, then we say that x, y **satisfy our relation**. Using this formulation, we can restate our conditions for a partial ordering relation in the following form. For all $x, y, z \in A$:

PO 1. *We have* $(x, x) \in R$.

PO 2. *If* $(x, y) \in R$ *and* $(y, z) \in R$ *then* $(x, z) \in R$.

PO 3. *If* $(x, y) \in R$ *and* $(y, x) \in R$ *then* $x = y$.

The notation we used previously is, however, much easier to use, and having shown how this notation can be explained only in terms of sets, we shall continue to use it as before.

Let A be a partially ordered set, and B a subset. Then we can define a partial ordering on B by defining $x \leqq y$ for $x, y \in B$ to hold if and only if $x \leqq y$ in A. In other words, if $R \subset A \times A$ is the subset of $A \times A$ defining our relation of partial ordering in A, we let $R_0 = R \cap (B \times B)$, and then R_0 defines a relation of partial ordering in B. We shall say that R_0 is the partial ordering on B **induced** by R, or is the **restriction** to B of the partial ordering of A.

Let S be a partially ordered set. By a **least** element of S (or a **smallest** element) one means an element $a \in S$ such that $a \leqq x$ for all $x \in S$. Similarly, by a **greatest element** one means an element b such that $x \leqq b$ for all $x \in S$.

By a **maximal element** m of S one means an element such that if $x \in S$ and $x \geqq m$, then $x = m$. Note that a maximal element need not be a greatest element. There may be many maximal elements in S, whereas if a greatest element exists, then it is unique (proof?).

Let S be a partially ordered set. We shall say that S is **totally ordered** if given $x, y \in S$ we have necessarily $x \leqq y$ or $y \leqq x$.

Example 4. The integers $\mathbf{Z}$ are totally ordered by the usual ordering. So are the real numbers.

Let S be a partially ordered set, and T a subset. An **upper bound** of T (in S) is an element $b \in S$ such that $x \leqq b$ for all $x \in T$. A **least upper bound** of T in S is an upper bound b such that, if c is another upper bound, then $b \leqq c$. We shall say that S is **inductively ordered** if every non-empty totally ordered subset has an upper bound.

We shall say that S is **strictly inductively ordered** if every non-empty totally ordered subset has a least upper bound.

In Examples 1, 2, 3 in each case, the set is strictly inductively ordered. To prove this, let us take Example 1. Let T be a non-empty totally ordered subset of the set of subgroups of G. This means that if $H, H' \in T$, then $H \subset H'$ or $H' \subset H$. Let U be the union of all sets in T. Then:

(1) U is a subgroup. *Proof*: If $x, y \in U$, there exist subgroups $H, H' \in T$ such that $x \in H$ and $y \in H'$. If, say, $H \subset H'$, then both $x, y \in H'$ and hence $xy \in H'$. Hence $xy \in U$. Also, $x^{-1} \in H'$, so $x^{-1} \in U$. Hence U is a subgroup.

(2) U is an upper bound for each element of T. *Proof*: Every $H \in T$ is contained in U, so $H \leqq U$ for all $H \in T$.

(3) U is a least upper bound for T. *Proof*: Any subgroup of G which contains all the subgroups $H \in T$ must then contain their union U.

The proof that the sets in Examples 2, 3 are strictly inductively ordered is entirely similar.

We can now state the axiom mentioned at the beginning of the section.

Zorn's lemma. *Let S be a non-empty inductively ordered set. Then there exists a maximal element in S.*

We shall see by two examples how one applies Zorn's lemma.

Theorem 2.1. *Let R be a commutative ring with unit element $1 \neq 0$. Then there exists a maximal ideal in R.*

(Recall that a maximal ideal is an ideal M such that $M \neq R$, and if J is an ideal such that $M \subset J \subset R$, then $J = M$ or $J = R$.)

Proof. Let S be the set of proper ideals of R, that is ideals J such that $J \neq R$. Then S is not empty, because the zero ideal is in S. Furthermore, S is inductively ordered by inclusion. To see this, let T be a non-empty totally ordered subset of S. Let U be the union of all ideals in T. Then U is an ideal (the proof being similar to the proof we gave before concerning Example 1). The crucial thing here, however, is that U is not equal to R. Indeed, if $U = R$, then $1 \in U$, and hence there is some ideal $J \in T$ such that $1 \in J$ because U is the union of such ideals J. This is impossible since S is a set of *proper* ideals. Hence U is in S, and is obviously an upper bound for T (even a least upper bound), so S is inductively ordered, and the theorem follows by Zorn's lemma.

Let V be a non-zero vector space over a field K. Let $\{v_i\}_{i \in I}$ be a family of elements of V. If $\{a_i\}_{i \in I}$ is a family of elements of K, such that $a_i = 0$ for all but a finite number of indices i, then we can form the sum

$$\sum_{i \in I} a_i v_i.$$

If $i_1, \ldots, i_n$ are those indices for which $a_i \neq 0$, then the above sum is defined to be

$$a_{i_1} v_{i_1} + \cdots + a_{i_n} v_{i_n}.$$

We shall say that family $\{v_i\}_{i \in I}$ is **linearly independent** if, whenever we have a family $\{a_i\}_{i \in I}$ with $a_i \in K$, all but a finite number of which are 0, and

$$\sum_{i \in I} a_i v_i = 0,$$

then all $a_i = 0$. For simplicity, we shall abbreviate "all but a finite number" by "almost all." We say that a family $\{v_i\}_{i \in I}$ of elements of V **generates** V if every element $v \in V$ can be written in the form

$$v = \sum_{i \in I} a_i v_i$$

for some family $\{a_i\}_{i \in I}$ of elements of K, almost all a_i being 0. A family $\{v_i\}_{i \in I}$ which is linearly independent and generates V is called a **basis** of V.

If U is a subset of V, we may view U as a family, indexed by its own elements. Thus if for each $v \in U$ we are given an element $a_v \in K$, almost all $a_v = 0$, we can form the sum

$$\sum_{v \in U} a_v v.$$

In this way, we can define what it means for a subset of V to generate V and to be linearly independent. We can define a basis of V to be a subset of V which generates V and is linearly independent.

Theorem 2.2. *Let V be a non-zero vector space over the field K. Then there exists a basis of V.*

Proof. Let S be the set of linearly independent subsets of V. Then S is not empty, because for any $v \in V$, $v \neq 0$, the set $\{v\}$ is linearly independent. If B, B' are elements of S, we define $B \leqq B'$ if $B \subset B'$. Then S is partially ordered, and is inductively ordered, because if T is a totally ordered subset of S, then

$$U = \bigcup_{B \in T} B$$

is an upper bound for T in S. It is easily checked that U is linearly independent. Let M be a maximal element of S, by Zorn's lemma. Let $v \in V$. Since M is maximal, if $v \notin M$, the set $M \cup \{v\}$ is not linearly independent. Hence there exist elements $a_w \in K$ ($w \in M$) and $b \in K$ not all 0, but almost all 0, such that

$$bv + \sum_{w \in M} a_w w = 0.$$

If $b = 0$, then we contradict the fact that M is linearly independent. Hence $b \neq 0$, and

$$v = \sum_{w \in M} -b^{-1} a_w w$$

is a linear combination of elements of M. If $v \in M$, then trivially, v is a linear combination of elements of M. Hence M generates V, and is therefore the desired basis of V.

Remark. Zorn's lemma is not psychologically completely satisfactory as an axiom, because its statement is too involved, and one does not visualize easily the existence of the maximal element asserted in that statement. It can be shown that Zorn's lemma is implied by the following statement, known as the **axiom of choice**:

Let $\{S_i\}_{i \in I}$ be a family of sets, and assume that each S_i is not empty. Then there exists a family of elements $\{x_i\}_{i \in I}$ with each $x_i \in S_i$.

For a proof of the implication, see for instance Appendix 2 of my *Algebra*.

X, §2. EXERCISES

1. Write out in detail the proof that the sets of Examples 2, 3 are inductively ordered.

2. In the proof of Theorem 2.2, write out in detail the proof of the statement: "It is easily checked that U is linearly independent."

3. Let R be a ring and E a finitely generated module over R, i.e. a module with a finite number of generators $v_1, \ldots, v_n$. Assume that E is not the zero module. Show that E has a maximal submodule, i.e. a submodule $M \neq E$ such that if N is a submodule, $M \subset N \subset E$, then $M = N$ or $N = E$.

4. Let R be a commutative ring and S a subset of R, S not empty, and $0 \notin S$. Show that there exists an ideal M whose intersection with S is empty, and is maximal with respect to this property. We then say that M is a maximal ideal not meeting S.

5. Let A, B be two non-empty sets. Show that there exists an injective map of A into B, or there exists a bijective map of a subset of A onto B. [*Hint*: Use Zorn's lemma on the family of injective maps of subsets of A into B.]

X, §3. CARDINAL NUMBERS

Let A, B be sets. We shall say that the **cardinality of A is the same as the cardinality of B**, and write

$$\mathrm{card}(A) = \mathrm{card}(B),$$

if there exists a bijection of A onto B.

We say $\mathrm{card}(A) \leqq \mathrm{card}(B)$ if there exists an injective mapping (injection) $f: A \to B$. We also write $\mathrm{card}(B) \geqq \mathrm{card}(A)$ in this case. It is clear that if $\mathrm{card}(A) \leqq \mathrm{card}(B)$ and $\mathrm{card}(B) \leqq \mathrm{card}(C)$, then

$$\mathrm{card}(A) \leqq \mathrm{card}(C).$$

This amounts to saying that a composite of injective mappings is injective. Similarly, if $\mathrm{card}(A) = \mathrm{card}(B)$ and $\mathrm{card}(B) = \mathrm{card}(C)$ then

$$\mathrm{card}(A) = \mathrm{card}(C).$$

This amounts to saying that a composite of bijective mappings is bijective. We clearly have $\mathrm{card}(A) = \mathrm{card}(A)$.

Finally, Exercise 5 of §2 shows:

Let A, B be non-empty sets. Then we have

$$\mathrm{card}(A) \leqq \mathrm{card}(B) \quad \textit{or} \quad \mathrm{card}(B) \leqq \mathrm{card}(A).$$

We shall first discuss denumerable sets. A set D is called **denumerable** if there exists a bijection of D with the positive integers, and such a bijection is called an **enumeration** of the set D.

Theorem 3.1. *Any infinite subset of a denumerable set is denumerable.*

One proves this easily by induction. (We sketch the proof: It suffices to prove that any infinite subset of the positive integers is denumerable. Let $D = D_1$ be such a subset. Then D_1 has a least element a_1. Suppose inductively that we have defined D_n for an integer $n \geqq 1$. Let D_{n+1} be the set of all elements of D_n which are greater than the least element of D_n. Let a_n be the least element of D_n. Then we get an injective mapping

$$n \mapsto a_n$$

of $\mathbf{Z}^+$ into D, and one sees at once that this map is surjective.)

Theorem 3.2. *Let D be a denumerable set. Then $D \times D$ is denumerable.*

Proof. It suffices to prove that $\mathbf{Z}^+ \times \mathbf{Z}^+$ is denumerable. Consider the map

$$(m, n) \mapsto 2^m 3^n.$$

It is an injective map of $\mathbf{Z}^+ \times \mathbf{Z}^+$ into $\mathbf{Z}^+$, and hence $\mathbf{Z}^+ \times \mathbf{Z}^+$ has the same cardinality as an infinite subset of $\mathbf{Z}^+$, whence $\mathbf{Z}^+ \times \mathbf{Z}^+$ is denumerable, as was to be shown.

In this proof, we have used the factorization of integers. One can also give a proof without using that fact. The idea for such a proof is illustrated in the following diagram:

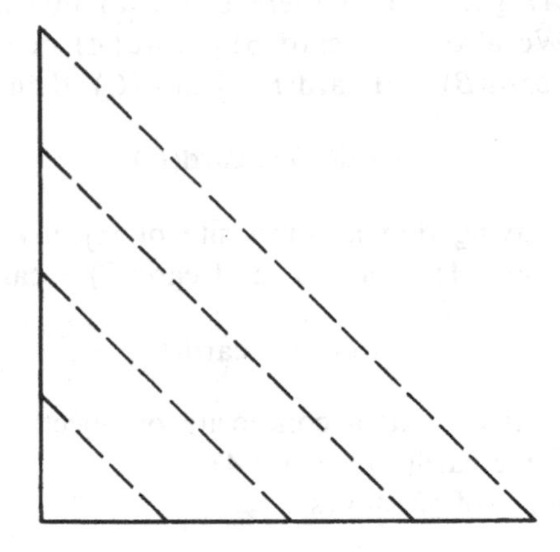

We must define a bijection $\mathbf{Z}^+ \to \mathbf{Z}^+ \times \mathbf{Z}^+$. We map 1 on $(1, 1)$. Inductively, suppose that we have defined an injective map

$$f: \{1, \dots, n\} \to \mathbf{Z}^+ \times \mathbf{Z}^+.$$

We wish to define $f(n + 1)$.

If $f(n) = (1, k)$ then we let $f(n + 1) = (k + 1, 1)$.
If $f(n) = (r, k)$ with $r \neq 1$ then we let $f(n + 1) = (r - 1, k + 1)$.

It is then routinely checked that we obtain an injection of $\{1, \dots, n + 1\}$ into $\mathbf{Z}^+ \times \mathbf{Z}^+$. By induction, we obtain a map of $\mathbf{Z}^+$ into $\mathbf{Z}^+ \times \mathbf{Z}^+$ which is also routinely verified to be a bijection. In the diagram, our map f can be described as follows. We start in the corner $(1, 1)$, and then move towards the inside of the quadrant, starting on the horizontal axis, moving diagonally leftwards until we hit the vertical axis, and then starting on the horizontal axis one step further, repeating the process. Geometrically, it is then clear that our path goes through every point (i, j) of $\mathbf{Z}^+ \times \mathbf{Z}^+$.

Corollary 3.3. *For every positive integer n, the product* $D \times \cdots \times D$ *taken n times is denumerable.*

Proof. Induction.

Corollary 3.4. *Let* $\{D_1, D_2, \ldots\}$ *be a sequence of denumerable sets, also written* $\{D_i\}_{i \in \mathbf{Z}^+}$. *Then the union*

$$U = \bigcup_{i=1}^{\infty} D_i$$

is denumerable.

Proof. For each i we have an enumeration of the elements of D_i, say

$$D_i = \{a_{i1}, a_{i2}, \ldots\}.$$

Then the map

$$(i, j) \mapsto a_{ij}$$

is a map from $\mathbf{Z}^+ \times \mathbf{Z}^+$ into U, and is in fact surjective. Let

$$f: \mathbf{Z}^+ \times \mathbf{Z}^+ \to U$$

be this map. For each $a \in U$ there exists an element $x \in \mathbf{Z}^+ \times \mathbf{Z}^+$ such that $f(x) = a$, and we can write this element x in the form x_a. The association $a \mapsto x_a$ is an injection of U into $\mathbf{Z}^+ \times \mathbf{Z}^+$, and we can now apply the theorem to conclude the proof.

In the preceding proof, we used a special case of a cardinality statement which it is useful to state in general:

Let $f: A \to B$ *be a surjective map of a set A onto a set B. Then*

$$\operatorname{card}(B) \leqq \operatorname{card}(A).$$

This is easily seen, because for each $y \in B$ there exists an element $x \in A$, denoted by x_y, such that $f(x_y) = y$. Then the association $y \mapsto x_y$ is an injective mapping of B into A, whence by definition,

$$\operatorname{card}(B) \leqq \operatorname{card}(A).$$

In dealing with arbitrary cardinalities, one needs a theorem which is somewhat less trivial than in the denumerable case.

Theorem 3.5 (Schroeder–Bernstein). *Let A, B be sets, and suppose that* $\operatorname{card}(A) \leqq \operatorname{card}(B)$, *and* $\operatorname{card}(B) \leqq \operatorname{card}(A)$. *Then*

$$\operatorname{card}(A) = \operatorname{card}(B).$$

Proof. Let $f: A \to B$ and $g: B \to A$ be injective maps. We separate A into two disjoint sets A_1 and A_2. We let A_1 consist of all $x \in A$ such that, when we lift back x by a succession of inverse maps,

$$x, \quad g^{-1}(x), \quad f^{-1} \circ g^{-1}(x), \quad g^{-1} \circ f^{-1} \circ g^{-1}(x), \ldots,$$

then at some stage we reach an element of A which cannot be lifted back to B by g^{-1}. We let A_2 be the complement of A_1, in other words, the set of $x \in A$ which can be lifted back indefinitely, or such that we get stopped in B (i.e. reach an element of B which has no inverse image in A by f^{-1}). Then $A = A_1 \cup A_2$. We shall define a bijection h of A onto B.

If $x \in A_1$, we define $h(x) = f(x)$.
If $x \in A_2$, we define $h(x) = g^{-1}(x) =$ unique element $y \in B$ such that $g(y) = x$.

Then trivially, h is injective. We must prove that h is surjective. Let $b \in B$. If, when we try to lift back b by a succession of maps

$$\cdots \circ f^{-1} \circ g^{-1} \circ f^{-1} \circ g^{-1} \circ f^{-1}(b)$$

we can lift back indefinitely, or if we get stopped in B, then $g(b)$ belongs to A_2 and consequently $b = h(g(b))$, so b lies in the image of h. On the other hand, if we cannot lift back b indefinitely, and get stopped in A, then $f^{-1}(b)$ is defined (i.e. b is in the image of f), and $f^{-1}(b)$ lies in A_1. In this case, $b = h(f^{-1}(b))$ is also in the image of h, as was to be shown.

Next we consider theorems concerning sums and products of cardinalities.

We shall reduce the study of cardinalities of products of arbitrary sets to the denumerable case, using Zorn's lemma. Note first that an infinite set A always contains a denumerable set. Indeed, since A is infinite, we can first select an element $a_1 \in A$, and the complement of $\{a_1\}$ is infinite. Inductively, if we have selected distinct elements $a_1, \ldots, a_n$ in A, the complement of $\{a_1, \ldots, a_n\}$ is infinite, and we can select a_{n+1} in this complement. In this way, we obtain a sequence of distinct elements of A, giving rise to a denumerable subset of A.

Let A be a set. By a **covering** of A one means a set Γ of subsets of A such that the union

$$\bigcup_{C \in \Gamma} C$$

of all the elements of Γ is equal to A. We shall say that Γ is a **disjoint covering** if whenever $C, C' \in \Gamma$, and $C \neq C'$, then the intersection of C and C' is empty.

Lemma 3.6. *Let A be an infinite set. Then there exists a disjoint covering of A by denumerable sets.*

Proof. Let S be the set whose elements are pairs (B, Γ) consisting of a subset B of A, and a disjoint covering of B by denumerable sets. Then S is not empty. Indeed, since A is infinite, A contains a denumerable set D, and the pair $(D, \{D\})$ is in S. If (B, Γ) and (B', Γ') are elements of S, we define

$$(B, \Gamma) \leqq (B', \Gamma')$$

to mean that $B \subset B'$, and $\Gamma \subset \Gamma'$. Let T be a totally ordered non-empty subset of S. We may write $T = \{(B_i, \Gamma_i)\}_{i \in I}$ for some indexing set I. Let

$$B = \bigcup_{i \in I} B_i \quad \text{and} \quad \Gamma = \bigcup_{i \in I} \Gamma_i.$$

If $C, C' \in \Gamma$, $C \neq C'$, then there exist some indices i, j such that $C \in \Gamma_i$ and $C' \in \Gamma_j$. Since T is totally ordered, we have, say,

$$(B_i, \Gamma_i) \leqq (B_j, \Gamma_j).$$

Hence in fact, C, C' are both elements of Γ_j, and hence C, C' have an empty intersection. On the other hand, if $x \in B$, then $x \in B_i$ for some i, and hence there is some $C \in \Gamma_i$ such that $x \in C$. Hence Γ is a disjoint covering of B. Since the elements of each Γ_i are denumerable subsets of A, it follows that Γ is a disjoint covering of B by denumerable sets, so (B, Γ) is in S, and is obviously an upper bound for T. Therefore S is inductively ordered.

Let (M, Δ) be a maximal element of S, by Zorn's lemma. Suppose that $M \neq A$. If the complement of M in A is infinite, then there exists a denumerable set D contained in this complement. Then

$$(M \cup D, \Delta \cup \{D\})$$

is a bigger pair than (M, Δ), contradicting the maximality of (M, Δ). Hence the complement of M in A is a finite set F. Let D_0 be an element of Δ. Let $D_1 = D_0 \cup F$. Then D_1 is denumerable. Let Δ_1 be the set consisting of all elements of Δ, except D_0, together with D_1. Then Δ_1 is a disjoint covering of A by denumerable sets, as was to be shown.

Theorem 3.7. *Let A be an infinite set, and let D be a denumerable set. Then*

$$\text{card}(A \times D) = \text{card}(A).$$

Proof. By the lemma, we can write

$$A = \bigcup_{i \in I} D_i$$

as a disjoint union of denumerable sets. Then

$$A \times D = \bigcup_{i \in I} (D_i \times D).$$

For each $i \in I$, there is a bijection of $D_i \times D$ on D_i by Theorem 3.2. Since the sets $D_i \times D$ are disjoint, we get in this way a bijection of $A \times D$ on A, as desired.

Corollary 3.8. *If F is a finite non-empty set, then*

$$\text{card}(A \times F) = \text{card}(A).$$

Proof. We have

$$\text{card}(A) \leqq \text{card}(A \times F) \leqq \text{card}(A \times D) = \text{card}(A).$$

We can then use Theorem 3.5 to get what we want.

Corollary 3.9. *Let A, B be non-empty sets, A infinite, and suppose $\text{card}(B) \leqq \text{card}(A)$. Then*

$$\text{card}(A \cup B) = \text{card}(A).$$

Proof. We can write $A \cup B = A \cup C$ for some subset C of B, such that C and A are disjoint. (We let C be the set of all elements of B which are not elements of A.) Then $\text{card}(C) \leqq \text{card}(A)$. We can then construct an injection of $A \cup C$ into the product

$$A \times \{1, 2\}$$

of A with a set consisting of 2 elements. Namely, we have a bijection of A with $A \times \{1\}$ in the obvious way, and also an injection of C into $A \times \{2\}$. Thus

$$\text{card}(A \cup C) \leqq \text{card}(A \times \{1, 2\}).$$

We conclude the proof by Corollary 3.8 and Theorem 3.5.

Theorem 3.10. *Let A be an infinite set. Then*

$$\text{card}(A \times A) = \text{card}(A).$$

Proof. Let S be the set consisting of pairs (B, f) where B is an infinite subset of A, and $f: B \to B \times B$ is a bijection of B onto $B \times B$. Then S is not empty because if D is a denumerable subset of A, we can always find a bijection of D onto $D \times D$. If (B, f) and (B', f') are in S, we define $(B, f) \leqq (B', f')$ to mean $B \subset B'$, and the restriction of f' to B is equal to f. Then S is partially ordered, and we contend that S is inductively ordered. Let T be a non-empty totally ordered subset of S, and say T consists of the pairs (B_i, f_i) for i in some indexing set I. Let

$$M = \bigcup_{i \in I} B_i.$$

We shall define a bijection $g: M \to M \times M$. If $x \in M$, then x lies in some B_i. We define $g(x) = f_i(x)$. This value $f_i(x)$ is independent of the choice of B_i in which x lies. Indeed, if $x \in B_j$ for some $j \in I$, then say

$$(B_i, f_i) \leqq (B_j, f_j).$$

By assumption, $B_i \subset B_j$, and $f_j(x) = f_i(x)$, so g is well defined. To show g is surjective, let $x, y \in M$ and $(x, y) \in M \times M$. Then $x \in B_i$ for some $i \in I$ and $y \in B_j$ for some $j \in I$. Again since T is totally ordered, say $(B_i, f_i) \leqq (B_j, f_j)$. Thus $B_i \subset B_j$, and $x, y \in B_j$. There exists an element $b \in B_j$ such that $f_j(b) = (x, y) \in B_j \times B_j$. By definition, $g(b) = (x, y)$, so g is surjective. We leave the proof that g is injective to the reader to conclude the proof that g is a bijection. We then see that (M, g) is an upper bound for T in S, and therefore that S is inductively ordered.

Let (M, g) be a maximal element of S, and let C be the complement of M in A. If $\text{card}(C) \leqq \text{card}(M)$, then

$$\text{card}(M) \leqq \text{card}(A) = \text{card}(M \cup C) = \text{card}(M)$$

by Corollary 3.9, and hence $\text{card}(M) = \text{card}(A)$ by Bernstein's theorem. Since $\text{card}(M) = \text{card}(M \times M)$, we are done with the proof in this case. If $\text{card}(M) \leqq \text{card}(C)$, then there exists a subset M_1 of C having the same cardinality as M. We shall prove this is not possible. We consider

$$\begin{aligned}(M \cup M_1) &\times (M \cup M_1) \\ &= (M \times M) \cup (M_1 \times M) \cup (M \times M_1) \cup (M_1 \times M_1).\end{aligned}$$

By the assumption on M and Corollary 3.9, the union of the last three sets in parentheses on the right of this equation has the same cardinality as M. Thus

$$(M \cup M_1) \times (M \cup M_1) = (M \times M) \cup M_2,$$

where M_2 is disjoint from $M \times M$, and has the same cardinality as M.

We now define a bijection

$$g_1: M \cup M_1 \to (M \cup M_1) \times (M \cup M_1).$$

We let $g_1(x) = g(x)$ if $x \in M$, and we let g_1 on M_1 be any bijection of M_1 on M_2. In this way we have extended g to $M \cup M_1$, and the pair $(M \cup M_1, g_1)$ is in S, contradicting the maximality of (M, g). The case $\operatorname{card}(M) \leqq \operatorname{card}(C)$ therefore cannot occur, and our theorem is proved.

Corollary 3.11. *If A is an infinite set, and $A^{(n)} = A \times \cdots \times A$ is the product taken n times, then*

$$\operatorname{card}(A^{(n)}) = \operatorname{card}(A).$$

Proof. Induction.

Corollary 3.12. *If $A_1, \ldots, A_n$ are non-empty infinite sets, and*

$$\operatorname{card}(A_i) \leqq \operatorname{card}(A_n)$$

for $i = 1, \ldots, n$, then

$$\operatorname{card}(A_1 \times \cdots \times A_n) = \operatorname{card}(A_n).$$

Proof. We have

$$\operatorname{card}(A_n) \leqq \operatorname{card}(A_1 \times \cdots \times A_n) \leqq \operatorname{card}(A_n \times \cdots \times A_n)$$

and we use Corollary 3.11 and the Schroeder–Bernstein theorem to conclude the proof.

Corollary 3.13. *Let A be an infinite set, and let Φ be the set of finite subsets of A. Then*

$$\operatorname{card}(\Phi) = \operatorname{card}(A).$$

Proof. Let Φ_n be the set of subsets of A having exactly n elements, for each integer $n = 1, 2, \ldots$. We first show that $\operatorname{card}(\Phi_n) \leqq \operatorname{card}(A)$. If F is an element of Φ_n, we order the elements of F in any way, say

$$F = \{x_1, \ldots, x_n\},$$

and we associate with F the element $(x_1, \ldots, x_n) \in A^{(n)}$,

$$F \mapsto (x_1, \ldots, x_n).$$

If G is another subset of A having n elements, say $G = \{y_1,\dots,y_n\}$, and $G \neq F$, then

$$(x_1,\dots,x_n) \neq (y_1,\dots,y_n).$$

Hence our map

$$F \mapsto (x_1,\dots,x_n)$$

of Φ_n into $A^{(n)}$ is injective. By Corollary 3.11, we conclude that

$$\operatorname{card}(\Phi_n) \leqq \operatorname{card}(A).$$

Now Φ is the disjoint union of the Φ_n for $n = 1, 2, \dots$ and it is an exercise to show that $\operatorname{card}(\Phi) \leqq \operatorname{card}(A)$ (cf. Exercise 1). Since

$$\operatorname{card}(A) \leqq \operatorname{card}(\Phi),$$

because in particular, $\operatorname{card}(\Phi_1) = \operatorname{card}(A)$, we see that our corollary is proved.

In the next theorem, we shall see that given a set, there always exists another set whose cardinality is bigger.

Theorem 3.14. *Let A be an infinite set, and T the set consisting of two elements $\{0, 1\}$. Let M be the set of all maps of A into T. Then*

$$\operatorname{card}(A) \leqq \operatorname{card}(M) \quad \textit{and} \quad \operatorname{card}(A) \neq \operatorname{card}(M).$$

Proof. For each $x \in A$ we let

$$f_x\colon A \to \{0, 1\}$$

be the map such that $f_x(x) = 1$ and $f_x(y) = 0$ if $y \neq x$. Then $x \mapsto f_x$ is obviously an injection of A into M, so that $\operatorname{card}(A) \leqq \operatorname{card}(M)$. Suppose that $\operatorname{card}(A) = \operatorname{card}(M)$. Let

$$x \mapsto g_x$$

be a bijection between A and M. We define a map $h\colon A \to \{0, 1\}$ by the rule

$$h(x) = 0 \quad \text{if } g_x(x) = 1,$$

$$h(x) = 1 \quad \text{if } g_x(x) = 0.$$

Then certainly $h \neq g_x$ for any x, and this contradicts the assumption that $x \mapsto g_x$ is a bijection, thereby proving Theorem 3.14.

Corollary 3.15. *Let A be an infinite set, and let S be the set of all subsets of A. Then* $\operatorname{card}(A) \leqq \operatorname{card}(S)$ *and* $\operatorname{card}(A) \neq \operatorname{card}(S)$.

Proof. We leave it as an exercise. [*Hint*: If B is a non-empty subset of A, use the characteristic function φ_B such that

$$\varphi_B(x) = 1 \quad \text{if} \quad x \in B,$$

$$\varphi_B(x) = 0 \quad \text{if} \quad x \notin B.$$

What can you say about the association $B \mapsto \varphi_B$?]

X, §3. EXERCISES

1. Prove the statement made in the proof of Corollary 3.13.

2. If A is an infinite set, and Φ_n is the set of subsets of A having exactly n elements, show that

$$\operatorname{card}(A) \leqq \operatorname{card}(\Phi_n)$$

for $n \geqq 1$.

3. Let A_i be infinite sets for $i = 1, 2, \ldots$ and assume that

$$\operatorname{card}(A_i) \leqq \operatorname{card}(A)$$

for some set A, and all i. Show that

$$\operatorname{card}\left(\bigcup_{i=1}^{\infty} A_i\right) \leqq \operatorname{card}(A).$$

4. Let K be a subfield of the complex numbers. Show that for each integer $n \geqq 1$, the cardinality of the set of extensions of K of degree n in $\mathbf{C}$ is $\leqq \operatorname{card}(K)$.

5. Let K be an infinite field, and E an algebraic extension of K. Show that $\operatorname{card}(E) = \operatorname{card}(K)$.

6. Finish the proof of Corollary 3.15.

7. If A, B are sets, denote by $M(A, B)$ the set of all maps of A into B. If B, B' are sets with the same cardinality, show that $M(A, B)$ and $M(A, B')$ have the same cardinality. If A, A' have the same cardinality, show that $M(A, B)$ and $M(A', B)$ have the same cardinality.

8. Let A be an infinite set and abbreviate $\operatorname{card}(A)$ by α. If B is an infinite set, abbreviate $\operatorname{card}(B)$ by β. Define $\alpha\beta$ to be $\operatorname{card}(A \times B)$. Let B' be a set disjoint from A such that $\operatorname{card}(B) = \operatorname{card}(B')$. Define $\alpha + \beta$ to be $\operatorname{card}(A \cup B')$. Denote by B^A the set of all maps of A into B, and denote $\operatorname{card}(B^A)$ by β^α.

Let C be an infinite set and abbreviate card(C) by γ. Prove the following statements:
(a) $\alpha(\beta + \gamma) = \alpha\beta + \alpha\gamma$ (b) $\alpha\beta = \beta\alpha$ (c) $\alpha^{\beta+\gamma} = \alpha^\beta\alpha^\gamma$ (d) $(\alpha^\beta)^\gamma = \alpha^{(\beta\gamma)}$.

9. Let K be an infinite field. Prove that there exists an algebraically closed field A containing K as a subfield, and algebraic over K. [*Hint*: Let Ω be a set of cardinality strictly greater than the cardinality of K, and containing K. Consider the set S of all pairs (E, φ) where E is a subset of Ω such that $K \subset E$, and φ denotes a law of addition and multiplication on E which makes E into a field such that K is a subfield, and E is algebraic over K. Define a partial ordering on S in an obvious way; show that S is inductively ordered, and that a maximal element is algebraic over K and algebraically closed. You will need Exercise 5 in the last step.]

10. Let K be an infinite field. Show that the field of rational functions $K(t)$ has the same cardinality as K.

11. Let J_n be the set of integers $\{1,\dots,n\}$. Let $\mathbf{Z}^+$ be the set of positive integers. Show that the following sets have the same cardinality:
(a) The set of all maps $M(\mathbf{Z}^+, J_n)$ with $n \geqq 2$.
(b) The set of all maps $M(\mathbf{Z}^+, J_2)$.
(c) The set of all real numbers x such that $0 \leqq x < 1$.
(d) The set of all real numbers.
[*Hint*: Use decimal expansions.]

12. Show that $M(\mathbf{Z}^+, \mathbf{Z}^+)$ has the same cardinality as the real numbers.

13. Prove that the sets $\mathbf{R}$, $M(\mathbf{Z}^+, \mathbf{R})$, $M(\mathbf{Z}^+, \mathbf{Z}^+)$ have the same cardinalities.

X, §4. WELL-ORDERING

A set A is said to be **well-ordered** if it is totally ordered, and if every non-empty subset B has a least element, i.e. an element $a \in B$ such that $a \leqq x$ for all $x \in B$.

Example 1. The set of positive integers $\mathbf{Z}^+$ is well-ordered. Any finite set can be well-ordered, and a denumerable set D can be well-ordered: Any bijection of D with $\mathbf{Z}^+$ will give rise to a well-ordering of D.

Example 2. Let D be a denumerable set which is well-ordered. Let b be an element of some set, and $b \notin D$. Let $A = D \cup \{b\}$. We define $x \leqq b$ for all $x \in D$. Then A is totally ordered, and is in fact well-ordered. *Proof*: Let B a non-empty subset of A. If B consists of b alone, then b is a least element of B. Otherwise, B contains some element $a \in D$. Then $B \cap D$ is not empty, and hence has a least element, which is obviously also a least element for B.

Example 3. Let D_1, D_2 be two denumerable sets, each one well-ordered, and assume that $D_1 \cap D_2$ is empty. Let $A = D_1 \cup D_2$. We define a total ordering in A by letting $x < y$ for all $x \in D_1$ and all $y \in D_2$. Using the same type of argument as in Example 2, we see that A is well-ordered.

Example 4. Proceeding inductively, given a sequence of disjoint denumerable sets $D_1, D_2, \ldots$ we let $A = \bigcup D_i$, and we can define a well-ordering on A by ordering each D_i like $\mathbf{Z}^+$, and then defining $x < y$ for $x \in D_i$ and $y \in D_{i+1}$. One may visualize this situation as follows:

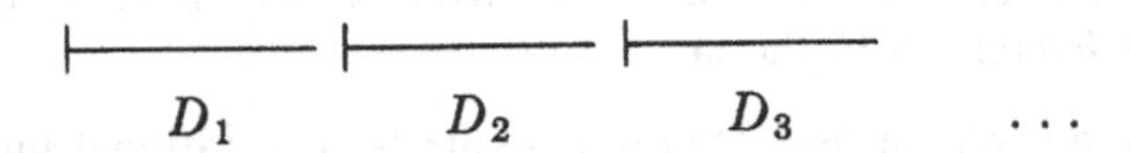

Theorem 4.1. *Every non-empty infinite set A can be well-ordered.*

Proof. Let S be the set of all pairs (X, R) where X is a subset of A, and R is a total ordering of X such that X is well-ordered. Then S is not empty, since given a denumerable subset D of A, we can always well-order it like the positive integers. If (X, R) and (Y, Q) are elements of S, we define $(X, R) \leqq (Y, Q)$ if $X \subset Y$, if the restriction of Q to X is equal to R, if X is the beginning segment of Y, and if every element $y \in Y$, $y \notin X$ is such that $x < y$ for all $x \in X$. Then S is partially ordered. To show that S is inductively ordered, let T be a totally ordered non-empty subset of S, say $T = \{(X_i, R_i)\}_{i \in I}$. Let

$$M = \bigcup_{i \in I} X_i.$$

Let $x, y \in M$. There exists $i, j \in I$ such that $x \in X_i$ and $y \in X_j$. Since T is totally ordered, say $(X_i, R_i) \leqq (X_j, R_j)$. Then both $x, y \in X_j$. We define $x \leqq y$ in M if $x \leqq y$ in X_j. This is easily seen to be independent of the choice of (X_j, R_j) such that $x, y \in X_j$, and it is then trivially verified that we have defined a total ordering on M, which we denote by (M, P). We contend that this total ordering on M is a well-ordering. To see this, let N be a non-empty subset of M. Let $x_0 \in N$. Then there exists some $i_0 \in I$ such that $x_0 \in X_{i_0}$. The subset $M \cap X_{i_0}$ is not empty. Let a be a least element. We contend that a is in fact a least element of N. Let $x \in N$. Then x lies in some X_i. Since T is totally ordered, we have

$$(X_i, R_i) \leqq (X_{i_0}, R_{i_0}) \quad \text{or} \quad (X_{i_0}, R_{i_0}) \leqq (X_i, R_i).$$

In the first case, $x \in X_i \subset X_{i_0}$ and hence $a \leqq x$. In the second case, if $x \notin X_{i_0}$ then by definition, $a < x$. This proves that (M, P) is a well-ordering.

We have therefore proved that S is inductively ordered. By Zorn's lemma, there exists a maximal element (M, P) of S. Then M is well-ordered, and all that remains to be shown is that $M = A$. Suppose $M \neq A$, and let z be an element of A and $z \notin M$. Let $M' = M \cup \{z\}$. We define a total ordering on M' by defining $x < z$ for all $x \in M$. Then M' is well-ordered, for let N be a totally ordered non-empty subset of M'. If $N \cap M$ is not empty, then $N \cap M$ has a least element a, which is obviously a least element for N. This contradicts the fact that M is maximal in S. Hence $M = A$, and our theorem is proved.

Remark. It is an elaborate matter to axiomatize the theory of sets beyond the point where we have carried it in the arguments of this chapter. Since all the arguments of the chapter are easily acceptable to working mathematicians, it is a reasonable policy to stop at this point without ever looking at the deeper foundations.

One may, however, be interested in these foundations for their own sake, as a matter of taste. We refer readers to technical books on the subject if they are so inclined.

Appendix

APP., §1. THE NATURAL NUMBERS

The purpose of this appendix is to show how the integers can be obtained axiomatically using only the terminology and elementary properties of sets. The rules of the game from now on allow us to use only sets and mappings.

We assume given once for all a set $\mathbf{N}$ called the set of **natural numbers**, and a map $\sigma: \mathbf{N} \to \mathbf{N}$, satisfying the following (Peano) axioms:

NN 1. *There is an element* $0 \in \mathbf{N}$.

NN 2. *We have* $\sigma(0) \neq 0$ *and if we let* $\mathbf{N}^+$ *denote the subset of* $\mathbf{N}$ *consisting of all* $n \in \mathbf{N}$, $n \neq 0$, *then the map* $x \mapsto \sigma(x)$ *is a bijection between* $\mathbf{N}$ *and* $\mathbf{N}^+$.

NN 3. *If* S *is a subset of* $\mathbf{N}$, *if* $0 \in S$, *and if* $\sigma(n)$ *lies in* S *whenever* n *lies in* S, *then* $S = \mathbf{N}$.

We often denote $\sigma(n)$ by n' and think of n' as the successor of n. The reader will recognize NN 3 as induction.

We denote $\sigma(0)$ by 1.

Our next task is to define addition between natural numbers.

Lemma 1.1. *Let* $f: \mathbf{N} \to \mathbf{N}$ *be maps such that*

$$f(0) = g(0) \qquad \text{and} \qquad \begin{cases} f(n') = f(n)', \\ g(n') = g(n)'. \end{cases}$$

Then $f = g$.

Proof. Let S be the subset of $\mathbf{N}$ consisting of all n such that

$$f(n) = g(n).$$

Then S obviously satisfies the hypotheses of induction, so $S = \mathbf{N}$, thereby proving the lemma.

For each $m \in \mathbf{N}$, we wish to define $m + n$ with $n \in \mathbf{N}$ such that

$$(1_m) \qquad m + 0 = m \qquad \text{and} \qquad m + n' = (m + n)' \qquad \text{for all} \quad n \in \mathbf{N}.$$

By Lemma 1.1, this is possible in only one way.

If $m = 0$, we define $0 + n = n$ for all $n \in \mathbf{N}$. Then (1_m) is obviously satisfied. Let T be the set of $m \in \mathbf{N}$ for which one can define $m + n$ for all $n \in \mathbf{N}$ in such a way that (1_m) is satisfied. Then $0 \in T$. Suppose $m \in T$. We define for all $n \in \mathbf{N}$,

$$m' + 0 = m' \qquad \text{and} \qquad m' + n = (m + n)'.$$

Then

$$m' + n' = (m + n')' = ((m + n)')' = (m' + n)'.$$

Hence $(1_{m'})$ is satisfied, so $m' \in T$. This proves that $T = \mathbf{N}$, and thus we have defined addition for all pairs (m, n) of natural numbers.

The properties of addition are easily proved.

Commutativity. Let S be the set of all natural numbers m such that

$$(2_m) \qquad m + n = n + m \qquad \text{for all} \quad n \in \mathbf{N}.$$

Then 0 is obviously in S, and if $m \in S$, then

$$m' + n = (m + n)' = (n + m)' = n + m',$$

thereby proving that $S = \mathbf{N}$, as desired.

Associativity. Let S be the set of natural numbers m such that

$$(3_m) \qquad (m + n) + k = m + (n + k) \qquad \text{for all} \quad n, k \in \mathbf{N}.$$

Then 0 is obviously in S. Suppose $m \in S$. Then

$$\begin{aligned}(m' + n) + k = (m + n)' + k, &= ((m + n) + k)' \\ &= (m + (n + k))' = m' + (n + k),\end{aligned}$$

thereby proving that $S = \mathbf{N}$, as desired.

Cancellation law. Let m be a natural number. We shall say that the **cancellation law holds for** m if for all $k, n \in \mathbf{N}$ satisfying $m + k = m + n$ we must have $k = n$. Let S be the set of m for which the cancellation law holds. Then obviously $0 \in S$, and if $m \in S$, then

$$m' + k = m' + n \quad \text{implies} \quad (m + k)' = (m + n)'.$$

Since the mapping $x \mapsto x'$ is injective, it follows that $m + k = m + n$, whence $k = n$. By induction, $S = \mathbf{N}$.

For multiplication, and other applications, we need to generalize Lemma 1.1.

Lemma 1.2. *Let S be a set, and $\varphi: S \to S$ a map of S into itself. Let f, g be maps of $\mathbf{N}$ into S. If*

$$f(0) = g(0) \quad \text{and} \quad \begin{cases} f(n') = \varphi \circ f(n), \\ g(n') = \varphi \circ g(n) \end{cases}$$

for all $n \in \mathbf{N}$, then $f = g$.

Proof. Trivial by induction.

For each natural number m, it follows from Lemma 1.2 that there is at most one way of defining a product mn satisfying

$$m0 = 0 \qquad \text{and} \qquad mn' = mn + m \qquad \text{for all} \quad n \in \mathbf{N}.$$

We in fact define the product this way in the same inductive manner that we did for addition, and then prove in a similar way that this product is *commutative, associative, and distributive*, that is

$$m(n + k) = mn + mk$$

for all $m, n, k \in \mathbf{N}$. We leave the details to the reader.

In this way, we obtain all the properties of a ring, *except* that $\mathbf{N}$ is not an additive group: We lack additive inverses. Note that 1 is a unit element for the multiplication, that is $1m = m$ for all $m \in \mathbf{N}$.

It is also easy to prove the *multiplicative cancellation law*, namely if $mk = mn$ and $m \neq 0$, then $k = n$. We also leave this to the reader. In particular, if $mn \neq 0$, then $m \neq 0$ and $n \neq 0$.

We recall that an **ordering** in a set X is a relation $x \leqq y$ between certain pairs (x, y) of elements of X, satisfying the conditions (for all $x, y, z \in X$):

PO 1. *We have $x \leqq x$.*

PO 2. *If $x \leqq y$ and $y \leqq z$, then $x \leqq z$.*

PO 3. *If $x \leqq y$ and $y \leqq x$, then $x = y$.*

The ordering is called a **total ordering** if given $x, y \in X$ we have $x \leqq y$ or $y \leqq x$. We write $x < y$ if $x \leqq y$ and $x \neq y$.

We can define an ordering in $\mathbf{N}$ by defining $n \leqq m$ if there exists $k \in \mathbf{N}$ such that $m = n + k$. The proof that this is an ordering is routine and left to the reader. *This is in fact a total ordering,* and we give the proof for that. Given a natural number m, let C_m be the set of $n \in \mathbf{N}$ such that $n \leqq m$ or $m \leqq n$. Then certainly $0 \in C_m$. Suppose that $n \in C_m$. If $n = m$, then $n' = m + 1$, so $m \leqq n'$. If $n < m$, then $m = n + k'$ for some $k \in \mathbf{N}$, so that

$$m = n + k' = (n + k)' = n' + k,$$

and $n' \leqq m$. If $m \leqq n$, then for some k, we have $n = m + k$, so that $n + 1 = m + k + 1$ and $m \leqq n + 1$. By induction, $C_m = \mathbf{N}$, thereby showing our ordering is total.

It is then easy to prove standard statements concerning inequalities, e.g.

$$m < n \quad \text{if and only if} \quad m + k < n + k \qquad \text{for some} \quad k \in \mathbf{N},$$

$$m < n \quad \text{if and only if} \quad mk < nk \qquad \text{for some} \quad k \in \mathbf{N},\ k \neq 0.$$

One can also replace "for some" by "for all" in these two assertions. The proofs are left to the reader. It is also easy to prove that if m, n are natural numbers and $m \leqq n \leqq m + 1$, then $m = n$ or $n = m + 1$. We leave the proof to the reader.

We now prove the first property of integers mentioned in Chapter I, §2, namely the well-ordering:

Every non-empty subset S of $\mathbf{N}$ has a least element.

To see this, let T be the subset of $\mathbf{N}$ consisting of all n such that $n \leqq x$ for all $x \in S$. Then $0 \in T$, and $T \neq \mathbf{N}$. Hence there exists $m \in T$ such that $m + 1 \notin T$ (by induction!). Then $m \in S$ (otherwise $m < x$ for all $x \in S$ which is impossible). It is then clear that m is the smallest element of S, as desired.

In Chapter IX, we assumed known the properties of finite cardinalities. We shall prove these here. For each natural number $n \neq 0$ let J_n be the set of natural numbers x such that $1 \leqq x \leqq n$.

If $n = 1$, then $J_n = \{1\}$, and there is only a single map of J_1 into itself. This map is obviously bijective. We recall that sets A, B are said to have the same cardinality if there is a bijection of A onto B. Since a composite of bijections is a bijection, it follows that if

$$\operatorname{card}(A) = \operatorname{card}(B) \qquad \text{and} \qquad \operatorname{card}(B) = \operatorname{card}(C),$$

then $\operatorname{card}(A) = \operatorname{card}(C)$.

Let m be a natural number $\geqq 1$ and let $k \in J_{m'}$. Then there is a bijection between

$$J_{m'} - \{k\} \quad \text{and} \quad J_m$$

defined in the obvious way: We let $f: J_{m'} - \{k\} \to J_m$ be such that

$$f: x \mapsto x \qquad \text{if} \quad x < k,$$
$$f: x \mapsto \sigma^{-1}(x) \qquad \text{if} \quad x > k.$$

We let $g: J_m \to J_{m'} - \{k\}$ be such that

$$g: x \mapsto x \qquad \text{if} \quad x < k,$$
$$g: x \mapsto \sigma(x) \qquad \text{if} \quad x \geqq k.$$

Then $f \circ g$ and $g \circ f$ are the respective identities, so f, g are bijections.

We conclude that for all natural numbers $m \geqq 1$, *if*

$$h: J_m \to J_m$$

is an injection, then h *is a bijection.*

Indeed, this is true for $m = 1$, and by induction, suppose the statement true for some $m \geqq 1$. Let

$$\varphi: J_{m'} \to J_{m'}$$

be an injection. Let $r \in J_{m'}$ and let $s = \varphi(r)$. Then we can define a map

$$\varphi_0: J_{m'} - \{r\} \to J_{m'} - \{s\}$$

by $x \mapsto \varphi(x)$. The cardinality of each set $J_{m'} - \{r\}$ and $J_{m'} - \{s\}$ is the same as the cardinality of J_m. By induction, it follows that φ_0 is a bijection, whence φ is a bijection, as desired.

We conclude that if $1 \leqq m < n$, *then a map*

$$f: J_n \to J_m$$

cannot be injective.

For otherwise by what we have seen,

$$f(J_m) = J_m,$$

and hence

$$f(n) = f(x)$$

for some x such that $1 \leqq x \leqq m$, so f is not injective.

Given a set A, we shall say that $\text{card}(A) = n$ (or the **cardinality of** A is n, or A has n elements) for a natural number $n \geqq 1$, if there is a bijection of A with J_n. By the above results, it follows that such a natural number n is uniquely determined by A. We also say that A has cardinality 0 if A is empty. We say that A is **finite** if A has cardinality n for some natural number n. It is then an exercise to prove the following statements:

If A, B are finite sets, and $A \cap B$ is empty, then

$$\text{card}(A) + \text{card}(B) = \text{card}(A \cup B).$$

Furthermore,

$$\text{card}(A)\,\text{card}(B) = \text{card}(A \times B).$$

We leave the proofs to the reader.

APP., §2. THE INTEGERS

Having the natural numbers, we wish to define the integers. We do this the way it is done in elementary school.

For each natural number $n \neq 0$ we select a new symbol denoted by $-n$, and we denote by $\mathbf{Z}$ the set consisting of the union of $\mathbf{N}$ and all the symbols $-n$ for $n \in \mathbf{N}$, $n \neq 0$. We must define addition in $\mathbf{Z}$. If $x, y \in \mathbf{N}$ we use the same addition as before. For all $x \in \mathbf{Z}$, we define

$$0 + x = x + 0 = x.$$

This is compatible with the addition defined in §1 when $x \in \mathbf{N}$.

Let $m, n \in \mathbf{N}$ and neither n nor $m = 0$. If $m = n + k$ with $k \in \mathbf{N}$ we define:

(a) $m + (-n) = (-n) + m = k$.
(b) $(-m) + n = n + (-m) = -k$ if $k \neq 0$, and $= 0$ if $k = 0$.
(c) $(-m) + (-n) = -(m + n)$.

Given $x, y \in \mathbf{Z}$, if not both x, y are natural numbers, then at least one of the situations (a), (b), (c) applies to their addition.

It is then tedious but routine to verify that $\mathbf{Z}$ is an additive group.

Next we define multiplication in $\mathbf{Z}$. If $x, y \in \mathbf{N}$ we use the same multiplication as before. For all $x \in \mathbf{Z}$ we define $0x = x0 = 0$.

Let $m, n \in \mathbf{N}$ and neither n nor $m = 0$. We define:

$$(-m)n = n(-m) = -(mn) \qquad \text{and} \qquad (-m)(-n) = mn.$$

Then it is routinely verified that **Z** is a commutative ring, and is in fact integral, its unit element being the element 1 in **N**. In this way we get the integers.

Observe that **Z** is an ordered ring in the sense of Chapter IX, §1 because the set of natural numbers $n \neq 0$ satisfies all the conditions given in that chapter, as one sees directly from our definitions of multiplication and addition.

APP., §3. INFINITE SETS

A set A is said to be **infinite** if it is not finite (and in particular, not empty).

We shall prove that *an infinite set A contains a denumerable subset.* For each nonempty subset T of A, let x_T be a chosen element of T. We prove by induction that for each positive integer n we can find uniquely determined elements $x_1,\ldots,x_n \in A$ such that $x_1 = x_A$ is the chosen element corresponding to the set A itself, and for each $k = 1,\ldots,n-1$, the element x_{k+1} is the chosen element in the complement of $\{x_1,\ldots,x_k\}$. When $n = 1$, this is obvious. Assume the statement proved for $n > 1$. Then we let x_{n+1} be the chosen element in the complement of $\{x_1,\ldots,x_n\}$. If $x_1,\ldots,x_n$ are already uniquely determined, so is x_{n+1}. This proves what we wanted. In particular, since the elements $x_1,\ldots,x_n$ are distinct for all n, it follows that the subset of A consisting of all elements x_n is a denumerable subset, as desired.

Index

D

E

F

G

Undergraduate Texts in Mathematics

Abbott: Understanding Analysis.
Anglin: Mathematics: A Concise History and Philosophy.
Readings in Mathematics.
Anglin/Lambek: The Heritage of Thales.
Readings in Mathematics.
Apostol: Introduction to Analytic Number Theory. Second edition.
Armstrong: Basic Topology.
Armstrong: Groups and Symmetry.
Axler: Linear Algebra Done Right. Second edition.
Beardon: Limits: A New Approach to Real Analysis.
Bak/Newman: Complex Analysis. Second edition.
Banchoff/Wermer: Linear Algebra Through Geometry. Second edition.
Berberian: A First Course in Real Analysis.
Bix: Conics and Cubics: A Concrete Introduction to Algebraic Curves.
Brémaud: An Introduction to Probabilistic Modeling.
Bressoud: Factorization and Primality Testing.
Bressoud: Second Year Calculus.
Readings in Mathematics.
Brickman: Mathematical Introduction to Linear Programming and Game Theory.
Browder: Mathematical Analysis: An Introduction.
Buchmann: Introduction to Cryptography.
Buskes/van Rooij: Topological Spaces: From Distance to Neighborhood.
Callahan: The Geometry of Spacetime: An Introduction to Special and General Relavitity.
Carter/van Brunt: The Lebesgue–Stieltjes Integral: A Practical Introduction.
Cederberg: A Course in Modern Geometries. Second edition.
Chambert-Loir: A Field Guide to Algebra
Childs: A Concrete Introduction to Higher Algebra. Second edition.
Chung/AitSahlia: Elementary Probability Theory: With Stochastic Processes and an Introduction to Mathematical Finance. Fourth edition.
Cox/Little/O'Shea: Ideals, Varieties, and Algorithms. Second edition.
Croom: Basic Concepts of Algebraic Topology.
Curtis: Linear Algebra: An Introductory Approach. Fourth edition.
Daepp/Gorkin: Reading, Writing, and Proving: A Closer Look at Mathematics.
Devlin: The Joy of Sets: Fundamentals of Contemporary Set Theory. Second edition.
Dixmier: General Topology.
Driver: Why Math?
Ebbinghaus/Flum/Thomas: Mathematical Logic. Second edition.
Edgar: Measure, Topology, and Fractal Geometry.
Elaydi: An Introduction to Difference Equations. Second edition.
Erdős/Surányi: Topics in the Theory of Numbers.
Estep: Practical Analysis in One Variable.
Exner: An Accompaniment to Higher Mathematics.
Exner: Inside Calculus.
Fine/Rosenberger: The Fundamental Theory of Algebra.
Fischer: Intermediate Real Analysis.
Flanigan/Kazdan: Calculus Two: Linear and Nonlinear Functions. Second edition.
Fleming: Functions of Several Variables. Second edition.
Foulds: Combinatorial Optimization for Undergraduates.
Foulds: Optimization Techniques: An Introduction.
Franklin: Methods of Mathematical Economics.

Frazier: An Introduction to Wavelets Through Linear Algebra
Gamelin: Complex Analysis.
Gordon: Discrete Probability.
Hairer/Wanner: Analysis by Its History. *Readings in Mathematics.*
Halmos: Finite-Dimensional Vector Spaces. Second edition.
Halmos: Naive Set Theory.
Hämmerlin/Hoffmann: Numerical Mathematics. *Readings in Mathematics.*
Harris/Hirst/Mossinghoff: Combinatorics and Graph Theory.
Hartshorne: Geometry: Euclid and Beyond.
Hijab: Introduction to Calculus and Classical Analysis.
Hilton/Holton/Pedersen: Mathematical Reflections: In a Room with Many Mirrors.
Hilton/Holton/Pedersen: Mathematical Vistas: From a Room with Many Windows.
Iooss/Joseph: Elementary Stability and Bifurcation Theory. Second edition.
Irving: Integers, Polynomials, and Rings: A Course in Algebra
Isaac: The Pleasures of Probability. *Readings in Mathematics.*
James: Topological and Uniform Spaces.
Jänich: Linear Algebra.
Jänich: Topology.
Jänich: Vector Analysis.
Kemeny/Snell: Finite Markov Chains.
Kinsey: Topology of Surfaces.
Klambauer: Aspects of Calculus.
Lang: A First Course in Calculus. Fifth edition.
Lang: Calculus of Several Variables. Third edition.
Lang: Introduction to Linear Algebra. Second edition.
Lang: Linear Algebra. Third edition.
Lang: Short Calculus: The Original Edition of "A First Course in Calculus."
Lang: Undergraduate Algebra. Third edition
Lang: Undergraduate Analysis.
Laubenbacher/Pengelley: Mathematical Expeditions.
Lax/Burstein/Lax: Calculus with Applications and Computing. Volume 1.
LeCuyer: College Mathematics with APL.
Lidl/Pilz: Applied Abstract Algebra. Second edition.
Logan: Applied Partial Differential Equations, Second edition.
Lovász/Pelikán/Vesztergombi: Discrete Mathematics.
Macki-Strauss: Introduction to Optimal Control Theory.
Malitz: Introduction to Mathematical Logic.
Marsden/Weinstein: Calculus I, II, III. Second edition.
Martin: Counting: The Art of Enumerative Combinatorics.
Martin: The Foundations of Geometry and the Non-Euclidean Plane.
Martin: Geometric Constructions.
Martin: Transformation Geometry: An Introduction to Symmetry.
Millman/Parker: Geometry: A Metric Approach with Models. Second edition.
Moschovakis: Notes on Set Theory.
Owen: A First Course in the Mathematical Foundations of Thermodynamics.
Palka: An Introduction to Complex Function Theory.
Pedrick: A First Course in Analysis.
Peressini/Sullivan/Uhl: The Mathematics of Nonlinear Programming.

Prenowitz/Jantosciak: Join Geometries.
Priestley: Calculus: A Liberal Art. Second edition.
Protter/Morrey: A First Course in Real Analysis. Second edition.
Protter/Morrey: Intermediate Calculus. Second edition.
Pugh: Real Mathematical Analysis.
Roman: An Introduction to Coding and Information Theory.
Roman: Introduction to the Mathematics of Finance: From Risk Management to Options Pricing.
Ross: Differential Equations: An Introduction with Mathematica®. Second edition.
Ross: Elementary Analysis: The Theory of Calculus.
Samuel: Projective Geometry. *Readings in Mathematics.*
Saxe: Beginning Functional Analysis
Scharlau/Opolka: From Fermat to Minkowski.
Schiff: The Laplace Transform: Theory and Applications.
Sethuraman: Rings, Fields, and Vector Spaces: An Approach to Geometric Constructability.
Sigler: Algebra.
Silverman/Tate: Rational Points on Elliptic Curves.
Simmonds: A Brief on Tensor Analysis. Second edition.
Singer: Geometry: Plane and Fancy.
Singer/Thorpe: Lecture Notes on Elementary Topology and Geometry.
Smith: Linear Algebra. Third edition.
Smith: Primer of Modern Analysis. Second edition.
Stanton/White: Constructive Combinatorics.
Stillwell: Elements of Algebra: Geometry, Numbers, Equations.
Stillwell: Elements of Number Theory.
Stillwell: Mathematics and Its History. Second edition.
Stillwell: Numbers and Geometry. *Readings in Mathematics.*
Strayer: Linear Programming and Its Applications.
Toth: Glimpses of Algebra and Geometry. Second Edition. *Readings in Mathematics.*
Troutman: Variational Calculus and Optimal Control. Second edition.
Valenza: Linear Algebra: An Introduction to Abstract Mathematics.
Whyburn/Duda: Dynamic Topology.
Wilson: Much Ado About Calculus.

Other Books by Lang Published by Springer

Math! Encounters with High School Students • The Beauty of Doing Mathematics • Geometry: A High School Course • Basic Mathematics • Short Calculus • A First Course in Calculus • Introduction to Linear Algebra • Calculus of Several Variables • Linear Algebra • Undergraduate Analysis • Undergraduate Algebra • Complex Analysis • Math Talks for Undergraduates • Algebra • Real and Functional Analysis • Introduction to Differentiable Manifolds • Fundamentals of Differential Geometry • Algebraic Number Theory • Cyclotomic Fields I and II • Introduction to Diophantine Approximations • $SL_2(\mathbf{R})$ • Spherical Inversion on $SL_n(\mathbf{R})$ (with Jay Jorgenson) • Elliptic Functions • Elliptic Curves: Diophantine Analysis • Introduction to Arakelov Theory • Riemann-Roch Algebra (with William Fulton) • Abelian Varieties • Introduction to Algebraic and Abelian Functions • Complex Multiplication • Introduction to Modular Forms • Modular Units (with Daniel Kubert) • Introduction to Complex Hyperbolic Spaces • Number Theory III • Survey on Diophantine Geometry • Collected Papers I-V, including the following: Introduction to Transcendental Numbers in volume I, Frobenius Distributions in GL2-Extensions (with Hale Trotter in volume II, Topics in Cohomology of Groups in volume IV, Basic Analysis of Regularized Series and Products (with Jay Jorgenson) in volume V and Explicit Formulas for Regularized Products and Series (with Jay Jorgenson) in volume V • THE FILE • CHALLENGES